LONDON MATHEMATICAL SOCIETY

Managing Editor: Professor M. Reid, Mathematics
University of Warwick, Coventry CV4 7AL, Unitec

The titles below are available from booksellers, or f
www.cambridge.org/mathematics

347 Surveys in contemporary mathematics, N. YOUNG & Y. CHOI (eds)
348 Transcendental dynamics and complex analysis, P.J. RIPPON & G.M. STALLARD (eds)
349 Model theory with applications to algebra and analysis I, Z. CHATZIDAKIS, D. MACPHERSON, A. PILLAY & A. WILKIE (eds)
350 Model theory with applications to algebra and analysis II, Z. CHATZIDAKIS, D. MACPHERSON, A. PILLAY & A. WILKIE (eds)
351 Finite von Neumann algebras and masas, A.M. SINCLAIR & R.R. SMITH
352 Number theory and polynomials, J. MCKEE & C. SMYTH (eds)
353 Trends in stochastic analysis, J. BLATH, P. MÖRTERS & M. SCHEUTZOW (eds)
354 Groups and analysis, K. TENT (ed)
355 Non-equilibrium statistical mechanics and turbulence, J. CARDY, G. FALKOVICH & K. GAWEDZKI
356 Elliptic curves and big Galois representations, D. DELBOURGO
357 Algebraic theory of differential equations, M.A.H. MACCALLUM & A.V. MIKHAILOV (eds)
358 Geometric and cohomological methods in group theory, M.R. BRIDSON, P.H. KROPHOLLER & I.J. LEARY (eds)
359 Moduli spaces and vector bundles, L. BRAMBILA-PAZ, S.B. BRADLOW, O. GARCÍA-PRADA & S. RAMANAN (eds)
360 Zariski geometries, B. ZILBER
361 Words: Notes on verbal width in groups, D. SEGAL
362 Differential tensor algebras and their module categories, R. BAUTISTA, L. SALMERÓN & R. ZUAZUA
363 Foundations of computational mathematics, Hong Kong 2008, F. CUCKER, A. PINKUS & M.J. TODD (eds)
364 Partial differential equations and fluid mechanics, J.C. ROBINSON & J.L. RODRIGO (eds)
365 Surveys in combinatorics 2009, S. HUCZYNSKA, J.D. MITCHELL & C.M. RONEY-DOUGAL (eds)
366 Highly oscillatory problems, B. ENGQUIST, A. FOKAS, E. HAIRER & A. ISERLES (eds)
367 Random matrices: High dimensional phenomena, G. BLOWER
368 Geometry of Riemann surfaces, F.P. GARDINER, G. GONZÁLEZ-DIEZ & C. KOUROUNIOTIS (eds)
369 Epidemics and rumours in complex networks, M. DRAIEF & L. MASSOULIÉ
370 Theory of p-adic distributions, S. ALBEVERIO, A.YU. KHRENNIKOV & V.M. SHELKOVICH
371 Conformal fractals, F. PRZYTYCKI & M. URBAŃSKI
372 Moonshine: The first quarter century and beyond, J. LEPOWSKY, J. MCKAY & M.P. TUITE (eds)
373 Smoothness, regularity and complete intersection, J. MAJADAS & A. G. RODICIO
374 Geometric analysis of hyperbolic differential equations: An introduction, S. ALINHAC
375 Triangulated categories, T. HOLM, P. JØRGENSEN & R. ROUQUIER (eds)
376 Permutation patterns, S. LINTON, N. RUŠKUC & V. VATTER (eds)
377 An introduction to Galois cohomology and its applications, G. BERHUY
378 Probability and mathematical genetics, N. H. BINGHAM & C. M. GOLDIE (eds)
379 Finite and algorithmic model theory, J. ESPARZA, C. MICHAUX & C. STEINHORN (eds)
380 Real and complex singularities, M. MANOEL, M.C. ROMERO FUSTER & C.T.C WALL (eds)
381 Symmetries and integrability of difference equations, D. LEVI, P. OLVER, Z. THOMOVA & P. WINTERNITZ (eds)
382 Forcing with random variables and proof complexity, J. KRAJÍČEK
383 Motivic integration and its interactions with model theory and non-Archimedean geometry I, R. CLUCKERS, J. NICAISE & J. SEBAG (eds)
384 Motivic integration and its interactions with model theory and non-Archimedean geometry II, R. CLUCKERS, J. NICAISE & J. SEBAG (eds)
385 Entropy of hidden Markov processes and connections to dynamical systems, B. MARCUS, K. PETERSEN & T. WEISSMAN (eds)
386 Independence-friendly logic, A.L. MANN, G. SANDU & M. SEVENSTER
387 Groups St Andrews 2009 in Bath I, C.M. CAMPBELL *et al* (eds)
388 Groups St Andrews 2009 in Bath II, C.M. CAMPBELL *et al* (eds)
389 Random fields on the sphere, D. MARINUCCI & G. PECCATI
390 Localization in periodic potentials, D.E. PELINOVSKY
391 Fusion systems in algebra and topology, M. ASCHBACHER, R. KESSAR & B. OLIVER
392 Surveys in combinatorics 2011, R. CHAPMAN (ed)
393 Non-abelian fundamental groups and Iwasawa theory, J. COATES *et al* (eds)
394 Variational problems in differential geometry, R. BIELAWSKI, K. HOUSTON & M. SPEIGHT (eds)
395 How groups grow, A. MANN
396 Arithmetic differential operators over the p-adic integers, C.C. RALPH & S.R. SIMANCA
397 Hyperbolic geometry and applications in quantum chaos and cosmology, J. BOLTE & F. STEINER (eds)
398 Mathematical models in contact mechanics, M. SOFONEA & A. MATEI
399 Circuit double cover of graphs, C.-Q. ZHANG
400 Dense sphere packings: a blueprint for formal proofs, T. HALES
401 A double Hall algebra approach to affine quantum Schur–Weyl theory, B. DENG, J. DU & Q. FU
402 Mathematical aspects of fluid mechanics, J.C. ROBINSON, J.L. RODRIGO & W. SADOWSKI (eds)
403 Foundations of computational mathematics, Budapest 2011, F. CUCKER, T. KRICK, A. PINKUS & A. SZANTO (eds)

404 Operator methods for boundary value problems, S. HASSI, H.S.V. DE SNOO & F.H. SZAFRANIEC (eds)
405 Torsors, étale homotopy and applications to rational points, A.N. SKOROBOGATOV (ed)
406 Appalachian set theory, J. CUMMINGS & E. SCHIMMERLING (eds)
407 The maximal subgroups of the low-dimensional finite classical groups, J.N. BRAY, D.F. HOLT & C.M. RONEY-DOUGAL
408 Complexity science: the Warwick master's course, R. BALL, V. KOLOKOLTSOV & R.S. MACKAY (eds)
409 Surveys in combinatorics 2013, S.R. BLACKBURN, S. GERKE & M. WILDON (eds)
410 Representation theory and harmonic analysis of wreath products of finite groups, T. CECCHERINI-SILBERSTEIN, F. SCARABOTTI & F. TOLLI
411 Moduli spaces, L. BRAMBILA-PAZ, O. GARCÍA-PRADA, P. NEWSTEAD & R.P. THOMAS (eds)
412 Automorphisms and equivalence relations in topological dynamics, D.B. ELLIS & R. ELLIS
413 Optimal transportation, Y. OLLIVIER, H. PAJOT & C. VILLANI (eds)
414 Automorphic forms and Galois representations I, F. DIAMOND, P.L. KASSAEI & M. KIM (eds)
415 Automorphic forms and Galois representations II, F. DIAMOND, P.L. KASSAEI & M. KIM (eds)
416 Reversibility in dynamics and group theory, A.G. O'FARRELL & I. SHORT
417 Recent advances in algebraic geometry, C.D. HACON, M. MUSTAŢĂ & M. POPA (eds)
418 The Bloch–Kato conjecture for the Riemann zeta function, J. COATES, A. RAGHURAM, A. SAIKIA & R. SUJATHA (eds)
419 The Cauchy problem for non-Lipschitz semi-linear parabolic partial differential equations, J.C. MEYER & D.J. NEEDHAM
420 Arithmetic and geometry, L. DIEULEFAIT *et al* (eds)
421 O-minimality and Diophantine geometry, G.O. JONES & A.J. WILKIE (eds)
422 Groups St Andrews 2013, C.M. CAMPBELL *et al* (eds)
423 Inequalities for graph eigenvalues, Z. STANIĆ
424 Surveys in combinatorics 2015, A. CZUMAJ *et al* (eds)
425 Geometry, topology and dynamics in negative curvature, C.S. ARAVINDA, F.T. FARRELL & J.-F. LAFONT (eds)
426 Lectures on the theory of water waves, T. BRIDGES, M. GROVES & D. NICHOLLS (eds)
427 Recent advances in Hodge theory, M. KERR & G. PEARLSTEIN (eds)
428 Geometry in a Fréchet context, C.T.J. DODSON, G. GALANIS & E. VASSILIOU
429 Sheaves and functions modulo p, L. TAELMAN
430 Recent progress in the theory of the Euler and Navier–Stokes equations, J.C. ROBINSON, J.L. RODRIGO, W. SADOWSKI & A. VIDAL-LÓPEZ (eds)
431 Harmonic and subharmonic function theory on the real hyperbolic ball, M. STOLL
432 Topics in graph automorphisms and reconstruction (2nd Edition), J. LAURI & R. SCAPELLATO
433 Regular and irregular holonomic D-modules, M. KASHIWARA & P. SCHAPIRA
434 Analytic semigroups and semilinear initial boundary value problems (2nd Edition), K. TAIRA
435 Graded rings and graded Grothendieck groups, R. HAZRAT
436 Groups, graphs and random walks, T. CECCHERINI-SILBERSTEIN, M. SALVATORI & E. SAVA-HUSS (eds)
437 Dynamics and analytic number theory, D. BADZIAHIN, A. GORODNIK & N. PEYERIMHOFF (eds)
438 Random walks and heat kernels on graphs, M.T. BARLOW
439 Evolution equations, K. AMMARI & S. GERBI (eds)
440 Surveys in combinatorics 2017, A. CLAESSON *et al* (eds)
441 Polynomials and the mod 2 Steenrod algebra I, G. WALKER & R.M.W. WOOD
442 Polynomials and the mod 2 Steenrod algebra II, G. WALKER & R.M.W. WOOD
443 Asymptotic analysis in general relativity, T. DAUDÉ, D. HÄFNER & J.-P. NICOLAS (eds)
444 Geometric and cohomological group theory, P.H. KROPHOLLER, I.J. LEARY, C. MARTÍNEZ-PÉREZ & B.E.A. NUCINKIS (eds)
445 Introduction to hidden semi-Markov models, J. VAN DER HOEK & R.J. ELLIOTT
446 Advances in two-dimensional homotopy and combinatorial group theory, W. METZLER & S. ROSEBROCK (eds)
447 New directions in locally compact groups, P.-E. CAPRACE & N. MONOD (eds)
448 Synthetic differential topology, M.C. BUNGE, F. GAGO & A.M. SAN LUIS
449 Permutation groups and cartesian decompositions, C.E. PRAEGER & C. SCHNEIDER
450 Partial differential equations arising from physics and geometry, M. BEN AYED *et al* (eds)
451 Topological methods in group theory, N. BROADDUS, M. DAVIS, J.-F. LAFONT & I. ORTIZ (eds)
452 Partial differential equations in fluid mechanics, C.L. FEFFERMAN, J.C. ROBINSON & J.L. RODRIGO (eds)
453 Stochastic stability of differential equations in abstract spaces, K. LIU
454 Beyond hyperbolicity, M. HAGEN, R. WEBB & H. WILTON (eds)
455 Groups St Andrews 2017 in Birmingham, C.M. CAMPBELL et al (eds)

London Mathematical Society Lecture Note Series: 453

Stochastic Stability of Differential Equations in Abstract Spaces

KAI LIU
Tianjin Normal University
and
University of Liverpool

CAMBRIDGE
UNIVERSITY PRESS

University Printing House, Cambridge CB2 8BS, United Kingdom

One Liberty Plaza, 20th Floor, New York, NY 10006, USA

477 Williamstown Road, Port Melbourne, VIC 3207, Australia

314–321, 3rd Floor, Plot 3, Splendor Forum, Jasola District Centre,
New Delhi – 110025, India

79 Anson Road, #06–04/06, Singapore 079906

Cambridge University Press is part of the University of Cambridge.

It furthers the University's mission by disseminating knowledge in the pursuit of education, learning, and research at the highest international levels of excellence.

www.cambridge.org
Information on this title: www.cambridge.org/9781108705172
DOI: 10.1017/9781108653039

First published 2019

Printed and bound in Great Britain by Clays Ltd, Elcograf S.p.A.

A catalogue record for this publication is available from the British Library.

Library of Congress Cataloging-in-Publication Data
Names: Liu, Kai, 1964– author.
Title: Stochastic stability of differential equations in abstract spaces /
Kai Liu (University of Liverpool).
Description: Cambridge ; New York, NY : Cambridge University Press, 2019. |
Series: London Mathematical Society lecture note series ; 453 |
Includes bibliographical references and index.
Identifiers: LCCN 2018049978 | ISBN 9781108705172 (pbk.)
Subjects: LCSH: Stochastic differential equations. | Differential equations, Linear. |
Differential equations, Nonlinear. | Algebraic spaces. | Stability. | Geometry, Algebraic.
Classification: LCC QA274.23 .L5825 2019 | DDC 515/.35–dc23
LC record available at https://lccn.loc.gov/2018049978

ISBN 978-1-108-70517-2 Paperback

Contents

	Preface	*page* vii
1	**Preliminaries**	1
1.1	Linear Operators, Semigroups, and Examples	1
1.2	Stochastic Processes and Martingales	16
1.3	Wiener Processes and Stochastic Integration	22
1.4	Stochastic Differential Equations	26
1.5	Definitions and Methods of Stochastic Stability	35
1.6	Notes and Comments	44
2	**Stability of Linear Stochastic Differential Equations**	46
2.1	Deterministic Linear Systems	46
2.2	Lyapunov Equations and Stochastic Stability	61
2.3	Systems with Boundary Noise	82
2.4	Exponentially Stable Stationary Solutions	88
2.5	Some Examples	92
2.6	Notes and Comments	95
3	**Stability of Nonlinear Stochastic Differential Equations**	98
3.1	An Extension of Linear Stability Criteria	98
3.2	Comparison Approach	105
3.3	Nonautonomous Stochastic Systems	108
3.4	Stability in Probability and Sample Path	122
3.5	Lyapunov Function Characterization	131
3.6	Two Applications	143
3.7	Invariant Measures and Ultimate Boundedness	150
3.8	Decay Rate	159
3.9	Stabilization of Systems by Noise	168
3.10	Notes and Comments	175

4 Stability of Stochastic Functional Differential Equations 178
4.1 Deterministic Systems 178
4.2 Linear Systems with Additive Noise 192
4.3 Linear Systems with Multiplicative Noise 196
4.4 Stability of Nonlinear Systems 206
4.5 Notes and Comments 223

5 Some Applications Related to Stochastic Stability 226
5.1 Applications in Mathematical Biology 226
5.2 Applications in Mathematical Physics 232
5.3 Applications in Stochastic Control 239
5.4 Notes and Comments 243

Appendix 245
A Proof of Theorem 4.1.5 245
B Proof of Proposition 4.1.7 247
C Proof of Proposition 4.1.10 248
D Proof of Proposition 4.1.14 250

References 252
Index 265

Preface

The aim of this book is to give a basic and systematic account of stability theory of stochastic differential equations in Hilbert spaces and its application to practical stochastic systems such as a stochastic partial differential equation (SPDE).

I have tried to organize the main content into an easily accessible monograph after a brief review of some preliminary material. I begin my account in Chapter 1 by recalling some notions and notations from the theory of stochastic differential equations in Hilbert spaces. Some fundamental concepts such as Q-Wiener process, stochastic integration, strong or mild solution will be reviewed carefully. Most theorems or propositions are stated without proofs here. However, I will present those proofs of results that are not available in the existing books and are to be found scattered in the literature.

Chapter 2 of this book is devoted to a detailed development of stability theory for linear stochastic evolution equations. The central part is a formulation of the characteristic conditions for mean square and almost sure exponential stability in terms of the Lyapunov type of equations.

In Chapter 3, I mainly focus on the development of a stability property for a wide class of nonlinear stochastic differential equations. This chapter contains basic theory and illustrative examples in connection with the stability behavior of nonlinear stochastic systems. In particular, I generalize those linear characteristic results from Chapter 2 in an appropriate way to obtain some nonlinear versions for semi-linear stochastic evolution equations. Motivated by the idea of reducing the stability problem of nonlinear stochastic systems to the corresponding one of linear systems, I develop the so-called Lyapunov function characterization method and explore the associated first-order approximation techniques. Other interesting topics such as non-exponential decay or stabilization by noise of systems are also considered.

Chapter 4 is a statement of stability theory for stochastic functional differential equations. Generally speaking, it is a natural idea to extend the so-called Lyapunov function theory in the previous chapters to develop a Lyapunov functional type of scheme for time delay stochastic systems although this program can frequently become frustrating. To avoid the underlying difficulty, I emphasize in linear cases the so-called fundamental solutions and associated lift-up methods. For nonlinear systems, I introduce various approaches such as a fixed-point theorem or Razumikhin function method to handle stochastic stability problems.

In Chapter 5, I present selected applications in which the choice of the material reflects my own personal preference. Here the treatment is somewhat sketchy and by no means the only way or even the most appropriate way. The purpose of my account is a desire to present some practical topics such as stochastic optimal control or stochastic population dynamics, beyond the main scheme of the book, but having a relation to stability theory of stochastic evolution equations. It is also hoped that the selected presentation here will stimulate further work in these and related fields.

Notes and Comments at the end of each chapter contain historical and related background material as well as references to the results discussed in that chapter. The pervading influence of a variety of authors' work in this book is obvious. I have drawn freely on their work and hopefully I can acknowledge my scientific debts to them by some remarks shown there. It should be emphasized that my choice of material in this book is highly subjective. In particular, this book is organized only to cover stability theory of stochastic models with white noise and a number of specific topics are not treated. For instance, I do not consider stochastic equations driven by a jump process or fractional Brownian motion. For time delay systems, I do not treat stochastic functional differential equations of infinite time lag or neutral type. Each of those subjects would require several additional chapters. On the other hand, the lengthy list of references at the end of the book is somewhat incomplete and only includes those titles which pertain directly to the content. The author wishes to apologize to those researchers whose work might have been overlooked.

The current volume is an outgrowth of the author's book *Stability of Infinite Dimensional Stochastic Differential Equations with Applications* published by Chapman & Hall/CRC in 2006. Since its publication, many new results have appeared and significant progress has been made in this exciting research field. Most ingredients in the author's old version are maintained as the skeleton of the present volume. But they are reorganized and integrated with much material developed in the last decade to reflect new developments and update the existing results.

I would like to thank all those who helped in the realization of this book through encouragement, advice, or scientific exchanges: Jianhai Bao, Joris Bierkens, Tomás Caraballo, María Garrido-Atienza, Anna Kwiecińska, Vidyadhar Mandrekar, Xuerong Mao, Bohdan Maslowski, Michael Scheutzow, Takeshi Taniguchi, Aubrey Truman, Feng-Yu Wang, Wei Wang, George Yin, and Chenggui Yuan. I am extremely grateful to Professor Peter Giblin for his language comments which have led to a significant improvement of my presentation. My special thanks go to Professor Pao-Liu Chow and Professor Jerzy Zabczyk whose warm assistance and concern has provided lasting support at various stages of my academic career. Last but certainly not least, I want to thank my wife and daughter, Lihong and Annie, who had to put up with me being even more "random" during the preparation of this book than usual.

1
Preliminaries

We begin by recalling some basic definitions and results, especially those from functional analysis, partial differential equations (PDEs), probability, and stability theories. We review basic notions and notations from deterministic systems and recall important results from stochastic differential equations. We introduce two notions of solutions, mild and strong, for infinite-dimensional stochastic differential equations and consider the existence and uniqueness of solutions under suitable assumptions. We introduce and clarify various definitions of stochastic stability in Hilbert spaces, which are a natural generalization of deterministic stability concepts. To present the proofs of all the results here would require preparatory background material that would significantly increase both the size and scope of this book. Therefore, we adopt the approach of omitting those proofs, which are treated in detail in well-known standard textbooks such as Da Prato and Zabczyk [53], Pazy [187], and Yosida [224]. However, those proofs will be presented that are not available in the existing books and are to be found scattered in the literature, or that discuss ideas specially relevant to our purpose.

1.1 Linear Operators, Semigroups, and Examples

Throughout this book, the sets of nonnegative integers, positive integers, real numbers, and complex numbers are denoted by $\mathbb{N}$, $\mathbb{N}_+$, $\mathbb{R}$, and $\mathbb{C}$, respectively. Also, $\mathbb{R}_+$ denotes the set of all nonnegative real numbers and $\mathbb{R}^n$ denotes the n-dimensional real vector space equipped with the usual Euclidean norm $\|\cdot\|_{\mathbb{R}^n}$, $n \geq 1$. For any $\lambda \in \mathbb{C}$, the symbols $Re\,\lambda$ and $Im\,\lambda$ denote the real and imaginary parts of λ, respectively. Given a set E, the symbol $\mathbf{1}_E$ denotes the characteristic function of E, i.e., $\mathbf{1}_E(x) = 1$ if $x \in E$ and $\mathbf{1}_E(x) = 0$ if $x \notin E$.

A *Banach space* $(X, \|\cdot\|_X)$, real or complex, is a complete normed linear space over $\mathbb{R}$ or $\mathbb{C}$. If the norm $\|\cdot\|_X$ is induced by an inner product $\langle\cdot,\cdot\rangle_X$, then X is called a *Hilbert space*. In this book, we always take the inner product $\langle\cdot,\cdot\rangle_X$ of X to be linear in the first entry and conjugate-linear in the second. We say that a sequence $\{x_n\}_{n\geq 1} \subseteq X$ *(strongly) converges* to $x \in X$ if $\lim_{n\to\infty} \|x_n - x\|_X = 0$. If X contains n linearly independent vectors, but every system of $n+1$ vectors in X is linearly dependent, then X is called an *n-dimensional space*, denoted by $\dim X = n$. Otherwise, the space X is said to be infinite dimensional. We say that X is *separable* if there exists a countable set $S \subseteq X$ such that $\overline{S} = X$, where $\overline{S}$ is the closure of S in X. For a Hilbert space X, a collection $\{e_i\}_{i\geq 1}$ of elements in X is called an *orthonormal set* if $\langle e_i, e_i\rangle_X = 1$ for all i, and $\langle e_i, e_j\rangle_X = 0$ if $i \neq j$. If S is an orthonormal set and no other orthonormal set contains S as a proper subset, then S is called an *orthonormal basis* for X. A Hilbert space is separable if and only if it has a countable orthonormal basis $\{e_i\}$, $i = 1, 2, \ldots$

A typical example of Banach spaces is the so-called Sobolev space, which plays an important role in PDE theory. Let $\mathcal{O}$ be a nonempty domain of $\mathbb{R}^n$, and m be a positive integer. For $1 \leq p < \infty$ we denote by $W^{m,p}(\mathcal{O}; X)$ the set of all elements $y \in L^p(\mathcal{O}; X)$ such that y and its distributional derivatives $\partial^\alpha y$ of order $|\alpha| \leq m$ are in $L^p(\mathcal{O}; X)$, where

$$\partial^\alpha = \frac{\partial^{|\alpha|}}{\partial {x_1}^{\alpha_1} \cdots \partial {x_n}^{\alpha_n}} \quad \text{and} \quad |\alpha| = \sum_{i=1}^{n} \alpha_i.$$

Then $W^{m,p}(\mathcal{O}; X)$ is a Banach space under the norm

$$\|y\|_{m,p} = \left(\int_{\mathcal{O}} \sum_{|\alpha|\leq m} \|\partial^\alpha y(x)\|_X^p dx \right)^{1/p}, \qquad y \in W^{m,p}(\mathcal{O}; X).$$

On the other hand, we denote by $C^m(\mathcal{O}; X)$ the set of all m-times continuously differentiable vectors in $\mathcal{O}$, and by $C_0^m(\mathcal{O}; X)$ the subspace of $C^m(\mathcal{O}; X)$ consisting of those vectors that have compact supports in $\mathcal{O}$. Another important Banach space $W_0^{m,p}(\mathcal{O}; X)$ is defined as the completion of $C_0^\infty(\mathcal{O}; X)$ in the metric of $W^{m,p}(\mathcal{O}; X)$.

In general, the spaces $W^{m,p}(\mathcal{O}; X)$ and $W_0^{m,p}(\mathcal{O}; X)$ do not coincide for bounded $\mathcal{O}$. However, it is true that

$$W^{m,p}(\mathbb{R}^n, \mathbb{R}) = W_0^{m,p}(\mathbb{R}^n, \mathbb{R}).$$

The case $p=2$ is special since the spaces $W^{m,2}(\mathcal{O}; X)$, $W_0^{m,2}(\mathcal{O}; X)$ (frequently written as $H^m(\mathcal{O}; X)$, $H_0^m(\mathcal{O}; X)$) are Hilbert spaces if X is a Hilbert space under the scalar product

$$\langle y, z\rangle_{m,2} = \int_{\Omega} \sum_{|\alpha|\le m} \langle \partial^{\alpha} y(x), \partial^{\alpha} z(x)\rangle_X dx.$$

Let X and Y be two Banach spaces and $\mathscr{D}(A)$ a subspace of X. A map $A\colon \mathscr{D}(A) \subseteq X \to Y$ is called a *linear operator* if the following relation holds:

$$A(\alpha x + \beta y) = \alpha Ax + \beta Ay \qquad \text{for any} \quad x,\ y \in \mathscr{D}(A),\ \ \alpha,\ \beta \in \mathbb{R} \text{ or } \mathbb{C}.$$

The subspace $\mathscr{D}(A)$ is called the *domain* of A. If A maps any bounded subsets of $\mathscr{D}(A)$ into bounded subsets of Y, we say that A is a *bounded linear operator*. We denote by $\mathscr{L}(X, Y)$ the set of all bounded linear operators A from X to Y with $\mathscr{D}(A) = X$. It may be shown that $\mathscr{L}(X, Y)$ is a Banach space under the operator norm $\|\cdot\|_{\mathscr{L}(X,Y)}$, or simply $\|\cdot\|$, given by

$$\|A\| := \sup_{\|x\|_X \le 1} \|Ax\|_Y = \sup_{\|x\|_X = 1} \|Ax\|_Y \qquad \text{for any} \qquad A \in \mathscr{L}(X, Y).$$

For simplicity, we frequently write $\mathscr{L}(X)$ for $\mathscr{L}(X, X)$.

For any linear operator $A\colon \mathscr{D}(A) \subseteq X \to Y$, we define $\mathscr{K}(A) = \{x \in \mathscr{D}(A)\colon Ax = 0\}$ and $\mathscr{R}(A) = \{Ax\colon x \in \mathscr{D}(A)\}$. They are called the *kernel* and *range* spaces of A, respectively.

Theorem 1.1.1 *Let X and Y be two Banach spaces. Then the following results hold:*

(i) (Open Mapping Theorem) *$A \in \mathscr{L}(X, Y)$ and $\mathscr{R}(A) = Y$ imply that for any open set $E \subseteq X$, the set $A(E)$ is open in Y.*
(ii) (Inverse Mapping Theorem) *$A \in \mathscr{L}(X, Y)$ with $\mathscr{R}(A) = Y$ and $\mathscr{K}(A) = \{0\}$ imply that the inverse operator A^{-1} exists and $A^{-1} \in \mathscr{L}(Y, X)$.*
(iii) (Principle of Uniform Boundedness) *$\Sigma \subseteq \mathscr{L}(X, Y)$ and $\sup_{A\in\Sigma} \|Ax\|_Y < \infty$ for each $x \in X$ imply that $\sup_{A\in\Sigma} \|A\| < \infty$.*

Let $Y = K$ where $K = \mathbb{R}$ or $\mathbb{C}$. Any $f \in \mathscr{L}(X, K)$ is called a *bounded linear functional* on X. In the sequel, we put $X^* = \mathscr{L}(X, K)$, which is a Banach space under the norm $\|\cdot\|_{X^*}$ and call X^* the *dual space* of X. Quite often, we write $f(x)$ for any $f \in X^*$, $x \in X$ by $\langle\langle x, f\rangle\rangle_{X,X^*}$, and the symbol $\langle\langle \cdot, \cdot\rangle\rangle_{X,X^*}$ is referred to as the *duality pair* between X and X^*. The following theorem assures the existence of nontrivial bounded linear functionals on any Banach space.

Theorem 1.1.2 (Hahn–Banach Theorem) *Let X be a Banach space and X_0 a subspace of X. Let $f_0 \in X_0^*$, then there exists an extension $f \in X^*$ of f_0 such that $\|f\|_{X^*} = \|f_0\|_{X_0^*}$.*

Since X^* is a Banach space, we may also talk about the dual space of X^*, i.e., $X^{**} := (X^*)^*$. It is known that for any $x \in X$, by defining

$$x^{**}(f) = f(x) = \langle\langle x, f\rangle\rangle_{X,X^*} \quad \text{for any} \quad f \in X^*, \tag{1.1.1}$$

we have $x^{**} \in X^{**}$ and $\|x\|_X = \|x^{**}\|_{X^{**}}$. Thus, the map $x \to x^{**}$ from X into X^{**} is linear and injective and preserves the norm so that X is embeddable into X^{**}. If we regard x exactly the same as x^{**}, then $X \subset X^{**}$. In general, the strict inclusion may hold, a fact that naturally leads to the following definition.

Definition 1.1.3 A Banach space X is said to be *reflexive* if $X = X^{**}$. Precisely, for any $x^{**} \in X^{**}$, there exists an $x \in X$ such that (1.1.1) holds.

The most important class of reflexive spaces are Hilbert spaces, a fact that is justified by the following theorem.

Theorem 1.1.4 (Riesz Representation Theorem) *Let X be a Hilbert space, then $X^* = X$. More precisely, for any $f \in X^*$, there exists a unique element $y \in X$ such that*

$$f(x) = \langle x, y\rangle_X \quad \textit{for any} \quad x \in X, \tag{1.1.2}$$

and conversely, for any $y \in X$, by defining f as in (1.1.2), one has $f \in X^*$. It clearly makes sense to write $\langle\cdot,\cdot\rangle_X$ for $\langle\langle\cdot,\cdot\rangle\rangle_{X,X^*}$ on this occasion.

Closed linear operators, generally unbounded, frequently appear in applications, notably in connection with partial differential equations.

Definition 1.1.5 Let X and Y be two Banach spaces. A linear operator $A\colon \mathscr{D}(A) \subseteq X \to Y$ is said to be *closed* if whenever

$$x_n \in \mathscr{D}(A), \;\; n \geq 1, \;\; \text{and} \;\; \lim_{n\to\infty} x_n = x, \quad \lim_{n\to\infty} Ax_n = y,$$

it follows that $x \in \mathscr{D}(A)$ and $Ax = y$.

For a closed linear operator $A\colon \mathscr{D}(A) \subseteq X \to X$, it can be shown that the domain $\mathscr{D}(A)$ is a Banach space under the graph norm $\|x\|_{\mathscr{D}(A)} := \|x\|_X + \|Ax\|_X$, $x \in \mathscr{D}(A)$. It is easy to see that any bounded linear operator having a closed domain is closed. The converse statement can be true in the following sense.

Theorem 1.1.6 (Closed Graph Theorem) *Suppose that $A\colon \mathscr{D}(A) \subseteq X \to Y$ is a closed linear operator. If $\mathscr{D}(A)$ is closed in X, then operator A is bounded.*

In general, it is difficult to prove that an operator is closed. The next theorem states that if this operator is the algebraic inverse of a bounded linear operator, then it is closed.

Theorem 1.1.7 *Assume that X and Y are Banach spaces and let A be a linear operator from X to Y. If A is invertible with $A^{-1} \in \mathscr{L}(Y,X)$, then A is a closed linear operator.*

Let X and Y be two Banach spaces and a linear operator $A\colon \mathscr{D}(A) \subseteq X \to Y$ is called *densely defined* if $\overline{\mathscr{D}(A)} = X$. If A is densely defined, we may define *Banach space adjoint operator* $A'\colon \mathscr{D}(A') \subseteq Y^* \to X^*$ of A in the following manner. Let

$$\mathscr{D}(A') = \left\{y^* \in Y^*\colon y^*A \text{ is continuous on } \mathscr{D}(A)\right\}.$$

The linear operator $A'\colon \mathscr{D}(A') \subseteq Y^* \to X^*$ is defined by

$$\langle\langle x, A'y^*\rangle\rangle_{X,X^*} = \langle\langle Ax, y^*\rangle\rangle_{Y,Y^*} \quad \text{for any} \quad y^* \in \mathscr{D}(A'),\ x \in \mathscr{D}(A).$$

It turns out that A' is uniquely defined and closed and map $A \to A'$ is linear.

Now let us consider the case where A is a densely defined linear operator on a Hilbert space X. Then the Banach space adjoint A' of A is a mapping from X^* into itself. Let $\iota : X \to X^*$ be the map that assigns, for each $x \in X$, the bounded linear functional $\langle \cdot, x\rangle_X$ in X^*. Then ι is a linear isometry, which is surjective by the Riesz Representation Theorem. Now define a map $A^*\colon X \to X$ by

$$A^* = \iota^{-1}A'\iota.$$

Then $A^*\colon X \to X$ satisfies

$$\langle Ay, x\rangle_X = (\iota x)(Ay) = (A'\iota x)(y) = \langle y, \iota^{-1}A'\iota x\rangle_X = \langle y, A^*x\rangle_X$$

for any $y \in \mathscr{D}(A)$, $x \in \mathscr{D}(A^*)$, and A^* is called the *Hilbert space adjoint*, or simply *adjoint*, of A. In general, $A^* \neq A'$. However, if X is a real Hilbert space, then $A^* = A'$.

Definition 1.1.8 Let X be a Hilbert space. A densely defined linear operator $A\colon \mathscr{D}(A) \subseteq X \to X$ is *symmetric* if for all $x, y \in \mathscr{D}(A)$, $\langle Ax, y\rangle_X = \langle x, Ay\rangle_X$. A symmetric operator A is called *self-adjoint* if $\mathscr{D}(A^*) = \mathscr{D}(A)$.

All bounded and symmetric operators are self-adjoint. It may be shown that the adjoint of a densely defined linear operator on a Hilbert space X is closed, and so is every self-adjoint operator. A linear operator A on the Hilbert space X is called *nonnegative*, denoted by $A \geq 0$, if $\langle Ax, x\rangle_X \geq 0$ for all $x \in \mathscr{D}(A)$. It is called *positive* if $\langle Ax, x\rangle_X > 0$ for all non zero $x \in \mathscr{D}(A)$ and *coercive* if

$\langle Ax, x\rangle_X \geq c\|x\|_X^2$ for some $c > 0$ and all $x \in \mathscr{D}(A)$. We denote the spaces of all nonnegative, positive, and coercive operators by $\mathscr{L}^+(X)$, $\mathscr{L}_0^+(X)$, and $\mathscr{L}_c^+(X)$, respectively. A linear operator B is called the *square root* of A if $B^2 = A$.

Theorem 1.1.9 *Let A be a linear operator on the Hilbert space X. If A is self-adjoint and nonnegative, then it has a unique square root, denoted by* $A^{1/2}$*, which is self-adjoint and nonnegative such that* $\mathscr{D}(A) \subset \mathscr{D}(A^{1/2})$*. Furthermore, if A is positive, so is* $A^{1/2}$.

Theorem 1.1.10 *Suppose that A is self-adjoint and nonnegative on the Hilbert space X. Then A is coercive if and only if it has a bounded inverse* $A^{-1} \in \mathscr{L}(X)$*. In this case,* A^{-1} *is self-adjoint and nonnegative.*

In the family of all bounded linear operators, there is a subclass, called compact operators, which are in many ways analogous to linear operators in finite-dimensional spaces.

Definition 1.1.11 Let X and Y be two Banach spaces. An operator $A \in \mathscr{L}(X, Y)$ is *compact* if for any bounded sequence $\{x_n\}_{n\geq 1}$ in X, the sequence $\{Ax_n\}_{n\geq 1}$ has a convergent subsequence in Y.

Let X be a separable Hilbert space and $\{e_i\}_{i=1}^{\infty}$ an orthonormal basis. Then for any nonnegative operator $A \in \mathscr{L}(X)$, we define $Tr(A) = \sum_{i=1}^{\infty}\langle e_i, Ae_i\rangle_X$. The number $Tr(A)$ is called the *trace* of A and is independent of the orthonormal basis chosen. An operator $A \in \mathscr{L}(X)$ is called *trace class* if $Tr(|A|) < \infty$, where $|A| = (A^*A)^{1/2}$. If we endow the trace norm $\|A\|_1 := Tr(|A|)$ for any trace class operator A, then the associated family $\mathscr{L}_1(X)$ of all trace class operators forms a Banach space. An operator $A \in \mathscr{L}(X)$ is called *Hilbert–Schmidt* if $Tr(A^*A) < \infty$. The norm corresponding to a Hilbert–Schmidt inner product is $\|A\|_2 := (Tr(A^*A))^{1/2}$ under which all the Hilbert–Schmidt operators form a Hilbert space $\mathscr{L}_2(X)$. It is easy to show that the following inclusions hold and they are all proper when X is infinite dimensional:

$$\{\text{trace class}\} \subset \{\text{Hilbert–Schmidt}\} \subset \{\text{compact}\}.$$

An operator $A \in \mathscr{L}(X)$ is said to have *finite trace* if the series

$$\sum_{i=1}^{\infty}\langle e_i, Ae_i\rangle_X < \infty \tag{1.1.3}$$

for any orthonormal basis $\{e_i\}_{i\geq 1}$ in X. In general, it is not true that

$$\sum_{i=1}^{\infty} |\langle e_i, Ae_i\rangle_X| < \infty \tag{1.1.4}$$

for some orthonormal basis implies that $A \in \mathscr{L}_1(X)$. However, for a trace class operator A the sum in (1.1.3) is absolutely convergent and independent of the choice of the orthonormal basis. In particular, for a nonnegative operator $A \in \mathscr{L}(X)$, the concept of a trace class operator coincides with that of an operator having finite trace.

Let $A\colon \mathscr{D}(A) \subseteq X \to X$ be a linear operator on a Banach space X. The *resolvent set* $\rho(A)$ of A is the set of all complex numbers $\lambda \in \mathbb{C}$ such that $(\lambda I - A)^{-1}$ exists and $(\lambda I - A)^{-1} \in \mathscr{L}(X)$, where I is the identity operator on X. For $\lambda \in \rho(A)$, we write $R(\lambda, A) = (\lambda I - A)^{-1}$ and call it the *resolvent operator* of A. The *spectrum* of A is defined to be $\sigma(A) = \mathbb{C} \setminus \rho(A)$. It may be shown that the resolvent set $\rho(A)$ is open in $\mathbb{C}$.

Definition 1.1.12 Let A be a linear operator on Banach space X. Define

(i) $\sigma_p(A) = \{\lambda \in \mathbb{C}\colon \lambda I - A$ is not injective$\}$, and $\sigma_p(A)$ is called the *point spectrum* of A. Moreover, each $\lambda \in \sigma_p(A)$ is called the *eigenvalue*, and each nonzero $x \in \mathscr{D}(A)$ satisfying $(\lambda I - A)x = 0$ is called the *eigenvector* of A corresponding to λ.
(ii) $\sigma_c(A) = \{\lambda \in \mathbb{C}\colon \lambda I - A$ is injective, $\mathscr{R}(\lambda I - A) \neq X$ and $\overline{\mathscr{R}(\lambda I - A)} = X\}$, and $\sigma_c(A)$ is called the *continuous spectrum* of A.
(iii) $\sigma_r(A) = \{\lambda \in \mathbb{C}\colon \lambda I - A$ is injective and $\overline{\mathscr{R}(\lambda I - A)} \neq X\}$, and $\sigma_r(A)$ is called the *residual spectrum* of A.

From this definition, it is immediate that $\sigma_p(A)$, $\sigma_c(A)$, and $\sigma_r(A)$ are mutually exclusive and their union is $\sigma(A)$. If A is self-adjoint, we have $\sigma_r(A) = \emptyset$. Note that if $\dim X < \infty$, all the linear operators A on X are compact and in this case $\sigma(A) = \sigma_p(A)$, a fact that is extendable to any compact operators in infinite-dimensional spaces.

Theorem 1.1.13 *Let X be a Banach space with $\dim X = \infty$. If $A \in \mathscr{L}(X)$ is compact, then one and only one of the following cases holds:*

(i) $\sigma(A) = \{0\}$;
(ii) $\sigma(A) = \{0, \lambda_1, \ldots, \lambda_n\}$ *where for each* $1 \leq k \leq n$, $\lambda_k \neq 0$ *and* λ_k *is an eigenvalue of* A;
(iii) $\sigma(A) = \{0, \lambda_1, \lambda_2, \ldots\}$ *where for each* $k \geq 1$, $\lambda_k \neq 0$ *and* λ_k *is an eigenvalue of* A *with* $\lim_{k\to\infty} \lambda_k = 0$.

In this book, we shall employ the theory of linear semigroups, which usually allows a uniform treatment of many systems such as some parabolic, hyperbolic, and delay equations.

Definition 1.1.14 A *strongly continuous or C_0-semigroup* $S(t) \in \mathscr{L}(X)$, $t \geq 0$, on a Banach space X is a family of bounded linear operators $S(t)\colon X \to X$, $t \geq 0$, satisfying the following:

(i) $S(0)x = x$ for all $x \in X$;
(ii) $S(t+s) = S(t)S(s)$ for all $t, s \geq 0$;
(iii) $S(t)$ is strongly continuous, i.e., for any $x \in X$, $S(\cdot)x\colon [0,\infty) \to X$ is continuous.

For any C_0-semigroup $S(t)$ on X, there exist constants $M \geq 1$ and $\mu \in \mathbb{R}$ such that

$$\|S(t)\| \leq Me^{\mu t}, \qquad t \geq 0. \tag{1.1.5}$$

In particular, the semigroup $S(t)$ is called *(uniformly) bounded* if $\mu = 0$ and *exponentially stable* if $\mu < 0$. The semigroup $S(t)$, $t \geq 0$, is called *eventually norm continuous* if the map $t \to S(t)$ is continuous from (r,∞) to $\mathscr{L}(X)$ for some $r > 0$. In particular, $S(t)$, $t \geq 0$, is simply called *norm continuous* if the map $t \to S(t)$ is continuous from $(0,\infty)$ to $\mathscr{L}(X)$. If $M = 1$ in (1.1.5), the semigroup $S(t)$, $t \geq 0$, is called a *pseudo contraction* C_0-semigroup, and if further $\mu = 0$, it is called a *contraction* C_0-semigroup.

In association with the C_0-semigroup $S(t)$, we may define a linear operator $A\colon \mathscr{D}(A) \subseteq X \to X$ by

$$\mathscr{D}(A) = \left\{ x \in X\colon \lim_{t\downarrow 0} \frac{S(t)x - x}{t} \text{ exists in } X \right\},$$

$$Ax = \lim_{t\downarrow 0} \frac{S(t)x - x}{t}, \quad x \in \mathscr{D}(A).$$

The operator A is called the *infinitesimal generator*, or simply *generator*, of the semigroup $\{S(t)\}_{t\geq 0}$, which is frequently written as e^{tA}, $t \geq 0$, in this book. It may be shown that A is densely defined and closed.

For an arbitrary C_0-semigroup e^{tA}, $t \geq 0$, the following theorem gives a characterization of its generator A.

Theorem 1.1.15 (Hille–Yosida Theorem) *Let X be a Banach space and $A\colon \mathscr{D}(A) \subseteq X \to X$ be a linear operator. Then the following are equivalent:*

(i) A generates a C_0-semigroup e^{tA}, $t \geq 0$, on X such that (1.1.5) holds for some $M \geq 1$ and $\mu \in \mathbb{R}$.

(ii) *A is densely defined, closed, and there exist constants* $\mu \in \mathbb{R}$, $M \geq 1$ *such that* $\rho(A) \supset \{\lambda \in \mathbb{C} \colon Re\,\lambda > \mu\}$ *and*

$$\|R(\lambda, A)^n\| \leq \frac{M}{(Re\,\lambda - \mu)^n} \quad \text{for any} \quad n \in \mathbb{N}_+, \ Re\,\lambda > \mu. \tag{1.1.6}$$

In general, it is not easy to verify (1.1.6) for each $n \in \mathbb{N}_+$. We can give, however, a simple characterization of linear operators that generate pseudo contraction C_0-semigroups.

Definition 1.1.16 A linear operator $A \colon \mathscr{D}(A) \subset X \to X$ on a Banach space X is called *dissipative* if

$$\|(\lambda I - A)x\|_X \geq \lambda \|x\|_X \quad \text{for all} \quad x \in \mathscr{D}(A) \quad \text{and} \quad \lambda > 0.$$

Theorem 1.1.17 (Lumer and Phillips Theorem) *Let* $A \colon \mathscr{D}(A) \subset X \to X$ *be a linear operator defined on* X. *Then* A *is the generator of a contraction* C_0*-semigroup on* X *if and only if*

(i) *A is a closed linear operator with dense domain in* X;
(ii) *A and its adjoint operator* A' *are dissipative.*

If X is a Hilbert space, the conditions in Theorem 1.1.17 may be simplified. In particular, we have the following proposition, which is a consequence of Theorem 1.1.17.

Proposition 1.1.18 *Let* A *be a closed, densely defined linear operator on a Hilbert space* X. *There exists a real number* $\alpha \in \mathbb{R}$ *such that*

$$Re\,\langle x, Ax\rangle_X \leq \alpha \|x\|_X^2 \quad \text{for all} \quad x \in \mathscr{D}(A), \tag{1.1.7}$$

and

$$Re\,\langle x, A^* x\rangle_X \leq \alpha \|x\|_X^2 \quad \text{for all} \quad x \in \mathscr{D}(A^*), \tag{1.1.8}$$

if and only if A *generates a pseudo contraction* C_0*-semigroup* e^{tA}, $t \geq 0$, *satisfying*

$$\|e^{tA}\| \leq e^{\alpha t} \quad \text{for all} \quad t \geq 0. \tag{1.1.9}$$

We state some properties of C_0-semigroups and their generators.

Proposition 1.1.19 *Let* e^{tA}, $t \geq 0$, *be a* C_0*-semigroup on a Banach space* X *and* $A_n = nAR(n, A) \in \mathscr{L}(X)$, $n \in \rho(A)$, *called the Yosida approximation of* A. *Then*

$$\lim_{n\to\infty} \|A_n x - Ax\|_X = 0 \quad \textit{for any} \quad x \in \mathscr{D}(A),$$

and

$$\lim_{n\to\infty} \sup_{t\in[0,T]} \|e^{tA_n}x - e^{tA}x\|_X = 0 \quad \textit{for any} \quad x \in X, \ \ T \geq 0.$$

Proposition 1.1.20 *For the generator A of a C_0-semigroup e^{tA}, $t \geq 0$, on a Banach space X,*

(i) if $x \in \mathscr{D}(A)$, then $e^{tA}x \in \mathscr{D}(A)$ and

$$\frac{d}{dt}e^{tA}x = e^{tA}Ax = Ae^{tA}x \quad \textit{for all} \quad t \geq 0;$$

(ii) for every $t \geq 0$ and $x \in X$,

$$\int_0^t e^{sA}x ds \in \mathscr{D}(A) \qquad \textit{and} \qquad A\int_0^t e^{sA}x ds = e^{tA}x - x.$$

Let X be a Banach space and consider the following deterministic linear Cauchy problem on X,

$$\begin{cases} \dfrac{dy(t)}{dt} = Ay(t), & t \geq 0, \\ y(0) = y_0 \in X, \end{cases} \tag{1.1.10}$$

where A is a linear operator that generates a C_0-semigroup e^{tA}, $t \geq 0$, on X. If $y_0 \in \mathscr{D}(A)$, we have by Proposition 1.1.20 that $e^{tA}y_0 \in \mathscr{D}(A)$ and

$$\frac{d}{dt}(e^{tA}y_0) = Ae^{tA}y_0, \qquad t \geq 0. \tag{1.1.11}$$

Hence, $y(t) = e^{tA}y_0$, $t \geq 0$, is a solution of the differential equation (1.1.10). If $y_0 \notin \mathscr{D}(A)$, the equality (1.1.11) may not be meaningful. However, for any $y_0 \in X$ it does make sense to define $y(t) = e^{tA}y_0$, $t \geq 0$, which is called a *mild solution* of (1.1.10). Quite a few partial differential equations can be formulated in the form (1.1.10).

Example 1.1.21 Let $\{\lambda_i\}$ be a sequence of complex numbers and $\{e_i\}$, $i \in \mathbb{N}_+$, be an orthonormal basis in a separable Hilbert space H. We define on H an operator A by

$$Ax = \sum_{i=1}^{\infty} \lambda_i \langle x, e_i\rangle_H e_i, \qquad x \in \mathscr{D}(A),$$

with its domain

$$\mathscr{D}(A) = \left\{ x \in H \colon \sum_{i=1}^{\infty} |\lambda_i \langle x, e_i\rangle_H|^2 < \infty \right\}.$$

It can be shown that A is a closed, densely defined linear operator and $\lambda I - A$ is invertible if and only if $\inf_{i\geq 1} |\lambda_i - \lambda| > 0$. Moreover, it is true by virtue of the Hille–Yosida Theorem that A generates a C_0-semigroup e^{tA}, $t \geq 0$, if $\sup_{i\geq 1}\{Re\,\lambda_i\} < \infty$, and in this case, we have

$$e^{tA}x = \sum_{i=1}^{\infty} e^{\lambda_i t}\langle x, e_i\rangle_H e_i, \qquad x \in H, \qquad t \geq 0.$$

Moreover, if $\lambda_i \in \mathbb{R}$ for each $i \geq 1$, then A is a self-adjoint operator on H.

As a special case, we could take A to be the classical Laplace operator $\Delta = \partial^2/\partial x_1^2 + \cdots + \partial^2/\partial x_N^2$ on some open bounded set $\mathcal{O} \subset \mathbb{R}^N$ with zero boundary conditions on a smooth boundary $\partial\mathcal{O}$ such that

$$\mathscr{D}(A) = H^2(\mathcal{O}) \cap H_0^1(\mathcal{O}).$$

In particular, if $N = 1$ and $\mathcal{O} = (0,1)$, we may have $e_i(x) = \sqrt{2}\sin(i\pi x)$, $x \in (0,1)$, $\lambda_i = -i^2\pi^2$, $i \geq 1$. As soon as locating the eigenfunctions and eigenvalues of Δ, one can give in terms of semigroup $e^{t\Delta}$, $t \geq 0$, the solution of the partial differential equation

$$\begin{cases} \dfrac{\partial y(t,x)}{\partial t} = \Delta y(t,x) \quad \text{in } \mathcal{O}, \ \ t \geq 0, \\ y(t,x)|_{\partial\mathcal{O}} = 0, \quad t \geq 0; \ \ y(0,x) = y_0(x) \in L^2(\mathcal{O}). \end{cases}$$

Example 1.1.22 Let A be a self-adjoint, nonnegative operator on a Hilbert space H such that the coercive condition holds:

$$\langle Ax, x\rangle_H \geq \beta\|x\|_H^2, \qquad \forall x \in \mathscr{D}(A), \quad \beta > 0.$$

Then, by virtue of Theorem 1.1.10, A has a bounded inverse A^{-1} that is self-adjoint and nonnegative. Moreover, we know by virtue of Theorem 1.1.9 that both the square root operators $A^{1/2}$ and $A^{-1/2}$ are well defined. Let $B \in \mathscr{L}(\mathscr{D}(A^{1/2}), H)$ be a self-adjoint operator on H with $\mathscr{D}(A^{1/2}) \subset \mathscr{D}(B)$. Assume that there exists a number $\alpha \in \mathbb{R}$ such that

$$\langle x, Bx\rangle_H \leq \alpha\|x\|_H^2, \qquad x \in \mathscr{D}(B).$$

We are interested in the following abstract wave equation on H,

$$\begin{cases} \dfrac{d^2u(t)}{dt^2} + Au(t) = B\dfrac{du(t)}{dt}, \qquad t \geq 0, \\ u(0) = u_0 \in H, \quad \dfrac{du}{dt}(0) = u_1 \in H. \end{cases} \tag{1.1.12}$$

To formulate (1.1.12) as a first-order equation in the form (1.1.10), we introduce a space $\mathcal{H} = \mathscr{D}(A^{1/2}) \times H$, equipped with a mapping $\langle\cdot,\cdot\rangle_{\mathcal{H}}: \mathcal{H} \times$

$\mathcal{H} \to \mathbb{C}$,

$$\langle y, \tilde{y} \rangle_{\mathcal{H}} := \langle A^{1/2} y_1, A^{1/2} \tilde{y}_1 \rangle_H + \langle y_2, \tilde{y}_2 \rangle_H,$$

where

$$y = \begin{pmatrix} y_1 \\ y_2 \end{pmatrix}, \qquad \tilde{y} = \begin{pmatrix} \tilde{y}_1 \\ \tilde{y}_2 \end{pmatrix} \in \mathcal{H}.$$

It turns out that $\mathcal{H}$ is a Hilbert space under the inner product $\langle \cdot, \cdot \rangle_{\mathcal{H}}$. Define two linear operators on $\mathcal{H}$

$$\mathcal{A} = \begin{pmatrix} 0 & I \\ -A & B \end{pmatrix} \quad \text{with domain} \quad \mathscr{D}(\mathcal{A}) = \mathscr{D}(A) \times \mathscr{D}(A^{1/2}),$$

and

$$\mathcal{B} = \begin{pmatrix} A^{-1}B & -A^{-1} \\ I & 0 \end{pmatrix},$$

then $\mathcal{B}$ is a bounded linear operator on $\mathcal{H}$ with the range $\mathscr{R}(\mathcal{B}) = \mathscr{D}(\mathcal{A})$ and $\mathcal{B}$ is the inverse of operator $\mathcal{A}$. This implies by Theorem 1.1.7 that $\mathcal{A}$ is a closed operator. Hence, (1.1.12) may be rewritten as a first-order differential equation on $\mathcal{H}$,

$$\begin{cases} \dfrac{dy(t)}{dt} = \mathcal{A}y(t), & t \geq 0, \\ y(0) = y_0 \in \mathcal{H}, \end{cases} \tag{1.1.13}$$

where

$$y(t) = \begin{pmatrix} u(t) \\ du(t)/dt \end{pmatrix}, \qquad y_0 = \begin{pmatrix} u_0 \\ u_1 \end{pmatrix}.$$

On the other hand, it is straightforward to show that

$$Re\,\langle \mathcal{A}y, y \rangle_{\mathcal{H}} = Re\,\langle Ay_1, y_2 \rangle_H + Re\,\langle -Ay_1 + By_2, y_2 \rangle_H \leq \alpha \|y_2\|_H^2 \leq |\alpha| \|y\|_{\mathcal{H}}^2$$

for any $y \in \mathscr{D}(A) \times \mathscr{D}(A^{1/2})$. Similarly, the adjoint of $\mathcal{A}$ is easily shown to be

$$\mathcal{A}^* \begin{pmatrix} y_1 \\ y_2 \end{pmatrix} = \begin{pmatrix} 0 & -I \\ A & B \end{pmatrix} \begin{pmatrix} y_1 \\ y_2 \end{pmatrix}, \qquad \mathscr{D}(\mathcal{A}^*) = \mathscr{D}(\mathcal{A}),$$

which immediately yields $Re\,\langle \mathcal{A}^* y, y \rangle_{\mathcal{H}} \leq |\alpha| \|y\|_{\mathcal{A}}^2$ for any $y \in \mathscr{D}(\mathcal{A}^*)$. Therefore, by virtue of Proposition 1.1.18, it follows that $\mathcal{A}$ generates a C_0-semigroup $e^{t\mathcal{A}}$, $t \geq 0$, on $\mathcal{H}$.

As a typical example, we define for $t \geq 0$, $x \in (0, 1)$,

$$Au(t,x) = -u''_{xx}(t,x), \qquad \mathscr{D}(A) = H^2(0,1) \cap H_0^1(0,1),$$
$$\text{and} \qquad Bu(t,x) = \alpha u'_x(t,x).$$

It is easy to verify that all the preceding conditions are satisfied, and the partial differential equation that gives the abstract version (1.1.13) is

$$\begin{cases} \dfrac{\partial^2 u(t,x)}{\partial t^2} - \dfrac{\partial^2 u(t,x)}{\partial x^2} = \alpha \dfrac{\partial^2 u(t,x)}{\partial x \partial t}, \quad t \geq 0, \quad \alpha \in \mathbb{R}, \\ u(t,0) = u(t,1) = 0, \quad u(0,x) = u_0(x) \in H^2(0,1) \cap H_0^1(0,1), \quad t \geq 0, \\ \dfrac{\partial u}{\partial t}(0,x) = u_1(x) \in H_0^1(0,1). \end{cases}$$

Example 1.1.23 Consider a retarded differential equation in $\mathbb{C}^n$ of the form

$$\begin{cases} dy(t) = A_0 y(t)dt + A_1 y(t-r)dt, \quad t \geq 0, \\ y(0) = \phi_0 \in \mathbb{C}^n, \quad y(t) = \phi_1(t) \in L^2([-r,0];\mathbb{C}^n), \quad -r \leq t \leq 0, \end{cases} \tag{1.1.14}$$

where $A_0, A_1 \in \mathscr{L}(\mathbb{C}^n)$.

We wish to formulate (1.1.14) into an abstract linear differential equation on a proper Hilbert space. To this end, we introduce a product Hilbert space $\mathcal{H} = \mathbb{C}^n \times L^2([-r,0];\mathbb{C}^n)$, equipped with the usual inner product, and meanwhile a linear operator $\mathcal{A}$ on $\mathcal{H}$ by

$$\mathcal{A}\Phi = \left(A_0\phi_0 + A_1\phi_1(-r), \frac{d\phi_1}{d\theta}(\theta) \right) \quad \text{for} \quad \phi = (\phi_0,\phi_1) \in \mathscr{D}(\mathcal{A}),$$

with its domain

$$\mathscr{D}(\mathcal{A}) = \left\{ \phi = (\phi_0,\phi_1) \in \mathcal{H} \colon \phi_1 \in W^{1,2}([-r,0];\mathbb{C}^n),\ \phi_1(0) = \phi_0 \right\}.$$

It may be shown (see Appendix B) that $\mathcal{A}$ generates a C_0-semigroup $e^{t\mathcal{A}}$, $t \geq 0$, on $\mathcal{H}$ and the equation (1.1.14) becomes a Cauchy problem without delay on $\mathcal{H}$,

$$\begin{cases} dY(t) = \mathcal{A}Y(t)dt, \quad t \geq 0, \\ Y(0) = (\phi_0,\phi_1) \in \mathcal{H}, \end{cases} \tag{1.1.15}$$

where $Y(t) := (y(t), y_t)$, $t \geq 0$, and $y_t(\theta) := y(t+\theta)$, $\theta \in [-r,0]$, is the so-called *lift-up* system of (1.1.14).

Last, we review some specific types of C_0-semigroups with delicate properties.

Definition 1.1.24 Let e^{tA}, $t \geq 0$, be a C_0-semigroup on a Banach space X with its generator $A \colon \mathscr{D}(A) \subset X \to X$.

(i) The semigroup e^{tA}, $t \geq 0$, is called *eventually compact* if there exists $r > 0$ such that $e^{tA} \in \mathscr{L}(X)$ is compact for any $t \in (r,\infty)$. Particularly, if e^{tA} is compact for all $t \in (0, \infty)$, this semigroup is simply called *compact*.

(ii) The semigroup e^{tA}, $t \geq 0$, is called *analytic* if it admits an extension e^{zA} on $z \in \Delta_\theta := \{z \in \mathbb{C}\colon |\arg z| < \theta\}$ for some $\theta \in (0, \pi]$, such that $z \to e^{zA}$ is analytic on Δ_θ and satisfies the following:

(a) $e^{(z_1+z_2)A} = e^{z_1 A} e^{z_2 A}$ for any $z_1, z_2 \in \Delta_\theta$;
(b) $\lim_{\Delta_{\bar{\theta}} \ni z \to 0} \|e^{zA}x - x\|_X = 0$ for all $x \in X$ and $0 < \bar{\theta} < \theta$.

Theorem 1.1.25 *Assume that A generates a C_0-semigroup e^{tA}, $t \geq 0$, on a Banach space X. For some $r \geq 0$, the semigroup e^{tA} is compact at any $t \in (r, \infty)$ if and only if e^{tA} is norm continuous on (r, ∞) and the operator $R(\lambda, A)e^{rA}$ is compact for some (thus, all) $\lambda \in \rho(A)$.*

Theorem 1.1.26 *Assume that A generates a C_0-semigroup e^{tA}, $t \geq 0$, which is eventually compact on a Banach space X, then the spectrum of A consists of isolated eigenvalues, i.e., for any numbers m and M, there are only a finite number of eigenvalues of A in the strip*

$$\{\lambda \in \mathbb{C}\colon m \leq Re\,\lambda \leq M\}.$$

For analytic semigroups, we have the following characterization.

Theorem 1.1.27 *Let e^{tA}, $t \geq 0$, be a C_0-semigroup on a Banach space X with generator A. The following statements are equivalent.*

(i) The semigroup e^{tA}, $t \geq 0$, is analytic.
(ii) There exist constants $M > 0$ and $L \geq 0$ such that

$$\|AR(\lambda, A)^{n+1}\| \leq M/n\lambda^n \quad \textit{for all} \quad \lambda > nL, \; n = 1, 2, \ldots$$

(iii) The semigroup e^{tA} is differentiable for $t > 0$, i.e., for every $x \in X$, the mapping $t \to e^{tA}x$ is differentiable for $t > 0$, and there exist constants $M > 0$ and $\mu > 0$ such that

$$\|Ae^{tA}\| \leq \frac{M}{t} e^{\mu t} \quad \textit{for} \quad t > 0.$$

In general, it is hard to check (ii) for every $n \in \mathbb{N}_+$. The following theorem is much more easily verified and thus quite useful in application.

Theorem 1.1.28 *Let e^{tA}, $t \geq 0$, be a C_0-semigroup with generator A on X. The semigroup e^{tA}, $t \geq 0$, is analytic if and only if there exist $M > 0$ and $\mu \in \mathbb{R}$ such that*

$$\rho(A) \supset \{\lambda\colon Re\,\lambda \geq \mu\} \quad \textit{and} \quad \|R(\lambda, A)\| \leq \frac{M}{1 + |\lambda|} \quad \textit{for all} \quad Re\,\lambda \geq \mu.$$

Assume that A generates an exponentially stable analytic semigroup e^{tA}, $t \geq 0$, on X. Then $i\mathbb{R} \subset \rho(A)$ and for any $\alpha \in (0, 1)$, the integral

$$(-A)^{-\alpha} := \frac{\sin \alpha\pi}{\pi} \int_0^\infty t^{-\alpha} R(t, A) dt$$

is well defined, which is a bounded linear operator $(-A)^{-\alpha} \in \mathscr{L}(X)$. It may be shown that this operator $(-A)^{-\alpha}$ is injective, a fact that leads to the following definition:

$$(-A)^\alpha := \begin{cases} [(-A)^{-\alpha}]^{-1} & \text{if} \quad 0 < \alpha < 1, \\ I & \text{if} \quad \alpha = 0. \end{cases}$$

The operator $(-A)^\alpha$ with domain $\mathscr{D}((-A)^\alpha)$, $\alpha \in [0, 1)$, is called a *fractional power* of $-A$. Further, we have a relation

$$\mathscr{D}((-A)^\beta) \subset \mathscr{D}((-A)^\alpha) \subset X, \qquad 0 \leq \alpha \leq \beta < 1,$$

and there exists a number $C_\alpha > 0$ such that

$$\|(-A)^\alpha e^{tA}\| \leq C_\alpha t^{-\alpha} \quad \text{for each } t > 0.$$

In finite-dimensional spaces, it is well known that the spectrum relation $\sigma(e^{tA})\backslash\{0\} = e^{t\sigma(A)}$ holds for each $t \geq 0$ between C_0-semigroup e^{tA}, $t \geq 0$, and its generator A. A partial result remains valid in infinite dimensions, which is the content of the following *spectral mapping theorem*.

Theorem 1.1.29 *Let e^{tA}, $t \geq 0$, be a C_0-semigroup with generator A on a Banach space X. Then*

$$e^{t\sigma(A)} \subset \sigma(e^{tA})\backslash\{0\} \quad \textit{for all} \quad t \geq 0. \tag{1.1.16}$$

In general, the strict inclusion in (1.1.16) may hold, although this is not the case for norm continuous semigroups in which compact, differentiable, and analytic semigroups are typical examples.

Theorem 1.1.30 *Let e^{tA}, $t \geq 0$, be a C_0-semigroup that is (eventually) norm continuous on the Banach space X. Then*

$$\sigma(e^{tA})\backslash\{0\} = e^{t\sigma(A)} \quad \textit{for each} \quad t \geq 0.$$

For an arbitrary C_0-semigroup e^{tA}, $t \geq 0$, it is not generally true that the adjoint of e^{tA} is a C_0-semigroup since the mapping $e^{tA} \to (e^{tA})'$ does not necessarily preserve the strong continuity of e^{tA}. But this could be true if the underlying space X is a Hilbert space.

Proposition 1.1.31 *Suppose that X is a Hilbert space and e^{tA}, $t \geq 0$, is a C_0-semigroup on X. Then $(e^{tA})^*$, $t \geq 0$, is a C_0-semigroup on X with its infinitesimal generator A^*, i.e., $(e^{tA})^* = e^{tA^*}$ for $t \geq 0$.*

1.2 Stochastic Processes and Martingales

A *measurable space* is a pair $(\Omega, \mathscr{F})$ where Ω is a set and $\mathscr{F}$ is a σ-field, also called a σ-algebra, of subsets of Ω. This means that the family $\mathscr{F}$ contains Ω and is closed under the operation of taking complements and countable unions of its elements. If $(\Omega, \mathscr{F})$ and $(S, \mathscr{S})$ are two measurable spaces, then a mapping ξ from Ω into S such that the set $\{\omega \in \Omega\colon \xi(\omega) \in A\} = \{\xi \in A\}$ belongs to $\mathscr{F}$ for arbitrary $A \in \mathscr{S}$ is called *measurable* from $(\Omega, \mathscr{F})$ into $(S, \mathscr{S})$. In this book, we shall only be concerned with the case where S is a complete metric space. Thus, we always set $\mathscr{S} = \mathscr{B}(S)$, the Borel σ-field of S, which is the smallest σ-field containing all closed (or open) subsets of S.

A *probability measure* $\mathbb{P}$ on a measurable space $(\Omega, \mathscr{F})$ is a σ-additive function from $\mathscr{F}$ into $[0,1]$ such that $\mathbb{P}(\Omega) = 1$. The triplet $(\Omega, \mathscr{F}, \mathbb{P})$ is called a *probability space*. If $(\Omega, \mathscr{F}, \mathbb{P})$ is a probability space, we set

$$\overline{\mathscr{F}} = \{A \subset \Omega\colon \exists B, C \in \mathscr{F};\ B \subset A \subset C,\ \mathbb{P}(B) = \mathbb{P}(C)\}.$$

Then it may be shown that $\overline{\mathscr{F}}$ is a σ-field, called the *completion* of $\mathscr{F}$. If $\mathscr{F} = \overline{\mathscr{F}}$, the probability space $(\Omega, \mathscr{F}, \mathbb{P})$ is said to be *complete*. Unless otherwise stated, completeness of $(\Omega, \mathscr{F}, \mathbb{P})$ will always be assumed in this book.

Let $(\Omega, \mathscr{F}, \mathbb{P})$ denote a complete probability space. A family $\{\mathscr{F}_t\}$, $t \geq 0$, for which each $\mathscr{F}_t$ is a sub-σ-field of $\mathscr{F}$ and forms an increasing family of σ-fields, is called a *filtration* of $\mathscr{F}$. With this $\{\mathscr{F}_t\}_{t\geq 0}$, one can associate another filtration by setting σ-fields $\mathscr{F}_{t+} = \bigcap_{s>t} \mathscr{F}_s$ for $t \geq 0$. We say that the filtration $\{\mathscr{F}_t\}_{t\geq 0}$ is *normal* or satisfies *the usual conditions* if $\mathscr{F}_{t+} = \mathscr{F}_t$ for each $t \geq 0$ and $\mathscr{F}_0$ contains all $\mathbb{P}$-null sets in $\mathscr{F}$.

If ξ is a measurable mapping from $(\Omega, \mathscr{F})$ into $(S, \mathscr{B}(S))$ or an S-valued *random variable* and $\mathbb{P}$ a probability measure on $(\Omega, \mathscr{F})$, then we will denote by $\mathbb{D}_\xi(\cdot)$ the image of $\mathbb{P}$ under the mapping ξ:

$$\mathbb{D}_\xi(A) = \mathbb{P}\{\omega \in \Omega\colon \xi(\omega) \in A\}, \quad \forall A \in \mathscr{B}(S).$$

This is a probability measure on $(S, \mathscr{B}(S))$, which is called the *distribution* or *law* of ξ.

Definition 1.2.1 Let $\{\mathscr{F}_t\}_{t\geq 0}$ be a filtration of $\mathscr{F}$. A mapping $\tau : \Omega \to [0, \infty]$ is called the *stopping time* with respect to $\{\mathscr{F}_t\}$, $t \geq 0$, if $\{\omega : \tau(\omega) \leq t\} \in \mathscr{F}_t$ for each $t \geq 0$. The σ-field of events prior to τ, denoted by $\mathscr{F}_\tau$, is defined as

$$\mathscr{F}_\tau = \big\{A \in \mathscr{F} : A \cap \{\tau \leq t\} \in \mathscr{F}_t \ \text{ for every } \ t \geq 0\big\}.$$

Now assume that $S = H$, a separable Hilbert space with norm $\|\cdot\|_H$, and ξ is an H-valued random variable on $(\Omega, \mathscr{F}, \mathbb{P})$. By a standard limit argument, we can define the integral $\int_\Omega \xi(\omega)\mathbb{P}(d\omega)$ of ξ with respect to probability measure $\mathbb{P}$, often denote it by $\mathbb{E}(\xi)$. The integral defined in this way is a Bochner type of integral, which is frequently called the *expectation* or *mean* of ξ in this book. We denote by $L^p(\Omega, \mathscr{F}, \mathbb{P}; H)$, $p \in [1,\infty)$, the set of all equivalence classes of H-valued random variables with respect to the equivalence relation of almost sure equality. Then one can verify that $L^p(\Omega, \mathscr{F}, \mathbb{P}; H)$, $p \in [1,\infty)$, equipped with the norm

$$\|\xi\|_p = (\mathbb{E}\|\xi\|_H^p)^{1/p}, \qquad p \in [1, \infty), \qquad \xi \in L^p(\Omega, \mathscr{F}, \mathbb{P}; H),$$

is a Banach space. If Ω is an interval $[0, T]$, $\mathscr{F} = \mathscr{B}([0, T])$, $0 \leq T < \infty$, and $\mathbb{P}$ is the standard Lebesgue measure on $[0, T]$, we also write $L^p([0, T]; H)$, or more simply $L^p(0, T)$ when no confusion is possible.

Let K, H be two separable Hilbert spaces. A mapping $\Phi(\cdot)$ from Ω into $\mathscr{L}(K, H)$ is said to be *measurable* if for arbitrary $k \in K$, $\Phi(\cdot)k$ is measurable as a mapping from $(\Omega, \mathscr{F})$ into $(H, \mathscr{B}(H))$. Let $\mathscr{F}(\mathscr{L}(K, H))$ be the smallest σ-field of subsets of $\mathscr{L}(K, H)$ containing all sets of the form

$$\{\Phi \in \mathscr{L}(K, H) : \Phi k \in A\}, \quad k \in K, \quad A \in \mathscr{B}(H),$$

then $\Phi : \Omega \to \mathscr{L}(K, H)$ is a measurable mapping from $(\Omega, \mathscr{F})$ into the measurable space $(\mathscr{L}(K, H), \mathscr{F}(\mathscr{L}(K, H)))$. The mapping Φ is said to be *Bochner integrable* with respect to measure $\mathbb{P}$ if for arbitrary $k \in K$, the mapping $\Phi(\cdot)k$ is Bochner integrable and there exists a bounded linear operator $\Psi \in \mathscr{L}(K, H)$ such that

$$\int_\Omega \Phi(\omega)k\mathbb{P}(d\omega) = \Psi k, \quad \forall k \in K.$$

The operator Ψ is then denoted by $\Psi = \displaystyle\int_\Omega \Phi(\omega)\mathbb{P}(d\omega)$ and called the *Bochner integral* of Φ.

An arbitrary family $M = \{M(t)\}$, $t \geq 0$, of H-valued random variables defined on a probability space $(\Omega, \mathscr{F}, \mathbb{P})$ is called a *stochastic* or *random process*. Sometimes, we also write $M(t, \omega)$ or M_t in place of $M(t)$ for all $t \geq 0$. In the study of stochastic processes, we usually need additional regularities of M to proceed with our program. Specially, a process M is called *measurable* if the mapping $M(\cdot,\cdot)\colon \mathbb{R}_+ \times \Omega \to H$ is $\mathscr{B}(\mathbb{R}_+) \times \mathscr{F}$-measurable. Let $\{\mathscr{F}_t\}$, $t \ge 0$, be an increasing family of sub-σ-fields of $\mathscr{F}$. The process M is called $\{\mathscr{F}_t\}_{t\ge 0}$-*adapted* if each $M(t)$ is measurable with respect to $\mathscr{F}_t$, $t \ge 0$. Clearly, M is always $\{\mathscr{F}_t^M\}_{t\ge 0}$-adapted, where $\mathscr{F}_t^M := \sigma(M(s); 0 \le s \le t)$ is the family of the σ-fields generated by $M = \{M(t)\}_{t\ge 0}$. For any $\omega \in \Omega$, the function $M(\cdot,\omega)$ is called a *path* or *trajectory* of M. A stochastic process $N = \{N(t)\}$ is called a *modification* or *version* of $M = \{M(t)\}$ if

$$\mathbb{P}\{\omega \in \Omega\colon M(t,\omega) \ne N(t,\omega)\} = 0, \qquad \forall t \ge 0.$$

Given an H-valued process $M = \{M(t)\}$, $t \ge 0$, and a stopping time $\tau\colon \Omega \to \mathbb{R}_+$, it is desirable for many applications that the mapping $M_\tau\colon \Omega \to H$ defined by $M_{\tau(\omega)}(\omega) = M(\tau(\omega),\omega)$ is also measurable. This is generally not the case if M is only a measurable process. However, this could be true if we confine ourselves to a smaller class of stochastic processes, i.e., progressively measurable processes, defined as follows.

Definition 1.2.2 Suppose that $M = \{M(t)\}$, $t \ge 0$, is an H-valued process and $\{\mathscr{F}_t\}_{t\ge 0}$ is a filtration of $\mathscr{F}$. The process M is said to be *progressively measurable* with respect to $\{\mathscr{F}_t\}_{t\ge 0}$ if for every $t \ge 0$, the mapping

$$[0,t] \times \Omega \to H, \qquad (s,\omega) \to M(s,\omega),$$

is $\mathscr{B}([0,t]) \times \mathscr{F}_t$-measurable.

It is obvious that if M is progressively measurable with respect to $\{\mathscr{F}_t\}_{t\ge 0}$, then it must be both measurable and $\{\mathscr{F}_t\}_{t\ge 0}$-adapted. The following theorem provides the extent to which the converse is true.

Proposition 1.2.3 *Suppose that stochastic process $M = \{M(t)\}$, $t \ge 0$, is measurable and adapted to the filtration $\{\mathscr{F}_t\}_{t\ge 0}$. Then it has a progressively measurable modification.*

Theorem 1.2.4 *Let $M = \{M(t)\}$, $t \ge 0$, be an H-valued progressively measurable process with respect to $\{\mathscr{F}_t\}_{t\ge 0}$, and let τ be a finite stopping time. Then the random variable M_τ is $\mathscr{F}_\tau$-measurable.*

Let $\mathscr{G}$ be an arbitrary sub-σ-field of $\mathscr{F}$. We use $\mathbb{E}(\cdot \mid \mathscr{G})$ to denote the conditional expectation given $\mathscr{G}$. Let M be a stochastic process with state space H. Then it can be shown that there exists a function $\mathbb{P}(s, x, t, \Gamma)$ $(s < t,\ x \in H,\ \Gamma \in \mathscr{B}(H))$ associated with the process M such that

(a) for all (s, x, t), $\mathbb{P}(s, x, t, \cdot)$ is a probability measure on $\mathscr{B}(H)$;
(b) for each (s, t, Γ), $\mathbb{P}(s, \cdot, t, \Gamma)$ is $\mathscr{B}(H)$-measurable;
(c) $\mathbb{E}(\mathbf{1}_{\{M(t)\in\Gamma\}} \mid \mathscr{F}_s^M) = \mathbb{P}(s, x, t, \Gamma)|_{x=M(s)}$ almost surely.

An H-valued process $M = M(t)$, $t \ge 0$, defined on $(\Omega, \mathscr{F}, \mathbb{P})$ and adapted to the family $\{\mathscr{F}_t\}_{t\ge 0}$ is said to be a *Markov process* with respect to $\{\mathscr{F}_t\}_{t\ge 0}$ if the following property is satisfied: for all t, $s \ge 0$,

$$\mathbb{E}(f(M(t+s)) \mid \mathscr{F}_t) = \mathbb{E}(f(M(t+s)) \mid \sigma(M(t))) \qquad a.s. \tag{1.2.1}$$

for every bounded real-valued Borel function $f(\cdot)$ on H. In particular, if a relation of the form (1.2.1) continues to hold when the time t is replaced by a stopping time τ, we say that M has *strong Markov property* or M is a *strong Markov process*. If M is a Markov process with respect to $\mathscr{F}_t^M$, $t \ge 0$, we simply say that M is a Markov process. A function $\mathbb{P}(s, x, t, \Gamma)$ satisfying (a), (b), and (c) is called the *transition probability function* of the Markov process M if it further satisfies the following Chapman–Kolmogorov equation

$$\mathbb{P}(s, x, t, \Gamma) = \int_H \mathbb{P}(s, x, u, dy)\mathbb{P}(u, y, t, \Gamma) \tag{1.2.2}$$

for all $x \in H$, $\Gamma \in \mathscr{B}(H)$ and (s, u, t) such that $s \le u \le t$. The process $M(t)$, $t \ge 0$, is said to have *homogeneous* transition probability function if

$$\mathbb{P}(s,x,t,\Gamma) = \mathbb{P}(0,x,t-s,\Gamma) \quad \text{for all} \quad x \in H,\ \Gamma \in \mathscr{B}(H),\ s \le t.$$

In this case, we write $\mathbb{P}(x,t,\Gamma)$ for $\mathbb{P}(0,x,t,\Gamma)$ and the Chapman–Kolmogorov equation (1.2.2) now reduces to

$$\mathbb{P}(x, s+t, \Gamma) = \int_H \mathbb{P}(x, s, dy)\mathbb{P}(y, t, \Gamma) \quad \text{for every} \quad s, t \ge 0. \tag{1.2.3}$$

On the class of all bounded Borel-measurable functions $B_b(H)$ on H, we can define for any $t \ge 0$ that

$$\mathbb{P}_t f(x) = \int_H f(y)\mathbb{P}(t,x,dy), \qquad \forall f \in B_b(H). \tag{1.2.4}$$

Then by virtue of (1.2.3), we can establish the following semigroup property for the family $\mathbb{P}_t$, $t \ge 0$:

$$\mathbb{P}_{t+s} f = \mathbb{P}_t \mathbb{P}_s f \quad \text{for any} \quad s, t \ge 0, \tag{1.2.5}$$

which is, in essence, a restatement of (1.2.3). Let $C_b(H)$ be the class of all real-valued, bounded continuous functions on H.

Definition 1.2.5 Semigroup $\mathbb{P}_t$, $t \ge 0$, is said to have the *Feller property* if for arbitrary $f \in C_b(H)$ and $t \ge 0$, function $\mathbb{P}_t f(\cdot)$ is continuous. Further, $\mathbb{P}_t$ is said to have the *strongly Feller property* if for arbitrary $f \in B_b(H)$ and $t \ge 0$, the function $\mathbb{P}_t f(\cdot)$ is continuous.

Let H be a Hilbert space and $M = \{M(t)\}, t \geq 0$, be an H-valued stochastic process defined on $(\Omega, \mathscr{F}, \{\mathscr{F}_t\}_{t\geq 0}, \mathbb{P})$. If $\mathbb{E}\|M(t)\|_H < \infty$ for all $t \geq 0$, then M is called *integrable*. An integrable and adapted H-valued process $M(t)$, $t \geq 0$, is said to be a *martingale* with respect to $\{\mathscr{F}_t\}_{t\geq 0}$ if

$$\mathbb{E}(M(t) \mid \mathscr{F}_s) = M(s) \qquad \mathbb{P} - a.s. \tag{1.2.6}$$

for arbitrary $t \geq s \geq 0$. By the definition of conditional expectations, relation (1.2.6) is equivalent to the following statement

$$\int_F M(t)\mathbb{P}(d\omega) = \int_F M(s)\mathbb{P}(d\omega), \quad \forall F \in \mathscr{F}_s, \ s \leq t.$$

We also recall that a real-valued integrable and adapted process $M(t), t \geq 0$, is said to be a *submartingale* (resp. *supermartingale*) with respect to $\{\mathscr{F}_t\}_{t\geq 0}$ if

$$\mathbb{E}(M(t) \mid \mathscr{F}_s) \geq M(s), \ (\text{resp. } \mathbb{E}(M(t) \mid \mathscr{F}_s) \leq M(s)), \ \mathbb{P} - a.s.$$

for any $0 \leq s \leq t$. An H-valued stochastic process M is a *continuous martingale* if it is a martingale with almost surely continuous trajectories. An adapted process M is called a *local martingale* if there exists a sequence of stopping times τ_n such that $\tau_n \uparrow \infty$ and for each n, the stopped process $M(t \wedge \tau_n), t \geq 0$, is a martingale.

If $M(t), t \geq 0$, is an H-valued continuous martingale, then $\|M(t)\|_H^2, t \geq 0$, is a real-valued continuous submartingale. By the well-known Doob–Meyer decomposition, there exists a unique real-valued, nondecreasing process, denoted by $[M](t)$, with $[M](0) = 0$ such that $\|M(t)\|_H^2 - [M](t)$ is an $\mathscr{F}_t$-martingale. Recall the following strong law of large numbers for martingales, which is useful in stability analysis.

Proposition 1.2.6 *Let $M(t)$, $t \geq 0$, be an H-valued, continuous local martingale with $M(0) = 0$. If*

$$\overline{\lim_{t\to\infty}} \frac{[M](t)}{t} < \infty \qquad a.s.$$

then

$$\lim_{t\to\infty} \frac{M(t)}{t} = 0 \qquad a.s.$$

Let $[0, T]$, $0 \leq T < \infty$, be a subinterval of $[0, \infty)$. An H-valued stochastic process $M(t)$, $t \in [0, T]$, defined on $(\Omega, \mathscr{F}, \{\mathscr{F}_t\}_{t\in[0,T]}, \mathbb{P})$, is a *continuous L^p-martingale*, $p \geq 1$, with respect to $\{\mathscr{F}_t\}_{t\in[0,T]}$ if it is a martingale with almost surely continuous trajectories and satisfies, in addition, $\mathbb{E}\sup_{t\in[0,T]} \|M(t)\|_H^p < \infty$. Let us denote by $\mathcal{M}_T^p(H)$ the space of all

H-valued continuous L^p-martingales on $[0,T]$. By using Theorem 1.2.9 it is possible to show the following result.

Theorem 1.2.7 *For $p \geq 1$, the space $\mathcal{M}_T^p(H)$, equipped with the norm*

$$\|M\|_{\mathcal{M}_T^p(H)} = \left(\mathbb{E} \sup_{t\in[0,T]} \|M(t)\|_H^p\right)^{1/p}, \qquad \forall\, M \in \mathcal{M}_T^p(H),$$

is a Banach space.

An $\mathscr{L}_1(H)$-valued process V is said to be *nondecreasing* if operator $V(t)$, $t \in [0,T]$, is nonnegative, so denote it by $V(t) \geq 0$, i.e., for any $x \in H$ and $t \in [0,T]$, $\langle V(t)x, x\rangle_H \geq 0$ and $V(t) - V(s) \geq 0$ if $0 \leq s \leq t \leq T$. For any $M \in \mathcal{M}_T^2(H)$, an $\mathscr{L}_1(H)$-valued continuous, adapted, and nondecreasing process $V(t)$ with $V(0) = 0$ is called a *quadratic variation process* of M if for arbitrary $a, b \in H$, the process

$$\langle M(t), a\rangle_H \langle M(t), b\rangle_H - \langle V(t)a, b\rangle_H, \quad t \in [0,T],$$

is a continuous $\mathscr{F}_t$-martingale, $t \in [0,T]$. One can show that such a process $V(t)$, $t \in [0,T]$, is uniquely determined, thus denote it by $\langle\langle M\rangle\rangle(t)$, $t \in [0,T]$.

Theorem 1.2.8 *For arbitrary $M \in \mathcal{M}_T^2(H)$, there exists a unique nonnegative symmetric process $Q_M(t) \in \mathscr{L}_1(H)$, $t \in [0,T]$, such that*

$$\langle\langle M\rangle\rangle(t) = \int_0^t Q_M(s)d[M](s) \quad \text{for all} \quad t \in [0,T].$$

This process $M(t)$, $t \in [0,T]$, is called a $Q_M(t)$-martingale process.

In a similar manner, one can define the so-called *cross quadratic variation* for any $M \in \mathcal{M}_T^2(H)$, $N \in \mathcal{M}_T^2(H)$ as a unique continuous process $\langle\langle M, N\rangle\rangle$ of operators on H such that for arbitrary $a, b \in H$, the process

$$\langle M(t), a\rangle_H \langle N(t), b\rangle_H - \langle \langle\langle M, N\rangle\rangle(t)a, b\rangle_H, \qquad t \in [0,T],$$

is a continuous $\mathscr{F}_t$-martingale, $t \in [0,T]$.

As an immediate consequence of the classic maximal inequalities for real-valued submartingales, we have the following Doob's type of inequalities in Hilbert spaces.

Theorem 1.2.9 *Let $M(t)$, $t \in [0,T]$, be a continuous H-valued, L^p-martingale, $p \geq 1$. Then the following statements hold.*

(i) For $p \geq 1$ and any $\lambda > 0$,

$$\mathbb{P}\left\{\sup_{0\leq t\leq T} \|M(t)\|_H \geq \lambda\right\} \leq \lambda^{-p} \sup_{0\leq t\leq T} \mathbb{E}(\|M(t)\|_H^p). \qquad (1.2.7)$$

(ii) For $p > 1$,

$$\mathbb{E}\Big(\sup_{0\le t\le T} \|M(t)\|_H^p\Big) \le \Big(\frac{p}{p-1}\Big)^p \sup_{0\le t\le T} \mathbb{E}(\|M(t)\|_H^p). \tag{1.2.8}$$

(iii) For $p = 1$,

$$\mathbb{E}\Big(\sup_{0\le t\le T} \|M(t)\|_H\Big) \le 3\mathbb{E}\{Tr(\langle\langle M\rangle\rangle(t))\}^{1/2}. \tag{1.2.9}$$

1.3 Wiener Processes and Stochastic Integration

Let K be a separable Hilbert space with inner product $\langle\cdot,\cdot\rangle_K$. A probability measure $\mathcal{N}$ on $(K, \mathscr{B}(K))$ is called *Gaussian* if for arbitrary $u \in K$, there exist numbers $\mu \in \mathbb{R}$, $\sigma > 0$, such that

$$\mathcal{N}\{x \in K : \langle u, x\rangle_K \in A\} = N(\mu,\sigma)(A), \qquad A \in \mathscr{B}(\mathbb{R}),$$

where $N(\mu,\sigma)$ is the standard one-dimensional normal distribution with mean μ and variance σ. It is true that if $\mathcal{N}$ is Gaussian, there exist an element $m \in K$ and a nonnegative self-adjoint operator $Q \in \mathscr{L}_1(K)$ such that the characteristic function of $\mathcal{N}$ is given by

$$\int_K e^{i\langle\lambda,\, x\rangle_K}\mathcal{N}(dx) = e^{i\langle\lambda,\, m\rangle_K - \frac{1}{2}\langle Q\lambda,\, \lambda\rangle_K}, \qquad \lambda \in K.$$

Therefore, the measure $\mathcal{N}$ is uniquely determined by m and Q and denoted thus by $\mathcal{N}(m, Q)$. In particular, we call m the *mean* and Q the *covariance operator* of $\mathcal{N}(m, Q)$, respectively.

For a self-adjoint and nonnegative operator $Q \in \mathscr{L}(K)$, we assume, without loss of generality, that there exists an orthonormal basis $\{e_k\}_{k\ge 1}$ in K, and a bounded sequence of positive numbers λ_k such that

$$Qe_k = \lambda_k e_k, \qquad k = 1, 2, \ldots$$

A stochastic process W_t or $W(t)$, $t \ge 0$, is called a *Q-Wiener process* in K if

(i) $W(0) = 0$;
(ii) $W(t)$ has continuous trajectories;
(iii) $W(t)$ has independent increments;
(iv) $\mathbb{D}_{W(t)-W(s)} = \mathcal{N}(0, (t-s)Q)$ for all $t \ge s \ge 0$.

If $Tr(Q) = \sum_{k=1}^{\infty}\lambda_k < \infty$, then W is a genuine Wiener process that has continuous paths in K. It is possible that $Tr(Q) = \infty$, e.g., $Q = I$, and in this case, we call W a *cylindrical* Wiener process in K, which, in general, has continuous paths only in another Hilbert space larger than K. It is immediate that the quadratic variation of a Q-Wiener process with $Tr(Q) < \infty$ is given by $\langle\langle W\rangle\rangle(t) = tQ$, $t \geq 0$.

Assume that probability space $(\Omega, \mathscr{F}, \mathbb{P})$ is equipped with a normal filtration $\{\mathscr{F}_t\}_{t\geq 0}$. Let $W(t)$, $t \geq 0$, be a Q-Wiener process in K that is assumed to be adapted to $\{\mathscr{F}_t\}_{t\geq 0}$, and for every $t > s \geq 0$ the increments $W(t) - W(s)$ are independent of $\mathscr{F}_s$. Then $W(t)$, $t \geq 0$, is a continuous martingale relative to $\{\mathscr{F}_t\}_{t\geq 0}$, and W has the following representation:

$$W(t) = \sum_{i=1}^{\infty} \sqrt{\lambda_i} w_i(t) e_i, \qquad t \geq 0, \tag{1.3.1}$$

where $(\lambda_i > 0, i \in \mathbb{N}_+)$ are the eigenvalues of Q with their corresponding eigenvectors $(e_i, i \in \mathbb{N}_+)$, and $(w_i(t), i \in \mathbb{N}_+)$ is a group of independent standard real-valued Brownian motions. We introduce a subspace $K_Q = \mathscr{R}(Q^{1/2})$ of K, which is a Hilbert space endowed with the inner product

$$\langle u, v\rangle_{K_Q} = \langle Q^{-1/2}u, Q^{-1/2}v\rangle_K \quad \text{for any} \quad u,\ v \in K_Q.$$

Let $\mathscr{L}_2(K_Q, H)$ denote the space of all Hilbert–Schmidt operators from K_Q into H. Then $\mathscr{L}_2(K_Q, H)$ turns out to be a separable Hilbert space under the inner product

$$\langle L, P\rangle_{\mathscr{L}_2(K_Q,H)} = Tr[LQ^{1/2}(PQ^{1/2})^*] \qquad \text{for any} \quad L,\ P \in \mathscr{L}_2(K_Q, H).$$

For arbitrarily given $T \geq 0$, let $B(t, \omega)$, $t \in [0, T]$, be an $\mathscr{L}_2(K_Q, H)$-valued process. We define the following norm for arbitrary $t \in [0, T]$,

$$|B|_t := \left\{\mathbb{E}\int_0^t Tr\big[B(s)Q^{1/2}(B(s)Q^{1/2})^*\big]ds\right\}^{\frac{1}{2}}. \tag{1.3.2}$$

In particular, we denote all $\mathscr{L}_2(K_Q, H)$-valued measurable processes B, adapted to the filtration $\{\mathscr{F}_t\}_{t\in[0,T]}$, satisfying $|B|_T < \infty$ by $\mathcal{U}^2\big([0, T] \times \Omega; \mathscr{L}_2(K_Q, H)\big)$. Recall (see Da Prato and Zabczyk [53]) that the stochastic integral $\int_0^t B(s)dW(s) \in H$, $t \geq 0$, may be defined for all $B \in \mathcal{U}^2([0, T] \times \Omega; \mathscr{L}_2(K_Q, H))$ by

$$\int_0^t B(s)dW(s) = L^2 - \lim_{n\to\infty} \sum_{i=1}^{n} \int_0^t \sqrt{\lambda_i} B(s)e_i dw_i(s), \qquad t \in [0, T]. \tag{1.3.3}$$

It is worth mentioning that stochastic integral (1.3.3) may be generalized, as in finite-dimensional cases, to any $\mathscr{L}_2(K_Q, H)$-valued adapted process $B(\cdot)$ satisfying

$$\mathbb{P}\left\{\int_0^T \|B(s)\|^2_{\mathscr{L}_2(K_Q,H)}ds < \infty\right\} = 1.$$

By employing the definition of stochastic integral and a standard limiting procedure, we may establish some useful properties of stochastic integrals.

Proposition 1.3.1 *For arbitrary $T \geq 0$, assume that $B(\cdot) \in \mathcal{U}^2([0,T] \times \Omega; \mathscr{L}_2(K_Q,H))$. Then*

(i) the stochastic integral $\int_0^t B(s)dW(s)$ is a continuous, square integrable H-valued martingale on $[0,T]$. Moreover,

$$\mathbb{E}\left\|\int_0^t B(s)dW(s)\right\|_H^2 = |B|_t^2, \qquad t \in [0,T]; \tag{1.3.4}$$

(ii) the quadratic variation process of $\int_0^t B(s)dW(s)$ has the form

$$\left\langle\!\!\left\langle \int_0^\cdot B(s)dW(s)\right\rangle\!\!\right\rangle(t) = \int_0^t B(s)Q^{1/2}(B(s)Q^{1/2})^*ds, \qquad t \in [0,T].$$

Proposition 1.3.2 *Assume that B_1, $B_2 \in \mathcal{U}^2([0,T] \times \Omega; \mathscr{L}_2(K_Q,H))$. Then the covariance operators*

$$V(s,t) = Cov\left(\int_0^s B_1(u)dW(u), \int_0^t B_2(u)dW(u)\right), \qquad s,\ t \in [0,T],$$

are given by

$$V(s,t) = \mathbb{E}\int_0^{s\wedge t} B_1(u)Q^{1/2}(B_2(u)Q^{1/2})^*du \quad s,\ t \in [0,T].$$

Moreover, for any s, $t \in [0,T]$,

$$\mathbb{E}\left\langle \int_0^s B_1(u)dW(u), \int_0^t B_2(u)dW(u)\right\rangle_H = \mathbb{E}\int_0^{s\wedge t} Tr\big[B_1(u)Q^{1/2}(B_2(u)Q^{1/2})^*\big]du.$$

Assume that e^{tA}, $t \geq 0$, is a C_0-semigroup with its infinitesimal generator A on H. Suppose that $B \in \mathcal{U}^2([0,T] \times \Omega; \mathscr{L}_2(K_Q,H))$ is such a process that the stochastic integral

$$\int_0^t e^{(t-s)A}B(s)dW(s) =: W_A^B(t), \qquad t \in [0,T], \tag{1.3.5}$$

is well defined. This process $W_A^B(t)$ is called the *stochastic convolution* of B. In general, $W_A^B(t)$, $t \in [0,T]$, is no longer a martingale, a fact that makes

W_A^B fail to have decent properties. However, one can still expect some useful results to be valid for this process. For instance, we have a useful version of the following Burkholder–Davis–Gundy type of inequality for $W_A^B(t)$, $t \in [0, T]$.

Theorem 1.3.3 *Let $p > 2$, $T \geq 0$ and assume that process $B \in \mathcal{U}^2([0, T] \times \Omega; \mathscr{L}_2(K_Q, H))$ satisfies*

$$\mathbb{E}\left(\int_0^T \|B(s)\|^p_{\mathscr{L}_2(K_Q,H)} ds\right) < \infty.$$

Then there exists a number $C_{p,T} > 0$, depending on p and T, such that

$$\mathbb{E} \sup_{t\in[0,T]} \left\| \int_0^t e^{(t-s)A} B(s) dW(s) \right\|_H^p \leq C_{p,T} \cdot \mathbb{E}\left(\int_0^T \|B(s)\|^p_{\mathscr{L}_2(K_Q,H)} ds\right). \tag{1.3.6}$$

Note that in Theorem 1.3.3, there is a weak point on the condition $p > 2$ to secure the validness of (1.3.6) for any C_0-semigroup e^{tA}, $t \geq 0$, on H. An alternative version of this theorem is possible to cover the case $p = 2$, although we have to restrict at this moment the C_0-semigroup e^{tA}, $t \geq 0$, to a pseudocontraction one.

Theorem 1.3.4 *Let $p \geq 2$ and $T \geq 0$. Assume that A generates a pseudocontraction C_0-semigroup e^{tA}, $t \geq 0$, and $B \in \mathcal{U}^2([0, T] \times \Omega; \mathscr{L}_2(K_Q, H))$. Then there exists a number $C_{p,T} > 0$, depending only on p and T, such that*

$$\mathbb{E}\left(\sup_{t\in[0,T]} \left\| \int_0^t e^{(t-s)A} B(s) dW(s) \right\|_H^p\right) \leq C_{p,T} \cdot \mathbb{E}\left(\int_0^T \|B(s)\|^2_{\mathscr{L}_2(K_Q,H)} ds\right)^{p/2}.$$

Moreover, if A generates a contraction C_0-semigroup, number $C_{p,T} > 0$ may be chosen to depend on p only.

The following stochastic version of the well-known Fubini theorem will be frequently used in this book.

Proposition 1.3.5 *Let $T \geq 0$ and*

$$B\colon [0, T] \times [0, T] \times \Omega \to \mathscr{L}_2(K_Q, H)$$

be measurable such that for each $s \in [0, T]$, $B(s, t)$ is $\{\mathscr{F}_t\}$-adapted, $t \in [0, T]$, and satisfies

$$\int_0^T \int_0^T \mathbb{E}\|B(s,t)\|^2_{\mathscr{L}_2(K_Q,H)} ds dt < \infty.$$

Then

$$\int_0^T \int_0^T B(s,t)dW(t)ds = \int_0^T \int_0^T B(s,t)dsdW(t) \qquad a.s. \tag{1.3.7}$$

Assume that $B \in \mathcal{U}^2([0,T]\times\Omega;\ \mathscr{L}_2(K_Q,H))$, and F is an H-valued, $\{\mathscr{F}_t\}$-adapted, Bochner integrable process on $[0,T]$. Then the following process

$$y(t) = y_0 + \int_0^t F(s)ds + \int_0^t B(s)dW(s), \qquad t \in [0,T], \qquad y_0 \in H, \tag{1.3.8}$$

is well defined. A function $\Lambda(t,x)\colon [0,T]\times H \to \mathbb{R}$ is called an *Itô functional* if Λ and its Fréchet partial derivatives Λ'_t, Λ'_x, Λ''_{xx} are continuous and bounded on any bounded subsets of $[0,T]\times H$.

Theorem 1.3.6 (Itô's formula) *Assume that* $\Lambda\colon [0,T]\times H \to \mathbb{R}$ *is an Itô functional. Then for all* $t \in [0,T]$, $\Lambda(t,y(t))$ *satisfies the following equality:*

$$\begin{aligned} d\Lambda(t,y(t)) = \Big\{ & \Lambda'_t(t,y(t)) + \langle \Lambda'_x(t,y(t)), F(t)\rangle_H \\ & + \frac{1}{2}Tr\big[\Lambda''_{xx}(t,y(t))B(t)Q^{1/2}(B(t)Q^{1/2})^*\big]\Big\}\, dt \\ & + \langle \Lambda'_x(t,y(t)), B(t)dW(t)\rangle_H. \end{aligned} \tag{1.3.9}$$

1.4 Stochastic Differential Equations

The theory of stochastic differential equations in Hilbert spaces is a natural generalization of finite-dimensional stochastic differential equations introduced by Itô and in a slightly different form by Gihman in the 1940s. The reader is referred to Da Prato and Zabczyk [53] for a systematic statement about this topic. On this occasion, we content ourselves with a presentation of how it is possible to formulate a standard stochastic partial differential equation as a stochastic differential equation in Hilbert spaces.

Let $\mathcal{O}$ be a bounded domain in $\mathbb{R}^n$, $n \in \mathbb{N}_+$, with smooth boundary $\partial\mathcal{O}$. Consider the following initial-boundary value problem for the randomly perturbed heat equation

$$\begin{cases} \dfrac{\partial y}{\partial t}(t,x) = \displaystyle\sum_{i=1}^n \dfrac{\partial^2 y}{\partial x_i^2}(t,x) + \dfrac{\partial}{\partial t}W(t,x), & t \ge 0, \quad x \in \mathcal{O}, \\ y(0,x) = y_0(x), \quad x \in \mathcal{O}; \quad y(t,x) = 0, & t \ge 0, \quad x \in \partial\mathcal{O}, \end{cases} \tag{1.4.1}$$

where $W(t,x)$ is a standard Wiener random field (see, e.g., [41]).

By analogy with partial differential equations, this stochastic partial differential equation (1.4.1) can be viewed in two different ways. One natural way is to consider its solution as a real-valued random field indexed by temporal and spatial variables t and x. In general, this approach uses complicated probability and calculus, and it will not be developed in this book. On the other hand, one can consider a solution of this equation as a stochastic process indexed by t with values in a proper space of functions of x, say, $L^2(\mathcal{O};\mathbb{R})$. In this manner, we can use advanced analysis to develop a stochastic process theory in an infinite-dimensional setting. For instance, we can write $\sum_{i=1}^n \partial^2/\partial x_i^2$ in (1.4.1) as an abstract operator, say A, from Sobolev space $H^2(\mathcal{O};\mathbb{R}) \cap H_0^1(\mathcal{O};\mathbb{R})$ into $L^2(\mathcal{O};\mathbb{R})$ or from $H_0^1(\mathcal{O};\mathbb{R})$ into $H^{-1}(\mathcal{O};\mathbb{R})$, the dual of $H^1(\mathcal{O};\mathbb{R})$, and let $W(t)$, $t \geq 0$, be a Wiener process in $L^2(\mathcal{O};\mathbb{R})$. Note that the Dirichlet boundary condition here is implicit in the fact that we look for solutions in $H_0^1(\mathcal{O};\mathbb{R})$. In other words, given an initial datum $y_0 \in L^2(\mathcal{O};\mathbb{R})$, we may reformulate (1.4.1) as a system in $L^2(\mathcal{O};\mathbb{R})$ or $H^{-1}(\mathcal{O};\mathbb{R})$ with the following form:

$$\begin{cases} dy(t) = Ay(t)dt + dW(t), & t \geq 0, \\ y(0) = y_0 \in L^2(\mathcal{O};\mathbb{R}). \end{cases}$$

In this book, we mainly adopt the latter viewpoint to establish a stochastic stability theory. We develop two formulations, i.e., semigroup and variational methods, to give a rigorous meaning to the solutions of abstract stochastic differential equations.

1.4.1 Semigroup Approach and Mild Solutions

Let $T \geq 0$ and consider the following semilinear stochastic system on a Hilbert space H,

$$\begin{cases} dy(t) = \big[Ay(t) + F(t, y(t))\big]dt + B(t, y(t))dW(t), & t \in [0, T], \\ y(0) = y_0 \in H, \end{cases} \tag{1.4.2}$$

where A is the infinitesimal generator of a C_0-semigroup e^{tA}, $t \geq 0$, of bounded linear operators on H. The coefficients $F(\cdot,\cdot)$ and $B(\cdot,\cdot)$ are two nonlinear measurable mappings from $[0, T] \times H$ into H and $\mathscr{L}_2(K_Q, H)$, respectively.

Definition 1.4.1 Let $T \geq 0$. An $\{\mathscr{F}_t\}_{t\geq 0}$-adapted stochastic process $y(t) \in H$, $t \in [0, T]$, defined on probability space $(\Omega, \mathscr{F}, \{\mathscr{F}_t\}_{t\geq 0}, \mathbb{P})$ is called a *mild solution* of (1.4.2) if it satisfies that

$$\mathbb{P}\left\{\int_0^T \|y(t)\|_H^2 dt < \infty\right\} = 1, \tag{1.4.3}$$

$$\mathbb{P}\left\{\int_0^T \left(\|F(t, y(t))\|_H + \|B(t, y(t))\|^2_{\mathscr{L}_2(K_Q, H)}\right)dt < \infty\right\} = 1, \tag{1.4.4}$$

and

$$\begin{aligned} y(t) = e^{tA}y_0 &+ \int_0^t e^{(t-s)A}F(s, y(s))ds \\ &+ \int_0^t e^{(t-s)A}B(s, y(s))dW(s), \qquad t \in [0, T], \end{aligned} \tag{1.4.5}$$

for $y_0 \in H$ almost surely.

By the standard Picard iteration procedure or a probabilistic fixed-point theorem type of argument, one can establish an existence and uniqueness theorem of mild solutions to (1.4.2) in the case that for any $y, z \in H$ and $t \in [0, T]$,

$$\begin{aligned} &\|F(t, y) - F(t, z)\|_H + \|B(t, y) - B(t, z)\|_{\mathscr{L}_2(K_Q, H)} \\ &\qquad\qquad\qquad\qquad \leq \alpha(T)\|y - z\|_H, \quad \alpha(T) > 0, \\ &\|F(t, y)\|_H + \|B(t, y)\|_{\mathscr{L}_2(K_Q, H)} \leq \beta(T)(1 + \|y\|_H), \quad \beta(T) > 0. \end{aligned} \tag{1.4.6}$$

Theorem 1.4.2 *Let $T \geq 0$, $p \geq 2$ and assume that condition (1.4.6) holds. Then there exists a unique mild solution $y \in C([0, T]; L^p(\Omega; H))$ to (1.4.2). If, in addition, $\mathbb{E}\|y_0\|_H^p < \infty$, $p > 2$, then the solution y satisfies*

$$\mathbb{E}\left(\sup_{0\leq t\leq T} \|y(t, y_0)\|_H^p\right) < \infty, \qquad p > 2.$$

As a direct application of semigroup theory, we have the following result straightaway.

Proposition 1.4.3 *For arbitrary $y_0 \in \mathscr{D}(A)$, assume that $y(t) \in \mathscr{D}(A)$, $t \in [0, T]$, is an $\{\mathscr{F}_t\}_{t\geq 0}$-adapted stochastic process satisfying (1.4.3), (1.4.4), and the equation*

$$y(t) = y_0 + \int_0^t (Ay(s) + F(s, y(s)))ds + \int_0^t B(s, y(s))dW(s), \qquad t \in [0, T], \tag{1.4.7}$$

then it is a mild solution of equation (1.4.2).

The process y satisfying (1.4.7) is called a solution of (1.4.2) in the strong sense, and a mild solution of (1.4.2) is not necessarily a solution in the strong sense. On the other hand, it is known that the stochastic convolution in (1.4.5) is no longer a martingale, implying that one cannot apply Itô's formula directly to mild solutions of (1.4.2), although it is possible to apply it to solutions of (1.4.2) in the strong sense. This consideration just suggests the usefulness of finding conditions under which a mild solution to (1.4.2) becomes a strong one.

Proposition 1.4.4 *Suppose that the following conditions hold:*

(1) $y_0 \in \mathscr{D}(A)$, $e^{(t-s)A}F(s,y) \in \mathscr{D}(A)$, $e^{(t-s)A}B(s,y)z \in \mathscr{D}(A)$ *for each* $y \in H$, $z \in K$, *and* $t \geq s$;

(2) $\left\|Ae^{(t-s)A}F(s,y)\right\|_H \leq f(t-s)\|y\|_H$, $y \in H$, *for some* $f \in L^1([0,T];\mathbb{R}_+)$;

(3) $\left\|Ae^{(t-s)A}B(s,y)\right\|_{\mathscr{L}_2(K_Q,H)} \leq g(t-s)\|y\|_H$, $y \in H$, *for some* $g \in L^2([0,T];\mathbb{R}_+)$.

Then for any mild solution $y(t)$, $t \in [0,T]$, *of (1.4.2), it is also a solution of (1.4.2) in the strong sense.*

Proof By the conditions (1), (2), and (3), it is easy to see that

$$\int_0^T \int_0^t \|Ae^{(t-r)A}F(r,y(r))\|_H dr dt < \infty \qquad a.s.$$

$$\int_0^T \int_0^t \|Ae^{(t-r)A}B(r,y(r))\|^2_{\mathscr{L}_2(K_Q,H)} dr dt < \infty \qquad a.s.$$

Thus by the classic Fubini's theorem and Proposition 1.1.20, we have

$$\begin{aligned}\int_0^t \int_0^s Ae^{(s-r)A}F(r,y(r))drds &= \int_0^t \int_r^t Ae^{(s-r)A}F(r,y(r))dsdr \\ &= \int_0^t e^{(t-r)A}F(r,y(r))dr - \int_0^t F(r,y(r))dr.\end{aligned} \tag{1.4.8}$$

Meanwhile, by virtue of Proposition 1.3.5,

$$\begin{aligned}\int_0^t \int_0^s Ae^{(s-r)A}B(r,y(r))dW(r)ds &= \int_0^t \int_r^t Ae^{(s-r)A}B(r,y(r))dsdW(r) \\ &= \int_0^t e^{(t-r)A}B(r,y(r))dW(r) \\ &\quad - \int_0^t B(r,y(r))dW(r).\end{aligned} \tag{1.4.9}$$

Hence, by the closedness of A, (1.4.8), and (1.4.9), it follows that $Ay(t) \in H$, $t \in [0,T]$, which is integrable almost surely and

$$\begin{aligned}\int_0^t Ay(s)ds &= e^{tA}y_0 - y_0 + \int_0^t e^{(t-r)A}F(r,y(r))dr - \int_0^t F(r,y(r))dr\\ &\quad + \int_0^t e^{(t-r)A}B(r,y(r))dW(r) - \int_0^t B(r,y(r))dW(r)\\ &= y(t) - y_0 - \int_0^t F(r,y(r))dr - \int_0^t B(r,y(r))dW(r).\end{aligned}$$

That is, y is also a solution of (1.4.2) in the strong sense. The proof is now complete. □

To employ Itô's formula in handling the mild solutions of (1.4.2), we introduce a Yosida approximating system of (1.4.2) in the following form:

$$\begin{cases} dy(t) = Ay(t)dt + R(n)F(t,y(t))dt + R(n)B(t,y(t))dW(t),\\ y(0) = R(n)y_0 \in \mathscr{D}(A), \end{cases} \tag{1.4.10}$$

where $n \in \rho(A)$, the resolvent set of A, $R(n) := nR(n,A)$ and $R(n,A) = (nI - A)^{-1}$ is the resolvent of A.

Proposition 1.4.5 *Let $T \geq 0$ and $p \geq 2$. Suppose that the nonlinear terms $F(\cdot,\cdot)$, $B(\cdot,\cdot)$ in (1.4.10) satisfy condition (1.4.6). Then, for each $n \in \rho(A)$, the equation (1.4.10) has a unique solution $y_n(t) \in \mathscr{D}(A)$ in the strong sense, which lies in $L^p(\Omega; C([0,T]; H))$. In addition, if $\mathbb{E}\|y_0\|_H^p < \infty$, $p > 2$, and we let y be the mild solution of (1.4.2), then we have*

$$\lim_{n\to\infty} \mathbb{E}\Big(\sup_{0\leq t\leq T} \|y_n(t) - y(t)\|_H^p\Big) = 0, \qquad p > 2. \tag{1.4.11}$$

Proof The existence of a unique mild solution $y_n \in C([0,T]; L^p(\Omega; H))$, $n \in \rho(A)$, of (1.4.10) is an immediate consequence of Theorem 1.4.2 through a probabilistic fixed-point theorem type of argument. The fact that $y_n \in L^p(\Omega; C([0,T]; H))$ and y_n is also a solution of (1.4.10) in the strong sense follows from Proposition 1.4.4 and the relation

$$AR(n) = nAR(n,A) = n - n^2R(n,A) \in \mathscr{L}(H), \qquad n \in \rho(A).$$

To prove the remainder of the proposition, let us suppose that $\mathbb{E}\|y_0\|_H^p < \infty$, $p > 2$, and consider for any $t \in [0,T]$,

$$\begin{aligned} y(t) - y_n(t) &= e^{tA}(y_0 - R(n)y_0)\\ &\quad + \int_0^t e^{(t-s)A}[F(s,y(s)) - R(n)F(s,y_n(s))]ds\\ &\quad + \int_0^t e^{(t-s)A}[B(s,y(s)) - R(n)B(s,y_n(s))]dW(s). \end{aligned} \tag{1.4.12}$$

Since $|a+b+c|^p \le 3^p(|a|^p+|b|^p+|c|^p)$ for any real numbers a, b, c, this yields, in addition to (1.4.12), that for any $T \ge 0$, $p > 2$,

$$
\begin{aligned}
&\mathbb{E}\sup_{0\le t\le T}\|y(t)-y_n(t)\|_H^p\\
&\quad\le 3^p\mathbb{E}\sup_{0\le t\le T}\left\|\int_0^t e^{(t-s)A}R(n)[F(s,y(s))-F(s,y_n(s))]ds\right\|_H^p\\
&\qquad+3^p\mathbb{E}\sup_{0\le t\le T}\left\|\int_0^t e^{(t-s)A}R(n)[B(s,y(s))-B(s,y_n(s))]dW(s)\right\|_H^p\\
&\qquad+3^p\Bigg\{\mathbb{E}\sup_{0\le t\le T}\Bigg\|e^{tA}(y_0-R(n)y_0)+\int_0^t e^{(t-s)A}[I-R(n)]F(s,y(s))ds\\
&\qquad+\int_0^t e^{(t-s)A}[I-R(n)]B(s,y(s))dW(s)\Bigg\|_H^p\Bigg\}\\
&\quad:=3^pI_1+3^pI_2+3^pI_3.
\end{aligned}
\tag{1.4.13}
$$

Note that by the Hille–Yosida theorem, $\|R(n)\| \le 2M$ for an $n \in \mathbb{N}_+$ large enough where $M \ge 1$ is the number given in (1.1.5). Condition (1.4.6) and Hölder's inequality imply that

$$
\begin{aligned}
I_1 &\le \mathbb{E}\sup_{0\le t\le T}\left(\int_0^t \left\|e^{(t-s)A}R(n)\big[F(s,y(s))-F(s,y_n(s))\big]\right\|_H ds\right)^p\\
&\le C_1(T)\mathbb{E}\sup_{0\le t\le T}\left\{\int_0^t \|F(s,y(s))-F(s,y_n(s))\|_H^p ds\right\}\\
&\le C_2(T)\mathbb{E}\int_0^T \sup_{0\le r\le s}\|y(r)-y_n(r)\|_H^p ds,
\end{aligned}
\tag{1.4.14}
$$

where $C_1(T)$, $C_2(T)$ are positive numbers, dependent on $T \ge 0$. In a similar way, by virtue of Theorem 1.3.3, for $n \in \mathbb{N}_+$ large enough there exists a real number $C_3(T) > 0$ such that

$$
\begin{aligned}
I_2 &\le \mathbb{E}\sup_{0\le t\le T}\left\|\int_0^t e^{(t-s)A}R(n)[B(s,y(s))-B(s,y_n(s))]dW(s)\right\|_H^p\\
&\le C_3(T)\mathbb{E}\int_0^T \sup_{0\le r\le s}\|y(r)-y_n(r)\|_H^p ds.
\end{aligned}
\tag{1.4.15}
$$

For the term I_3, it is easy to see that

$$I_3 \leq 3^p \Bigg\{ \mathbb{E} \sup_{0\leq t\leq T} \|e^{tA}(y_0 - R(n)y_0)\|_H^p + \mathbb{E} \sup_{0\leq t\leq T} \left\| \int_0^t e^{(t-s)A}[I - R(n)]F(s, y(s))ds \right\|_H^p + \mathbb{E} \sup_{0\leq t\leq T} \left\| \int_0^t e^{(t-s)A}[I - R(n)]B(s, y(s))dW(s) \right\|_H^p \Bigg\}. \tag{1.4.16}$$

We now estimate each term at the right-hand side of (1.4.16). By the Dominated Convergence Theorem and the fact that $R(n) \to I$ strongly as $n \to \infty$, it is easy to see that

$$\mathbb{E} \sup_{0\leq t\leq T} \|e^{tA}(y_0 - R(n)y_0)\|_H^p \leq C_4(T) \cdot \mathbb{E}\|y_0 - R(n)y_0\|_H^p \to 0, \quad n \to \infty,$$

where $C_4(T) > 0$ is some positive number. On the other hand, by using the Hölder inequality and Dominated Convergence Theorem, we get for some $C_5(T) > 0$ that

$$\begin{aligned} \mathbb{E} \sup_{0\leq t\leq T} \left\| \int_0^t e^{(t-s)A}[I - R(n)]F(s, y(s))ds \right\|_H^p &\leq C_5(T) \int_0^T \mathbb{E}\|[I - R(n)]F(s, y(s))\|_H^p ds \\ &\to 0 \quad \text{as} \quad n \to \infty. \end{aligned} \tag{1.4.17}$$

In a similar manner, by using Theorem 1.3.3 and the Dominated Convergence Theorem again, we have that for some $C_6(T) > 0$,

$$\begin{aligned} &\mathbb{E} \sup_{0\leq t\leq T} \left\| \int_0^t e^{(t-s)A}[I - R(n)]B(s, y(s))dW(s) \right\|_H^p \\ &\leq C_6(T) \int_0^T \mathbb{E}\|[I - R(n)]B(s, y(s))\|_{\mathscr{L}_2(K_Q, H)}^p ds \to 0 \quad \text{as} \quad n \to \infty. \end{aligned} \tag{1.4.18}$$

Combining (1.4.13) through (1.4.18), we thus have that there exist numbers $C(T) > 0$ and $\varepsilon(n) > 0$ such that

$$\mathbb{E} \sup_{0\leq t\leq T} \|y(t) - y_n(t)\|_H^p \leq C(T) \int_0^T \mathbb{E} \sup_{0\leq r\leq s} \|y(r) - y_n(r)\|_H^p ds + \varepsilon(n),$$

where $\lim_{n\to\infty} \varepsilon(n) = 0$. By the well-known Gronwall's inequality, we further deduce that

$$\mathbb{E} \sup_{0\leq t\leq T} \|y(t) - y_n(t)\|_H^p \leq \varepsilon(n)e^{C(T)T} \to 0, \quad \text{as} \quad n \to \infty. \tag{1.4.19}$$

The proof is thus complete. □

Corollary 1.4.6 *Let $y_0 \in H$ be an arbitrarily given nonrandom vector. Suppose that the nonlinear terms $F(\cdot,\cdot)$, $B(\cdot,\cdot)$ in (1.4.2) and (1.4.10) satisfy condition (1.4.6) for any $T \geq 0$. Then there exists a sequence $y_n(t) \in \mathscr{D}(A)$ of solutions to (1.4.10) in the strong sense, which lies in $L^p(\Omega; C([0,T]; H))$, $p > 2$, such that $y_n(t) \to y(t)$ almost surely as $n \to \infty$, uniformly on any compact set of $[0,\infty)$.*

Proof We may construct the desired sequence by a diagonal sequence trick. Indeed, by virtue of (1.4.11) there exists a positive integer sequence $\{n_1(i)\}$ in $\rho(A)$ such that $y_{n_1(i)}(t) \to y(t)$ almost surely as $i \to \infty$, uniformly with respect to $t \in [0,1]$. Now consider the sequence $y_{n_1(i)}(t)$. We can find a subsequence $y_{n_2(i)}(t)$ of $y_{n_1(i)}(t)$ such that $y_{n_2(i)}(t) \to y(t)$ almost surely as $i \to \infty$, uniformly with respect to $t \in [0,2]$. Proceeding inductively, we find successive subsequences $y_{n_m(i)}(t)$ such that (a) $y_{n_m(i)}(t)$ is a subsequence of $y_{n_{m-1}(i)}(t)$ and (b) $y_{n_m(i)} \to y(t)$ almost surely as $i \to \infty$, uniformly with respect to $t \in [0,m]$. To get a sequence converging for each m, one may take the diagonal sequence $\hat{n}(m) := \{n_m(m)\}$. Then the sequence $y_{\hat{n}(m)}(t)$, $y_{\hat{n}(m+1)}(t), \ldots$ is a subsequence of $y_{n_m(i)}(t)$ so that $y_{\hat{n}(i)}(t) \to y(t)$ almost surely as $i \to \infty$, uniformly with respect to $t \in [0,m]$ for each $m \in \mathbb{N}_+$. □

Remark 1.4.7 In general, it is not immediate to know from Theorem 1.4.2 that the mild solution of (1.4.2) has almost surely continuous paths. However, Corollary 1.4.6 permits a modification of any mild solution of (1.4.2) with continuous sample paths. Unless otherwise stated, we always assume that the mild solution of this kind of equation under investigation has continuous sample paths in the sequel.

1.4.2 Variational Approach and Strong Solutions

Let V be a reflexive Banach space which is densely and continuously embedded in a Hilbert space H. We identity H with its dual space H^* according to Theorem 1.1.4. Then we have the following relations:

$$V \hookrightarrow H \cong H^* \hookrightarrow V^*$$

where $\hookrightarrow$ denotes the injection. We denote the duality pair between V and V^* by $\langle\langle \cdot,\cdot \rangle\rangle_{V,V^*}$. Let K be a separable Hilbert space and assume that $W(t)$, $t \geq 0$, is a Q-Wiener process in K defined on some probability space $(\Omega, \mathscr{F}, \mathbb{P})$, equipped with a normal filtration $\{\mathscr{F}_t\}_{t\geq 0}$ with respect to which $\{W(t)\}_{t\geq 0}$ is a continuous martingale.

Let $T \geq 0$ and consider the following nonlinear stochastic differential equation on V^*:

$$\begin{cases} y(t) = y(0) + \int_0^t A(s, y(s))ds + \int_0^t B(s, y(s))dW(s), & t \in [0, T], \\ y(0) = y_0 \in H, \end{cases} \tag{1.4.20}$$

where $A\colon [0, T] \times V \to V^*$ and $B\colon [0, T] \times V \to \mathscr{L}_2(K_Q, H)$ are two families of nonlinear measurable functions (they may be random as well in an appropriate setting). For any $y_0 \in H$, an $\{\mathscr{F}_t\}_{t\geq 0}$-adapted, V-valued process y is said to be a *strong solution* of the equation (1.4.20) if $y \in L^p([0, T] \times \Omega; V)$ for some $p \geq 1$ and the equation (1.4.20) holds in V^* almost surely. In contrast with system (1.4.2), one remarkable feature of (1.4.20) is that both mappings A and B could be nonlinear here.

To obtain the existence and uniqueness of solutions to (1.4.20), we impose the following conditions on $A(\cdot, \cdot)$ and $B(\cdot, \cdot)$.

(a) (Coercivity) There exist numbers $p > 1, \alpha > 0, \lambda \in \mathbb{R}$ and function $\gamma \in L^1(0, T)$ such that for all $y \in V$ and $t \in [0, T]$,

$$2\langle\!\langle y, A(t, y)\rangle\!\rangle_{V,V^*} + \|B(t, y)\|^2_{\mathscr{L}_2(K_Q,H)} \leq -\alpha\|y\|^p_V + \lambda\|y\|^2_H + \gamma(t), \tag{1.4.21}$$

(b) (Boundedness) and there exists a function $\theta \in L^{\frac{p}{p-1}}(0, T)$ such that for all $y \in V, t \in [0, T]$,

$$\|A(t, y)\|_{V^*} \leq \theta(t) + c\|y\|^{p-1}_V \tag{1.4.22}$$

for some number $c > 0$.

(c) (Continuity) The map $s \in \mathbb{R} \to \langle\!\langle x, A(t, y + sz)\rangle\!\rangle_{V,V^*}$ is continuous for arbitrary $y, z, x \in V$ and $0 \leq t \leq T$.

(d) (Monotonicity) There exists a number $\mu \in \mathbb{R}$ such that for any $y, z \in V$, and $t \in [0, T]$,

$$\begin{aligned} &2\langle\!\langle y - z, A(t, y) - A(t, z)\rangle\!\rangle_{V,V^*} + \|B(t, y) - B(t, z)\|^2_{\mathscr{L}_2(K_Q,H)} \\ &\leq \mu\|y - z\|^2_H. \end{aligned} \tag{1.4.23}$$

Theorem 1.4.8 *Assume that $y_0 \in L^2(\Omega, \mathscr{F}_0, \mathbb{P}; H)$. Under the assumptions (a)–(d), equation (1.4.20) has a unique $\{\mathscr{F}_t\}_{t\geq 0}$-progressively measurable strong solution*

$$y \in L^2(\Omega; C([0, T]; H)) \cap L^2([0, T] \times \Omega; V) \quad \textit{for any} \quad T \geq 0,$$

which has strong Markov property and satisfies the following energy equation: for all $t \in [0, T]$,

$$\|y(t)\|_H^2 = \|y_0\|_H^2 + 2\int_0^t \langle\langle y(s), A(s, y(s))\rangle\rangle_{V,V^*} ds + 2\int_0^t \langle y(s), B(s, y(s))dW(s)\rangle_H + \int_0^t \|B(s, y(s))\|^2_{\mathscr{L}_2(K_Q,H)} ds. \tag{1.4.24}$$

Equality (1.4.24) is the usual Itô's formula for quadratic function $\Lambda(\cdot) = \|\cdot\|_H^2$. To extend this formula to more general function Λ, one need impose stronger conditions on Λ. A function $\Lambda: [0, T] \times H \to \mathbb{R}$ is called an *Itô type of functional* if it satisfies:

(i) Λ has locally bounded partial derivatives $\partial_t\Lambda$, $\partial_x\Lambda$ and $\partial^2_{xx}\Lambda$ on $[0, T] \times H$;
(ii) $\partial_t\Lambda$ and $\partial_x\Lambda$ are continuous in $[0, T] \times H$;
(iii) for any trace class operator P, the map $(t, x) \to Tr[\partial^2_{xx}\Lambda(t, x)P]$ is continuous on $[0, T] \times H$;
(iv) if $x \in V$, then $\partial_x\Lambda(t, x) \in V$ for any $t \in [0, T]$ and $\langle\langle \partial_x\Lambda(t, x), v^*\rangle\rangle_{V,V^*}$ is continuous in $t \in [0, T]$ for any $v^* \in V^*$. Moreover, there exists a number $M > 0$ such that

$$\|\partial_x\Lambda(t, x)\|_V \le M(1 + \|x\|_V), \qquad (t, x) \in [0, T] \times V. \tag{1.4.25}$$

Theorem 1.4.9 *Let $y \in L^2(\Omega; C([0, T]; H)) \cap L^2([0, T] \times \Omega; V)$, $T \ge 0$, be the strong solution of (1.4.20) with $y_0 \in L^2(\Omega, \mathscr{F}_0, \mathbb{P}; H)$. For any Itô type of functional Λ satisfying (i), (ii), (iii) and (iv) on $[0, T] \times H$, the following Itô's formula holds: for $t \in [0, T]$,*

$$\Lambda(t, y(t)) = \Lambda(0, y_0) + \int_0^t (\mathcal{L}\Lambda)(s, y(s))ds + \int_0^t \langle \partial_x\Lambda(s, y(s)), B(s, y(s))dW(s)\rangle_H,$$

where

$$(\mathcal{L}\Lambda)(s, y(s)) = \partial_s\Lambda(s, y(s)) + \langle\langle \partial_x\Lambda(s, y(s)), A(s, y(s))\rangle\rangle_{V,V^*} + \frac{1}{2}Tr\Big[\partial^2_{xx}\Lambda(s, y(s))B(s, y(s))Q^{1/2}(B(s, y(s))Q^{1/2})^*\Big]. \tag{1.4.26}$$

1.5 Definitions and Methods of Stochastic Stability

The term stability is one that has a variety of different meanings within mathematics. One often says that a system is stable if it is "continuous" with respect to initial conditions. Precisely, suppose that $y(t) = y(t, y_0)$, $t \geq 0$, is a solution to some differential equation on a Hilbert space H,

$$\begin{cases} dy(t) = f(t, y(t))dt, & t \geq 0, \\ y(0) = y_0 \in H, \end{cases} \tag{1.5.1}$$

where $f(\cdot, \cdot)$ is a properly given function. Let $\tilde{y}(t)$, $t \geq 0$, be a particular solution to (1.5.1) and the corresponding system is thought of as describing a process without perturbations. Those systems associated with other solutions $y(t)$ are regarded as perturbed ones. When we talk about stability of the solution $\tilde{y}(t)$, $t \geq 0$, it means that the norm $\|y(t) - \tilde{y}(t)\|_H$, $t \geq 0$, could be made smaller and smaller if the initial perturbation scale $\|y(0) - \tilde{y}(0)\|_H$ is sufficiently small.

Another notion of stability is that of asymptotic stability. Here we say an equation is stable if all of its solutions get close to some nice solution $\tilde{y}$, e.g., equilibrium solution, as time goes to infinity. In most situations, it is enough to consider asymptotic stability of the null solution for some relevant system. Indeed, let $z(t) = y(t) - \tilde{y}(t)$ in (1.5.1), then the equation (1.5.1) could be rewritten as

$$\begin{aligned} dz(t) &= dy(t) - d\tilde{y}(t) \\ &= [f(t, z(t) + \tilde{y}(t)) - f(t, \tilde{y}(t))]dt \\ &=: F(t, z(t))dt, \qquad t \geq 0, \end{aligned} \tag{1.5.2}$$

where $F(t, 0) = 0$, $t \geq 0$. Note that if $z(0) = z_0 = 0$, it is immediate that the null is the unique solution to system (1.5.2). Hence, this treatment could be thought of as considering asymptotic stability of this null solution.

Definition 1.5.1 The null solution of (1.5.2) is said to be *stable* if for arbitrarily given $\varepsilon > 0$, there exists $\delta = \delta(\varepsilon) > 0$ such that the relation $\|z_0\|_H < \delta$ implies

$$\|z(t, z_0)\|_H < \varepsilon \quad \text{for all} \quad t \geq 0. \tag{1.5.3}$$

Definition 1.5.2 The null solution of (1.5.2) is said to be *asymptotically stable* if it is stable and there exists $\delta > 0$ such that the relation $\|z_0\|_H < \delta$ implies

$$\lim_{t \to \infty} \|z(t, z_0)\|_H = 0. \tag{1.5.4}$$

For any $z_0 \in H$, if there exists $T(z_0) \geq 0$ such that (1.5.3) or (1.5.4) remains valid for all $t \geq T(z_0)$, then the null solution of (1.5.2) is said to have *global stability*. In addition to asymptotic stability, one might also want to know the rate of convergence, which leads to the following notion.

Definition 1.5.3 The null solution of (1.5.2) is said to be *(asymptotic) exponentially stable* if it is asymptotically stable and there exist numbers $M > 0$ and $\mu > 0$ such that

$$\|z(t,z_0)\|_H \leq M\|z_0\|_H e^{-\mu t} \qquad \text{for all} \qquad t \geq 0. \tag{1.5.5}$$

There are at least three times as many definitions for the stability of stochastic systems as there are for deterministic ones. This is certainly because in a stochastic setting, there exist three basic types of convergence: convergence in probability, convergence in mean, and convergence in sample paths. The preceding deterministic stability definitions can be translated into a stochastic setting by properly interpreting the notion of convergence.

Consider the following stochastic differential equation on the Hilbert space H,

$$\begin{cases} dy(t) = A(t,y(t))dt + B(t,y(t))dW(t), & t \geq 0, \\ y(0) = y_0 \in H, \end{cases} \tag{1.5.6}$$

where $y_0 \in H$ is a nonrandom vector; W is an infinite-dimensional Q-Wiener process; and A, B are families of measurable mappings with $A(t,0) = 0$, $B(t,0) = 0$ for any $t \geq 0$.

Definition 1.5.4 (Stability in Probability) The null solution of (1.5.6) is said to be *stable or strongly stable in probability* if for arbitrarily given $\varepsilon_1, \varepsilon_2 > 0$, there exists $\delta = \delta(\varepsilon_1,\varepsilon_2) > 0$ such that the relation $\|y_0\|_H < \delta$ implies

$$\mathbb{P}\big\{\|y(t,y_0)\|_H > \varepsilon_1\big\} < \varepsilon_2 \qquad \text{for all} \qquad t \geq 0, \tag{1.5.7}$$

or

$$\mathbb{P}\Big\{\sup_{t\geq 0}\|y(t,y_0)\|_H > \varepsilon_1\Big\} < \varepsilon_2.$$

Definition 1.5.5 (Asymptotic Stability in Probability) The null solution of (1.5.6) is said to have *asymptotic stability or strongly asymptotic stability in probability* if it is stable or strongly stable in probability and for each $\varepsilon > 0$, there exists $\delta = \delta(\varepsilon) > 0$ such that the relation $\|y_0\|_H < \delta$ implies

$$\lim_{t\to\infty} \mathbb{P}\big\{\|y(t,y_0)\|_H > \varepsilon\big\} = 0, \tag{1.5.8}$$

or

$$\lim_{T\to\infty} \mathbb{P}\Big\{\sup_{t\geq T}\|y(t,y_0)\|_H > \varepsilon\Big\} = 0.$$

Definition 1.5.6 (Stability in the pth Moment) The null solution of (1.5.6) is said to be *stable or strongly stable in the pth moment*, $p > 0$, if for arbitrarily given $\varepsilon > 0$, there exists $\delta = \delta(\varepsilon) > 0$ such that the relation $\|y_0\|_H < \delta$ implies

$$\mathbb{E}\|y(t, y_0)\|_H^p < \varepsilon \quad \text{for all} \quad t \geq 0, \tag{1.5.9}$$

or

$$\mathbb{E}\left\{\sup_{t\geq 0} \|y(t, y_0)\|_H^p\right\} < \varepsilon.$$

Definition 1.5.7 (Asymptotic Stability in the pth Moment) The null solution of (1.5.6) is said to have *asymptotic stability or strongly asymptotic stability in the pth moment*, $p > 0$, if it is stable or strongly stable in the pth moment and there exists $\delta > 0$ such that the relation $\|y_0\|_H < \delta$ implies

$$\lim_{t\to\infty} \mathbb{E}\|y(t, y_0)\|_H^p = 0, \tag{1.5.10}$$

or

$$\lim_{T\to\infty} \mathbb{E}\left\{\sup_{t\geq T} \|y(t, y_0)\|_H^p\right\} = 0.$$

Definition 1.5.8 (Pathwise Stability) The null solution of (1.5.6) is said to be *stable or strongly stable in sample paths* if for any $\varepsilon > 0$, there exists $\delta = \delta(\varepsilon) > 0$ such that the relation $\|y_0\|_H < \delta$ implies

$$\mathbb{P}\{\|y(t, y_0)\|_H > \varepsilon\} = 0 \quad \text{for all} \quad t \geq 0,$$

or

$$\mathbb{P}\left\{\sup_{t\geq 0} \|y(t, y_0)\|_H > \varepsilon\right\} = 0,$$

which means with probability one, all the paths of solutions are stable or strongly stable.

Definition 1.5.9 (Pathwise Asymptotic Stability) The null solution of (1.5.6) is said to have *asymptotic stability or strongly asymptotic stability in sample paths* if it is stable or strongly stable in probability and there exists $\delta > 0$ such that the relation $\|y_0\|_H < \delta$ implies

$$\mathbb{P}\left\{\lim_{t\to\infty} \|y(t, y_0)\|_H = 0\right\} = 1,$$

or

$$\mathbb{P}\left\{\lim_{T\to\infty} \sup_{t\geq T} \|y(t, y_0)\|_H = 0\right\} = 1.$$

For any $y_0 \in H$, if there exists $T(y_0) \geq 0$ such that the claims in Definitions 1.5.4 through 1.5.9 remain valid for all $t \geq T(y_0)$, then the system (1.5.6) is said to have its *global* stability, respectively.

In application, there exist various versions of stochastic stability that show explicitly the decay rate of systems. Most noteworthy is the pth moment or pathwise exponential stability. Let $\lambda\colon [0,\infty) \to (0,\infty)$ be a continuous function with $\lim_{t\to\infty} \lambda(t) = 0$.

Definition 1.5.10 For $p > 0$, the null solution of (1.5.6) is said to be *pth momently stable with rate λ* if for each $y_0 \in H$, there exists a number $M(y_0) > 0$ such that

$$\mathbb{E}\|y(t, y_0)\|_H^p \le M(y_0)\lambda(t) \qquad \text{for all} \quad t \ge 0. \tag{1.5.11}$$

Definition 1.5.11 The null solution of (1.5.6) is said to have *almost sure stability with rate λ* if for each $y_0 \in H$, there exists a random variable $M(y_0) > 0$ such that

$$\|y(t, y_0)\|_H \le M(y_0)\lambda(t) \qquad \text{for all} \quad t \ge 0 \quad \text{almost surely.} \tag{1.5.12}$$

In Definitions 1.5.10 and 1.5.11, if $\lambda(t) = e^{-\gamma t}$, $(1+t)^{-\gamma}$ or $(\ln(1+t))^{-\gamma}$, $t \ge 0$, for some constant $\gamma > 0$, the system (1.5.6) is said to have *exponential, polynomial, or logarithmic stability*, respectively. In general, if $\lambda\colon [0,\infty) \to (0,\infty)$ is a continuous function such that $\overline{\lim}_{t\to\infty} \lambda(t) < \infty$ and (1.5.11) or (1.5.12) holds, the system (1.5.6) is called *ultimately bounded* in the pth moment or almost sure sense.

Remark 1.5.12 The definition of the almost sure stability with rate $\lambda(t)$ can be equivalently stated in the following way: for each $y_0 \in H$, there exist a number $M(y_0) > 0$ and random time $T = T(y_0) \ge 0$ such that

$$\|y(t, y_0)\|_H \le M(y_0)\lambda(t) \qquad \text{for all} \quad t \ge T(y_0) \quad \text{almost surely.}$$

Remark 1.5.13 All the preceding stability definitions remain meaningful if we remove the condition $A(t,0) = 0$, $B(t,0) = 0$ in (1.5.6). This fact leads to a natural generalization of all the stability concepts. That is, we say in this case that the solution of system (1.5.6) has a *decay*, e.g., exponential decay.

Remark 1.5.14 For stochastic stability, it is enough in most cases to consider a nonrandom initial $y_0 \in H$. To illustrate this, suppose for the moment that y_0 is random and the null solution of (1.5.6) with nonrandom initial $x \in H$ is stable in probability, i.e., for arbitrarily given ε_1, $\varepsilon_2 > 0$, there exist $\delta = \delta(\varepsilon_1, \varepsilon_2) > 0$, $T = T(\varepsilon_1, \varepsilon_2) \ge 0$ such that if $\|x\|_H < \delta$, then

$$\mathbb{P}\big\{\omega\colon \|y(t,x)\|_H > \varepsilon_1\big\} < \varepsilon_2 \qquad \text{for all} \quad t \ge T.$$

Now suppose that $\|y_0(\omega)\|_H < \delta$ almost surely. Let $B_\delta = \{x \in H\colon \|x\|_H < \delta\}$ and define the law of $y_0(\omega)$ by

$$\mathbb{D}_{y_0}(A) = \mathbb{P}\{\omega \in \Omega\colon y_0(\omega) \in A\}, \qquad \forall A \in \mathscr{B}(H).$$

Then we have

$$\begin{aligned}\mathbb{P}\{\omega\colon \|y(t,y_0)\|_H > \varepsilon_1\} &= \int_{\{x\in H\colon \|x\|_H<\delta\}} \mathbb{P}\{\|y(t,x)\|_H > \varepsilon_1\}\mathbb{D}_{y_0}(dx)\\ &\le \int_{\{x\in H\colon \|x\|_H<\delta\}} \varepsilon_2 \mathbb{D}_{y_0}(dx)\\ &\le \varepsilon_2 \qquad \text{for all} \qquad t\ge T.\end{aligned}$$

Unless otherwise stated, we always assume in the sequel that the initial data of the system under consideration are nonrandom.

Remark 1.5.15 (Also see Example 6.7, pp. 225–226 in [103]) It is clear that strong stability of a stochastic system implies its stability. The converse statement is not true in general. To see this, let us consider a stochastic process on the circle with one unit radius,

$$\begin{cases} dy(t) = \left[-2\sin^2\frac{y(t)}{2} + \sin^3\frac{y(t)}{2}\cos\frac{y(t)}{2}\right]dt - 2\sin^2\frac{y(t)}{2}dw(t), \quad t\ge 0,\\ y(0) = y_0 < 0,\end{cases}$$

where $y(t)$ is the angle coordinate of a point on the circle and $w(t)$, $t\ge 0$, is a standard real Brownian motion.

It may be computed that the solution of this system is the process

$$y(t) = 2\text{arccot}\left(t + w(t) + \cot\frac{y_0}{2}\right), \qquad t\ge 0.$$

Since $t + w(t)\to\infty$ almost surely as $t\to\infty$, it is easy to see that the null solution is strongly unstable in sample paths. However, by a direct calculation one can show that the null solution is stable in the almost sure sense.

It is clear that (strong) stability in the pth moment of the null solution of (1.5.6) for any value of $p > 0$ implies its (strong) moment stability for every smaller value than p and (strong) stability in probability. On the other hand, one can easily show that the null solution could be the pth moment (strongly) stable for some $p > 0$ but not the qth moment (strongly) stable for $q > p$. The case most frequently discussed in the literature is (strong) moment stability with $p = 2$. We shall also refer to this case as (strong) *stability in mean square.*

Although some stability, for instance, Definition 1.5.7 or 1.5.10, does not appear to be as strong a restriction on systems as that given in Definition 1.5.4 or 1.5.5, there are significant implications in Definitions 1.5.7 and 1.5.10 for sample stability behavior. However, it is worth pointing out that stability of the moment alone does not always provide a satisfactory intuitive basis upon which to judge the stability characteristics of the systems of interest.

Example 1.5.16 Consider a simple one-dimensional linear Itô equation

$$dy(t) = ay(t)dt + by(t)dw(t), \qquad t \geq 0, \tag{1.5.13}$$

where $w(t)$, $t \geq 0$, is a standard one-dimensional Brownian motion, $y(0) = y_0 \in \mathbb{R}$, and a, b are real numbers.

A direct computation shows that the solution process $y(t)$, $t \geq 0$, is given by

$$y(t) = \exp\left\{bw(t) + (a - b^2/2)t\right\} y_0, \qquad t \geq 0. \tag{1.5.14}$$

Hence, by using the law of iterated logarithm for Brownian motion (cf. Revuz and Yor [197]), it is easy to deduce that the asymptotically exponential growth rate of solution y is given by

$$\overline{\lim_{t\to\infty}} \frac{\log|y(t)|}{t} = a - \frac{b^2}{2} \qquad a.s. \tag{1.5.15}$$

We then conclude that the null solution has global exponential stability in the almost sure sense if and only if $a < b^2/2$. On the other hand, using the standard exponential martingale properties for Brownian motion, it is also easy to see for any $n \in \mathbb{N}_+$ that

$$\mathbb{E}y(t)^n = y_0^n \cdot \exp\left\{(a - b^2/2)nt + \frac{b^2n^2}{2}t\right\}.$$

Hence, we conclude that the null solution has the global nth moment exponential stability if and only if $a < b^2(1 - n)/2$. Therefore, unlike deterministic systems, for $a < 0$, the first moment is exponentially stable, but higher moments are probably unstable. For $a < -b^2/2$, the first and second moments are exponentially stable, and higher moments are probably unstable, etc. It seems difficult to associate a physical meaning to the behavior of a system, knowing only that the first nth moments are stable and all higher moments are unstable. On the other hand, it is clear from (1.5.15) that the stability of sample trajectories are determined by the algebraic sign of $a - b^2/2$ only. It is interesting to note that for $a < b^2/2$, the sample path possesses almost surely asymptotic stability, but it is possible that all moments will diverge exponentially. Hence, we see in this example that unlike deterministic systems, even though stability in mean square implies almost sure stability, almost sure stability need not imply the moment stability.

Remark 1.5.17 If a system is almost surely (strong) asymptotically stable, then it is also (strong) asymptotically stable in probability. From the analogy of deterministic stability, it seems reasonable to assume in Definition 1.5.9 almost sure (strong) stability rather than (strong) stability in probability. However, it is worth pointing out this requirement is actually too strong. In fact, let us consider Example 1.5.16 again. By (1.5.14) and the properties of Brownian motion, it is easy to see that for no positive constant $\varepsilon > 0$ does there exist a number $\delta > 0$ such that almost all the sample trajectories of the solutions originating at $y_0 \neq 0$, $|y_0| < \delta$, remain in an ε-neighborhood of zero (i.e., not almost surely stable) even if the unperturbed term is very stable (i.e., $a < 0$) and $|b|$ is very small.

It is not always possible to get an explicit solution for a stochastic differential equation. Therefore, it is generally unrealistic to deal with stochastic stability problems in such a way as we did in Example 1.5.16. In the history of stability study, one of the most effective approaches is the so-called Lyapunov function method or Lyapunov's second (direct) method. To gain some insight into the main ideas of this method, let us analyze a simple situation.

Consider a nonnegative continuous function Λ on $\mathbb{R}^n$ with $\Lambda(0) = 0$ and $\Lambda(x) > 0$ for $x \neq 0$. Suppose that for some $\delta > 0$, the set $D_\delta = \{x \in \mathbb{R}^n : \Lambda(x) < \delta\}$ is bounded and $\Lambda(x)$ has continuous first-order derivatives in D_δ. Let $y(t) = y(t, y_0)$ be the unique solution of the initial value problem:

$$\begin{cases} dy(t) = f(y(t))dt, & t \geq 0, \\ y(0) = y_0 \in D_\delta \subset \mathbb{R}^n, \end{cases} \tag{1.5.16}$$

for a given function $f(\cdot) \in \mathbb{R}^n$ with $f(0) = 0$. Since $\Lambda(x)$ is continuous, the open set D_δ contains the origin and monotonically decreases to the singleton set $\{0\}$ as $\delta \downarrow 0$. If the total derivative $\dot{\Lambda}(y(t))$ of Λ, along the solution trajectory $y(t)$, satisfies

$$\dot{\Lambda}(y(t)) = f(y(t)) \cdot \left.\frac{d\Lambda(x)}{dx}\right|_{x=y(t)} = -k(y(t)) \leq 0, \qquad t \geq 0, \tag{1.5.17}$$

where $k(\cdot)$ is some nonnegative continuous function, then $\Lambda(y(t))$ is a nonincreasing function of t, i.e., $\Lambda(y_0) < \delta$ implies $\Lambda(y(t)) < \delta$ for all $t \geq 0$. In other words, $y_0 \in D_\delta$ implies that $y(t) \in D_\delta$ for all $t \geq 0$. This establishes the stability of the null solution to (1.5.16) in the sense of Lyapunov, and $\Lambda(x)$ is thus called a Lyapunov function of equation (1.5.16). If we further assume that $k(x) > 0$ for $x \in D_\delta \backslash \{0\}$, then $\Lambda(y(t))$, as a function of t, is strictly monotone decreasing. Moreover, we have from (1.5.17) that

$$0 < \Lambda(y_0) - \Lambda(y(t)) = \int_0^t k(y(s))ds < \infty \quad \text{for all} \quad t \in [0, \infty). \tag{1.5.18}$$

In this case, $\Lambda(y(t)) \to 0$ as $t \to \infty$ from (1.5.18) for sufficiently small $\delta > 0$ (otherwise, $\Lambda(y(t)) \geq \Lambda(y_0)$ for some sufficiently large $t > 0$). This further implies that $y(t) \to 0$ as $t \to \infty$, i.e., the null solution of system (1.5.16) is asymptotically stable.

It is possible to generalize the preceding Lyapunov function method to stochastic systems. For instance, let us consider a stochastic process $y(t) \in \mathbb{R}^n$, $t \geq 0$, on some probability space $(\Omega, \mathscr{F}, \mathbb{P})$. At present, it is not realistic to require that $\dot{\Lambda}(y(t,\omega)) \leq 0$ for all $\omega \in \Omega$. What one can expect for stability is that the time derivative of the expectation of $\Lambda(y(t))$, denote it by $\mathcal{L}\Lambda(\cdot)$, is nonpositive, where

$$\mathcal{L}\Lambda(y_0) := \lim_{t\to 0+} \frac{\mathbb{E}(\Lambda(y(t))) - \mathbb{E}\Lambda(y_0)}{t}, \qquad y_0 \in \mathbb{R}^n. \tag{1.5.19}$$

Here, the domain of $\mathcal{L}$ is defined as a family of those functions Λ for which (1.5.19) is well defined. This is a natural analogue of the total derivative of Λ along the process trajectory $y(t)$ to the deterministic case. Now suppose that there exists a Lyapunov function Λ satisfying the aforementioned conditions or

$$\mathcal{L}\Lambda(y_0) \leq 0, \quad y_0 \in \mathbb{R}^n,$$

then it is possible to show, usually under additional conditions such as a strong Markov property of y, that for any $t \geq s \geq 0$,

$$\mathbb{E}(\Lambda(y(t, y_0)) \mid \mathscr{F}_s^y) \leq \Lambda(y(s, y_0)) \qquad a.s.$$

This means that $\Lambda(y(t, y_0))$ is a nonnegative supermartingale, and by the well-known martingale convergence theorem, we may show that $\Lambda(y(t, y_0)) \to 0$, which further implies $y(t, y_0) \to 0$, almost surely as $t \to \infty$ and $\|y_0\|_{\mathbb{R}^n} \to 0$.

The Lyapunov function $\Lambda(\cdot)$ may be regarded as a generalized energy function of the system under investigation. The preceding argument illustrates the physical intuition that if the energy of a physical system is always decreasing near an equilibrium state, then the equilibrium state is stable.

Since Lyapunov's original work [159], the Lyapunov function method for stability has been extensively developed. The advantage of this method is that one can obtain considerable information about stability properties of a given system without being required to solve the system equation explicitly. The main drawback of this method is that there does not exist a general method to construct appropriate Lyapunov functions, especially for nonlinear systems. A stability criterion obtained in this manner, which usually provides only a sufficient condition, depends sensitively on the chosen Lyapunov function. In the remainder of this book, we shall mainly explore the Lyapunov function approach to establish a stochastic stability theory for infinite-dimensional stochastic differential equations.

1.6 Notes and Comments

All the material in Section 1.1 is standard, and the statement there is mainly based on Curtain and Zwart [49], Engel and Nagel [70], Kreyszig [112], Pazy [187], and Yosida [224]. The proof of Theorem 1.1.10 is sketched in Example A.4.2 in Curtain and Zwart [49]. A systematic presentation of the material in Sections 1.2 and 1.3 is given in Da Prato and Zabczyk [53]. Theorem 1.3.3 is presented in Da Prato and Zabczyk [53], and its version, Theorem 1.3.4, is established in Tubaro [212].

A systematic statement of the variational method for infinite-dimensional stochastic systems is presented by many authors such as Krylov and Rozovskii [113], Pardoux [184], and Prévôt and Röckner [190]. As for applications of semigroup approaches to infinite-dimensional stochastic systems, a comprehensive statement can be found in the existing literature such as Chow [41], Da Prato and Zabczyk [53], and Métivier [171]. Much material in Section 1.4.1 is taken from Ichikawa [91].

The stability of a real system is the ability of the system to resist an influence or disturbance unknown beforehand. The system is said to be stable if such a disturbance does not essentially change it. Indeed, an individual predictable process can be physically realized only if it is stable in the corresponding natural sense. For instance, we know from classical control theory that, before we can consider the design of a regulatory or tracking control system, we need to make sure that such a system is stable from input to output. For stability and the relevant Lyapunov function method of finite-dimensional deterministic systems, some systematic statements can be found in the literature, e.g., Hahn [82].

For a finite-dimensional stochastic system, there are two main techniques dealing with its stability properties. The first significantly extends Lyapunov's direct method for deterministic systems to a stochastic setting. The main ingredient here is a Lyapunov function, and as in the deterministic theory, a major difficulty in this method is to construct a suitable Lyapunov function to find the optimal stability condition for nonlinear stochastic systems. The earliest attempt to generalize the classic Lyapunov function method to stochastic stability goes back at least to Kats and Krasovskii [101]. During the initial development of the Lyapunov theory and method of stochastic stability, some confusion about the formulation of Lyapunov functions, their usefulness in application, and the relationship among the different concepts of stability existed. Kozin's survey [109] clarified some of the confusion and provided a good foundation for further development. Shortly, quite a few important works appeared, e.g., Kushner [114, 115] used martingale convergence techniques to develop a Lyapunov function theory for strong Markov processes and study related control problems, and Pinsky [188] introduced specific Lyapunov functions to handle stochastic stability of some Dirichlet problems in two dimensions. In the meanwhile, a comprehensive statement on stochastic stability theory was presented in Has'minskii [86] for diffusion processes given as the solutions of Itô's stochastic differential equations. For subsequent developments of this topic over the last several decades, the reader is referred to some informative monographs in this field such as Arnold [3], Khas'minskii [103], Kolmanovskii and Nosov [107], and Mao [165, 167], among others.

The other important development in finite-dimensional stochastic stability is the application of the so-called Lyapunov exponent method to stochastic systems. This is the stochastic counterpart of the notion of characteristic exponents introduced in Lyapunov's work on asymptotically exponential stability. Although the Lyapunov exponent method provides necessary and sufficient conditions for asymptotic (exponential) stability, this method needs to use sophisticated mathematical techniques, especially for nonlinear systems, and significant computational problems must be solved. Important studies of the Lyapunov exponent method, when applied to stochastic systems, have been made in such work as Arnold, Kliemann, and Oeljeklaus [7]; Arnold and Wihstutz [9]; Furstenberg [77, 78]; Khas'minskii [103]; Mohammed and Scheutzow [177, 178]; and Oseledec [182], among others.

Last, we mention some monographs that include chapters dealing with stability problems for finite-dimensional deterministic or stochastic systems with time delay: Hale [84], Kolmanovskii and Nosov [107], and Mao [165], among others.

2
Stability of Linear Stochastic Differential Equations

The purpose of this chapter is to establish stability results of systems defined by stochastic linear differential equations. We shall explore abstract stability, which is a random generalization of Lyapunov's classic work in finite-dimensional spaces. The characterizations of both mean square and almost sure exponential stability are established and applied to stochastic partial differential equations. Special attention is paid to the almost sure pathwise stability of trivial solutions, a case that can be considered most compatible with its deterministic counterpart. In some sense, it is this kind of stability that one really likes to have in applications.

2.1 Deterministic Linear Systems

We begin with a deterministic system where some linear unbounded operator A generates a stable C_0-semigroup $T(t)$, or e^{tA}, $t \geq 0$, of bounded linear operators on a Banach space X.

2.1.1 Stable Strongly Continuous Semigroups

First consider the following linear system on the standard n-dimensional Euclidean space $X = \mathbb{R}^n$:

$$\begin{cases} dy(t) = Ay(t)dt, & t \geq 0, \\ y(0) = y_0 \in \mathbb{R}^n, \end{cases} \tag{2.1.1}$$

where A is some $n \times n$ constant matrix. Clearly, for any $y_0 \in \mathbb{R}^n$ this equation has a unique solution that is explicitly given by

$$y(t, y_0) = e^{tA}y_0 = \sum_{n=0}^{\infty} \frac{t^n A^n}{n!} y_0, \qquad t \geq 0.$$

In particular, it is well known (see, e.g., Krasovskii [111]) that for this linear system, the following statements are equivalent:

(a) System (2.1.1) is asymptotically stable.
(b) System (2.1.1) is exponentially stable.

In an infinite-dimensional setting, the situation is quite different. Let X be a Banach space equipped with norm $\|\cdot\|_X$. Consider the following deterministic linear Cauchy problem on X,

$$\begin{cases} dy(t) = Ay(t)dt, & t \geq 0, \\ y(0) = y_0 \in X, \end{cases} \tag{2.1.2}$$

where A is a linear, generally unbounded, operator that generates a C_0-semigroup e^{tA}, $t \geq 0$, on X. In general, if $y_0 \notin \mathscr{D}(A)$, the initial value problem (2.1.2) does not have a classical solution. Instead, we may consider the mild solution $y(t) = e^{tA}y_0$, $t \geq 0$, of (2.1.2), and the time evolution of the C_0-semigroup e^{tA}, $t \geq 0$, totally determines stability of this system.

Definition 2.1.1 Let $T(t)$, $t \geq 0$, be a C_0-semigroup on Banach space X. It is called

(i) *weakly stable* if for every $x \in X$ and $y \in X^*$, $\langle\langle T(t)x, y\rangle\rangle_{X,X^*} \to 0$ as $t \to \infty$;
(ii) *asymptotically stable* if for every $x \in X$, $\|T(t)x\|_X \to 0$ as $t \to \infty$;
(iii) *(uniformly) stable* if $\lim_{t\to\infty} \|T(t)\| = 0$;
(iv) *exponentially stable* if there exist constants $M \geq 1$ and $\mu > 0$ such that $\|T(t)\| \leq Me^{-\mu t}$ for all $t \geq 0$;
(v) *L^p-stable* for some $1 \leq p < \infty$ if $\int_0^\infty \|T(t)x\|_X^p dt < \infty$ for any $x \in X$.

If X is finite dimensional, it can be proved (see, e.g., Krasovskii [111]) that all the stability definitions (i)–(v) are equivalent. In an infinite-dimensional setting, the following theorems and examples show, however, that (i) $\Leftarrow$ (ii) $\Leftarrow$ (iii) $\Leftrightarrow$ (iv) $\Leftrightarrow$ (v), while, in general, (i) $\not\Rightarrow$ (ii) $\not\Rightarrow$ (iii).

Firstly, note that, by definition, the implications (i) $\Leftarrow$ (ii) $\Leftarrow$ (iii) $\Leftarrow$ (iv) are immediate.

Theorem 2.1.2 *Let $T(t)$, $t \geq 0$, be a C_0-semigroup on Banach space X. The following statements are equivalent:*

(a) $T(t)$, $t \geq 0$, is exponentially stable;
(b) $\lim_{t\to\infty} \|T(t)\| = 0$;
(c) there exists some $t_0 > 0$ such that $\|T(t_0)\| < 1$.

Proof Clearly, it suffices to show that (c) $\Rightarrow$ (a). Suppose that (c) is true, then we have

$$\mu_0 := \inf_{t>0} \frac{\log \|T(t)\|}{t} < 0.$$

Let $w(t) = \log \|T(t)\|$, $t \geq 0$. It follows, by the semigroup property of $T(\cdot)$, that

$$w(t+s) \leq w(t) + w(s) \qquad \text{for all} \qquad s,\ t > 0.$$

Let $\varepsilon > 0$, then there exists a $\delta > 0$ such that

$$\mu_0 \leq \frac{w(\delta)}{\delta} \leq \mu_0 + \varepsilon.$$

Let $t > 0$ be arbitrarily given and n be the integer satisfying $t = n\delta + r$, $r \in [0, \delta)$. Then we have

$$\mu_0 \leq \frac{w(t)}{t} \leq \frac{nw(\delta)}{n\delta + r} + \frac{w(r)}{t}.$$

Hence, for t sufficiently large, or equivalently n sufficiently large, we have

$$\mu_0 \leq \frac{w(t)}{t} \leq \mu_0 + \varepsilon + \frac{w(r)}{t}.$$

As $\mu_0 < 0$ and ε can be sufficiently small, there exist $\mu > 0$ and $T_0 > 0$ such that

$$\|T(t)\| \leq e^{-\mu t} \qquad \text{for all} \qquad t \geq T_0.$$

On the other hand, since $\|T(t)\|$ is bounded on compact intervals according to (1.1.5), there exists an $M_0 \geq 1$ such that $\|T(t)\| \leq M_0$ for $t \in [0, T_0]$. Therefore, there are $M = M_0 e^{\mu T_0} \geq 1$ and $\mu > 0$ such that

$$\|T(t)\| \leq M e^{-\mu t}, \qquad \forall t \geq 0.$$

In other words, as claimed, (c) implies (a). The proof is thus complete. □

Example 2.1.3 Let l^2 be the Hilbert space of all square summable sequences with norm $\|a\|_{l^2}^2 = \sum_{i=1}^{\infty} a_i^2 < \infty$, $a = (a_1, \ldots, a_n, \ldots) \in l^2$. On the space l^2, we define a C_0-semigroup $T(t)$, $t \geq 0$, of bounded operators by

$$T(t)a = (e^{-t}a_1, e^{-t/2}a_2, \ldots, e^{-t/n}a_n, \ldots), \qquad t \geq 0,$$

for any $a = (a_1, a_2, \ldots, a_n, \ldots) \in l^2$. Then, it is easy to see that for each $t \in [0, \infty)$,

$$\|T(t)\| = \sup_{\|a\|_{l^2}=1} \|T(t)a\|_{l^2}$$
$$= \sup_{\|a\|_{l^2}=1} \left(\sum_{n=1}^{\infty} e^{-2t/n} a_n^2\right)^{1/2} \leq \sup_{\|a\|_{l^2}=1} \left(\sum_{n=1}^{\infty} a_n^2\right)^{1/2} = 1.$$

On the other hand, let $b_m := (\underbrace{0, 0, \ldots, 0}_{m-1}, 1, 0, 0, \ldots) \in l^2$, $m \in \mathbb{N}_+$. It follows by definition that

$$e^{-t/m} = \|T(t)b_m\|_{l^2} \leq \|T(t)\| \quad \text{for each} \quad m \in \mathbb{N}_+,$$

which, by passing on a limit, immediately yields

$$1 = \lim_{m\to\infty} e^{-t/m} \leq \|T(t)\|, \qquad t \in [0, \infty).$$

That is, for each $t \in [0, \infty)$, it is true that $\|T(t)\| = 1$. However, for each $a \in l^2$ we have by the Dominated Convergence Theorem that

$$\lim_{t\to\infty} \|T(t)a\|_{l^2}^2 = \lim_{t\to\infty} \sum_{n=1}^{\infty} e^{-2t/n} a_n^2 = \sum_{n=1}^{\infty} \lim_{t\to\infty} e^{-2t/n} a_n^2 = 0,$$

which means that $T(t)$, $t \geq 0$, is asymptotically stable, i.e., (ii) $\not\Rightarrow$ (iii).

Example 2.1.4 Let $T(t)$, $t \geq 0$ be the left translation semigroup on $X = L^2(\mathbb{R})$, i.e., for any function $f \in L^2(\mathbb{R})$, the semigroup $T(t)$, $t \geq 0$ is defined by

$$(T(t)f)(s) = f(s + t), \qquad s \in \mathbb{R}.$$

Then it is a C_0-semigroup of isometries, which is not asymptotically stable. However, for functions $f, g \in L^2(\mathbb{R})$ with compact supports and large $t > 0$, it follows that $T(t)f$ and g have disjoint supports. This fact yields that

$$\langle\langle T(t)f, g\rangle\rangle_{X,X^*} = \int_{-\infty}^{\infty} f(s + t)g(s)ds = 0.$$

For arbitrary $f, g \in L^2(\mathbb{R})$ and for each $n \in \mathbb{N}$, we can choose $f_n, g_n \in L^2(\mathbb{R})$ with compact supports such that $\|f - f_n\|_X \leq 1/n$ and $\|g - g_n\|_X \leq 1/n$. Then, when $t \geq 0$ is sufficiently large,

$$\begin{aligned}|\langle\langle T(t)f, g\rangle\rangle_{X,X^*}| &\leq |\langle\langle T(t)(f - f_n), g_n\rangle\rangle_{X,X^*}| + |\langle\langle T(t)f, g - g_n\rangle\rangle_{X,X^*}| \\ &\quad + |\langle\langle T(t)f_n, g_n\rangle\rangle_{X,X^*}| \\ &\leq \frac{1}{n}(\|g\|_X + 1 + \|f\|_X) + |\langle\langle T(t)f_n, g_n\rangle\rangle_{X,X^*}| \\ &\to 0 \ \text{ as } \ n \to \infty.\end{aligned}$$

Hence, $T(t)$, $t \geq 0$, is weakly stable, i.e., (i) $\not\Leftarrow$ (ii).

As for the property (v), it is easy to see that (iv) implies (v) for every $p \in [1, \infty)$. The following theorem shows that the converse implication (v) $\Rightarrow$ (iv) is true as well.

Theorem 2.1.5 *Let $T(t)$, $t \geq 0$, be a C_0-semigroup on X. Suppose that for some number $p \in [1, \infty)$,*

$$\int_0^\infty \|T(t)x\|_X^p dt < \infty \quad \textit{for every} \quad x \in X, \tag{2.1.3}$$

then there exist constants $M \geq 1$ and $\mu > 0$ such that $\|T(t)\| \leq Me^{-\mu t}$, $t \geq 0$.

Proof We first show that there exists a positive constant $C > 0$ such that for all $x \in X$,

$$\int_0^\infty \|T(t)x\|_X^p dt \leq C\|x\|_X^p. \tag{2.1.4}$$

To see this, let us define a family of operators $Q_n : X \to L^p(\mathbb{R}_+; X)$, $n \in \mathbb{N}$, by $Q_n x = \mathbf{1}_{[0,n]}(\cdot)T(\cdot)x$, $x \in X$. From (2.1.3), it follows immediately that for each $n \in \mathbb{N}$, Q_n is defined on all elements of X, and Q_n is linear and bounded. On the other hand, for each $x \in X$,

$$\sup_{n \in \mathbb{N}} \|Q_n x\|_{L^p(\mathbb{R}_+, X)} \leq \left(\int_0^\infty \|T(t)x\|_X^p \right)^{1/p} < \infty.$$

Hence, by the Principle of Uniform Boundedness, it follows that

$$\begin{aligned} \left(\int_0^\infty \|T(t)x\|_X^p dt \right)^{1/p} &= \sup_{n \in \mathbb{N}} \left(\int_0^n \|T(t)x\|_X^p dt \right)^{1/p} \\ &\leq \sup_{n \in \mathbb{N}} \|Q_n\| \|x\|_X < \infty, \qquad x \in X. \end{aligned} \tag{2.1.5}$$

That is, the inequality (2.1.4) with $C = (\sup_{n \in \mathbb{N}} \|Q_n\|)^p < \infty$ is valid. Let $M \geq 1$, $\mu > 0$ be two numbers such that $\|T(t)\| \leq Me^{\mu t}$ for all $t \geq 0$. Then, for all $x \in X$ and $t \geq 0$, we have

$$\begin{aligned} \frac{1 - e^{-p\mu t}}{p\mu} \|T(t)x\|_X^p &= \int_0^t e^{-p\mu s} \|T(t)x\|_X^p ds \\ &\leq \int_0^t e^{-p\mu s} \|T(s)\|^p \|T(t-s)x\|_X^p ds \\ &\leq M^p \int_0^t \|T(t-s)x\|_X^p ds \leq M^p C \|x\|_X^p. \end{aligned}$$

This implies that for all $t \in [1, \infty)$ and $x \in X$,

$$\|T(t)x\|_X^p \leq \frac{p\mu C M^p}{1 - e^{-p\mu t}} \|x\|_X^p \leq \frac{p\mu C M^p}{1 - e^{-p\mu}} \|x\|_X^p. \tag{2.1.6}$$

On the other hand, for any $t \in [0, 1]$, it is true that $\|T(t)\| \leq Me^{\mu}$. In addition to (2.1.6), this further implies that $\|T(t)\| \leq L$ for some constant $L > 0$ and all $t \geq 0$. Therefore, we have

$$t\|T(t)x\|_X^p = \int_0^t \|T(t)x\|_X^p ds \leq \int_0^t \|T(s)\|^p \|T(t-s)x\|_X^p ds \leq L^p C \|x\|_X^p$$

for all $x \in X$ and $t \geq 0$, and it follows that

$$\|T(t)\| \leq \frac{LC^{1/p}}{t^{1/p}}, \qquad t > 0,$$

which immediately implies $\|T(t_0)\| < 1$ for some $t_0 > 0$ sufficiently large. By virtue of Theorem 2.1.2, this implies the desired result and the proof is thus complete. □

Definition 2.1.6 Let A be the generator of a strongly continuous semigroup $T(t), t \geq 0$, on a Banach space X.

(i) The *growth bound* of A is defined by

$$\omega_g(A) := \inf\Big\{\mu \in \mathbb{R}: \text{there exists } M_\mu \geq 1 \text{ such that } \|T(t)\| \leq M_\mu e^{\mu t} \text{ for all } t \geq 0\Big\}.$$

(ii) The *spectral bound* of A is defined by

$$\omega_s(A) := \sup\{Re\,\lambda : \lambda \in \sigma(A)\},$$

where $\sigma(A)$ is the spectrum set of A.

From the well-known Hille–Yosida theorem, it is easy to see that

$$\omega_s(A) \leq \omega_g(A). \tag{2.1.7}$$

Although the equality in (2.1.7) is always valid when X is finite dimensional, the strict inequality may hold (cf. Example 4.2, chapter 4 in [187]) in an infinite-dimensional space. If this equality is valid, we would say that the *spectrum-determined growth condition* holds. In this case, knowledge of the spectrum of A may allow one to predict the limit behavior of solutions to (2.1.2). Hence, it is of importance to establish easily verified criteria under which a C_0-semigroup has the spectrum-determined growth condition.

Theorem 2.1.7 *Let $T(t)$, $t \geq 0$, be a C_0-semigroup with generator A on a Banach space X. If the following spectrum relation holds:*

$$\sigma(T(t))\backslash\{0\} \subset e^{t\sigma(A)} \quad \textit{for each} \quad t \geq 0, \tag{2.1.8}$$

then $\omega_g(A) \leq \omega_s(A)$ and the spectrum-determined growth condition holds. In particular, if the semigroup $T(t)$, $t \geq 0$, is eventually norm continuous, the associated spectrum-determined growth condition holds.

Proof It is well known (cf. theorem 2.1.6 in [49]) that

$$\omega_g(A) = \inf_{t>0} \frac{1}{t} \log \|T(t)\| = \lim_{t\to\infty} \frac{1}{t} \log \|T(t)\|.$$

For any fixed $t > 0$, the spectrum radius $r(T(t))$ is given (cf. theorem 7.5-5 in [112]) by

$$r(T(t)) = \lim_{n\to\infty} \|T(t)^n\|^{1/n} = \lim_{n\to\infty} \|T(nt)\|^{1/n}.$$

By taking the logarithm, we thus obtain

$$\log r(T(t)) = t \lim_{n\to\infty} \frac{\log \|T(nt)\|}{nt} = t\omega_g(A).$$

This yields $r(T(t)) = e^{t\omega_g(A)}$. However, relation (2.1.8) implies that $r(T(t)) \leq e^{t\omega_s(A)}$ for each $t \geq 0$, or equivalently, $e^{t\omega_g(A)} \leq e^{t\omega_s(A)}$, a fact that, in addition to $\omega_s(A) \leq \omega_g(A)$, shows the spectrum-determined growth condition.

If semigroup $T(t)$, $t \geq 0$, is eventually norm continuous, we know from Theorem 1.1.30 that (2.1.8) is true, so the desired result follows. The proof is thus complete. □

Although it is not generally true that asymptotic stability of C_0-semigroups implies their exponential stability in infinite-dimensional spaces, there is an important category of semigroups whose asymptotic stability is equivalent to exponential stability.

Theorem 2.1.8 *Suppose that for each $t > 0$, the C_0-semigroup $T(t)$ is compact, then the asymptotic stability of $T(t)$, $t \geq 0$, implies its exponential stability.*

Proof Firstly, note that if $T(t)$, $t \geq 0$, is asymptotically stable, then by the Principle of Uniform Boundedness there is a constant $M \geq 1$ such that $\|T(t)\| \leq M$ for all $t \geq 0$, which implies, in addition to (2.1.7), that

$$Re\,\lambda \leq 0 \quad \text{for all} \quad \lambda \in \sigma(A), \tag{2.1.9}$$

where A is the generator of $T(t)$, $t \geq 0$.

If $T(t)$ is asymptotically stable, then we claim that A has no eigenvalues on the imaginary axis. Indeed, if there exists $x \neq 0$ such that $Ax = ibx$ for some $b \in \mathbb{R}$, then by Theorem 2.1.7 and the compactness of $T(t)$, $T(t)x = e^{ibt}x$, $t > 0$, a fact that contradicts the assumption that $T(t)$ is asymptotically stable. By assumption, $T(t)$ is compact for $t > 0$, hence the spectrum $\sigma(A)$ consists only of eigenvalues, i.e., $\sigma(A) = \sigma_p(A)$ according to Theorems 1.1.25 and 1.1.26. Therefore, we have from (2.1.9) that

$$Re\,\lambda < 0 \qquad \text{for all} \qquad \lambda \in \sigma(A) = \sigma_p(A). \tag{2.1.10}$$

On the other hand, since $T(t)$ is compact for $t > 0$, we know by Theorems 2.1.7 and 1.1.26 that the spectrum-determined growth condition holds and A has only isolated eigenvalues. This fact, in addition to (2.1.10), implies the existence of constants $M \geq 1$ and $\mu > 0$ such that

$$\|T(t)\| \leq Me^{-\mu t}, \qquad \forall t \geq 0.$$

The proof is now complete. □

Next, we present a result about the equivalence between weak stability and asymptotic stability.

Theorem 2.1.9 *Suppose that $T(t)$, $t \geq 0$, is a weakly stable C_0-semigroup on Banach space X. If its infinitesimal generator A has compact resolvent for some λ, then $T(t)$, $t \geq 0$, is asymptotically stable.*

Proof Since a weakly convergent sequence in a Banach space is norm bounded (cf. p. 112 in [195]), for any $y \in X$ there exists a constant $M(y) > 0$ such that $\|T(n)y\|_X \leq M(y)$, uniformly for $n \in \mathbb{N}$. By virtue of the Principle of Uniform Boundedness, it follows that $\|T(n)\| \leq C$ for all $n \in \mathbb{N}$. We know that $\|T(t)\| \leq M$ for some $M > 0$ and all $t \in [0, 1]$. For any $t \in \mathbb{R}_+$, it is always possible to write $t = n + \delta$, where $n \in \mathbb{N}$ and $\delta \in [0, 1)$. Therefore, it follows that for any $t \in \mathbb{R}_+$,

$$\|T(t)\| = \|T(n)T(\delta)\| \leq \|T(n)\|\|T(\delta)\| \leq CM < \infty,$$

which shows that $T(t)$ is (uniformly) bounded for any $t \geq 0$.

On the other hand, by assumption there exists a $\lambda \in \mathbb{C}$ such that $(\lambda I - A)^{-1}$ is compact. For this λ, we show that for any $y \in X$, $(\lambda I - A)^{-1}T(t)y \to 0$ as $t \to \infty$. Indeed, suppose this is not true for some sequence; for simplicity, denote it by $\{n\}$, i.e., $\|(\lambda I - A)^{-1}T(n)y\|_X \geq \delta$ for some $\delta > 0$ when n is largely enough. For this sequence, since $T(n)y$ (thus, $(\lambda I - A)^{-1}T(n)y$) is weakly convergent to zero, it is norm bounded. By compactness of $(\lambda I - A)^{-1}$, this further implies the existence of a subsequence $\{n_k\}$ of $\{n\}$ such that

$(\lambda I - A)^{-1}T(n_k)y$ is strongly convergent, actually to zero as $k \to \infty$, which is a contradiction. This means that $(\lambda I - A)^{-1}T(t)y \to 0$ as $t \to \infty$. Now, for any $y \in X$, let $x \in (\lambda I - A)^{-1}y \in \mathscr{D}(A)$. Then

$$T(t)x = T(t)(\lambda I - A)^{-1}y = (\lambda I - A)^{-1}T(t)y \to 0 \quad \text{as} \quad t \to \infty, \tag{2.1.11}$$

which holds for any $x \in \mathscr{D}(A)$. Since $\mathscr{D}(A)$ is dense in X and $T(t)$ is uniformly bounded, (2.1.11) implies that $T(t)$ is asymptotically stable. □

2.1.2 Exponential Stability and Lyapunov Functions

In this section, we shall carry out a Lyapunov function type of program to generalize some stability results from finite- to infinite-dimensional spaces. At this point, we are especially interested in exponential stability as it shows explicitly the decay of the systems under consideration.

With regard to system (2.1.2), if the state space X is finite dimensional, there exist a few equivalent conditions for the null solution to be exponentially stable as mentioned in the beginning of this section. Some similar conditions can be also formulated, based on either properties of the spectrum of matrix A or the existence of an appropriate Lyapunov function. Precisely, recall the following results (cf. Krasovskii [111]).

Proposition 2.1.10 *Let X be finite dimensional, for instance, $\mathbb{R}^n$, $n \geq 1$. The null solution to (2.1.2) is exponentially stable if and only if one of the following conditions holds:*

(i) all eigenvalues of matrix A have negative real parts,

$$\sup\{Re\,\lambda:\ \det(\lambda I - A) = 0\} < 0; \tag{2.1.12}$$

(ii) there exists a nonnegative definite matrix, denote it by $P \geq 0$, such that the Lyapunov equation

$$PA + A^T P = -I \tag{2.1.13}$$

holds, where A^T is the transpose of A and I is the identity matrix. In the latter case, the function $\Lambda(x) = \langle Px, x\rangle_{\mathbb{R}^n}$, $x \in \mathbb{R}^n$, is a Lyapunov function of (2.1.2) in the sense that for solution $y(t)$, $t \geq 0$, to (2.1.2), the derivative of Λ along the trajectory $y(t)$ satisfies

$$\frac{d\Lambda(y(t))}{dt} = \langle \Lambda'(y(t)), Ay(t)\rangle_{\mathbb{R}^n} = -\|y(t)\|^2_{\mathbb{R}^n}, \qquad t \geq 0. \tag{2.1.14}$$

If space X is infinite dimensional, then Proposition 2.1.10 is only partially true. In this case, the condition (2.1.12), in general, does not imply exponential stability of this Cauchy problem unless some additional conditions are imposed on A. Note that on this occasion (2.1.12) is, of course, replaced by

$$\sup\{Re\,\lambda:\ \lambda \in \sigma(A)\} < 0, \tag{2.1.15}$$

where $\sigma(A)$ is the spectrum of operator A. This is certainly a consequence of the fact that linear operators in finite-dimensional spaces are always compact, and thus have only a finite number of point spectra. Although it is not generally true that (2.1.15) yields stable semigroups, a Lyapunov type of equation (2.1.13) or (2.1.14) has, nevertheless, its complete infinite-dimensional counterpart (see Theorems 2.1.11 and 2.1.12 and Remark 2.1.13).

Theorem 2.1.11 *Let $T(t)$, $t \geq 0$, be an exponentially stable C_0-semigroup on Banach space X. Then there exists a unique continuous mapping $\Lambda(\cdot)\colon X \to [0, \infty)$ such that*

(i) for each $x \in X$, the mapping $t \to \Lambda(T(t)x)\colon [0, \infty) \to [0, \infty)$ has the property that

$$\lim_{t\to\infty} \Lambda(T(t)x) = 0 \quad \text{and} \quad d\Lambda(T(t)x)/dt = -\|T(t)x\|_X^2; \tag{2.1.16}$$

(ii) there exists a constant $C > 0$ such that

$$\Lambda(x) \leq C\|x\|_X^2, \qquad \forall x \in X. \tag{2.1.17}$$

Conversely, if there exists a mapping $\Lambda(\cdot)\colon X \to [0, \infty)$ satisfying (i) and (ii), then the C_0-semigroup $T(t)$, $t \geq 0$, is exponentially stable.

Proof Suppose that $T(t)$, $t \geq 0$, is exponentially stable, i.e., $\|T(t)\| \leq Me^{-\mu t}$, $t \geq 0$, for some $M \geq 1$ and $\mu > 0$. Then the mapping Λ given by

$$\Lambda(x) = \int_0^\infty \|T(t)x\|_X^2 dt \tag{2.1.18}$$

is well defined for any $x \in X$. It is easy to see that $\Lambda(\cdot)$ satisfies an inequality of the form (2.1.17). Moreover, since $T(t)$ is exponentially stable and satisfies (2.1.17), it follows that

$$\Lambda(T(t)x) \leq C\|T(t)x\|_X^2 \leq CM^2e^{-2\mu t}\|x\|_X^2, \qquad x \in X,$$

where $C > 0$, $M \geq 1$, $\mu > 0$. Hence, $\Lambda(T(t)x) \to 0$ as $t \to \infty$ for any $x \in X$. On the other hand, from (2.1.18) we have that for $t \geq 0$ and $x \in X$,

$$\Lambda(T(t)x) = \int_0^\infty \|T(s)T(t)x\|_X^2 ds = \int_t^\infty \|T(s)x\|_X^2 ds,$$

and hence

$$d\Lambda(T(t)x)/dt = -\|T(t)x\|_X^2, \qquad t \geq 0, \quad x \in X,$$

which is exactly the second equality of (2.1.16). Uniqueness is a consequence of the equality

$$\Lambda(T(t)x) = \Lambda(x) + \int_0^t \frac{d\Lambda(T(s)x)}{ds} ds = \Lambda(x) - \int_0^t \|T(s)x\|_X^2 ds \quad (2.1.19)$$

and the fact that $\Lambda(T(t)x) \to 0$ as $t \to \infty$.

For the converse statement, note that we have (2.1.19) straightaway due to (2.1.16). Moreover, we have, by letting $t \to \infty$ in (2.1.19) and using (2.1.17), that

$$\int_0^\infty \|T(t)x\|_X^2 dt = \Lambda(x) \leq C\|x\|_X^2 < \infty, \qquad x \in X.$$

By virtue of Theorem 2.1.5, this implies immediately the exponential stability of $T(t)$, $t \geq 0$. The proof is thus complete. □

The mapping $\Lambda(\cdot)\colon X \to [0, \infty)$ in Theorem 2.1.11 is called a Lyapunov function of system (2.1.2). If X is a Hilbert space, one can obtain an infinite dimensional version of Proposition 2.1.10, (ii).

Theorem 2.1.12 *A C_0-semigroup $T(t)$ or e^{tA}, $t \geq 0$, on a Hilbert space H is exponentially stable if and only if for any self-adjoint operator $L \in \mathscr{L}^+(H)$, there exists a nonnegative, self-adjoint operator P_L, or simply, $P \in \mathscr{L}(H)$ such that the Lyapunov equation*

$$\langle Ax, Py\rangle_H + \langle Px, Ay\rangle_H = -\langle Lx, y\rangle_H \quad \textit{for any} \quad x,\ y \in \mathscr{D}(A), \quad (2.1.20)$$

holds. In this case, the Lyapunov function $\Lambda(\cdot) : H \to [0, \infty)$ is given by

$$\Lambda(x) = \langle P_I x, x\rangle_H, \qquad x \in H, \qquad (2.1.21)$$

where I is the identity operator on H.

Proof If $\|T(t)\| \leq Me^{-\mu t}$, $t \geq 0$, for some $M \geq 1$, $\mu > 0$, then it makes sense for each self-adjoint operator $L \in \mathscr{L}^+(H)$ to define a mapping P_L from H into itself by

$$P_L x = \int_0^\infty T^*(t)LT(t)x dt, \qquad x \in H, \qquad L \in \mathscr{L}^+(H).$$

It is easily seen that P_L is a self-adjoint operator on H and $P_L \geq 0$. Equation (2.1.20) follows from the fact that if x and y are both in $\mathscr{D}(A)$, then for any $t \geq 0$

$$
\begin{aligned}
&\langle AT(t)x, P_L T(t)y\rangle_H + \langle P_L T(t)x, AT(t)y\rangle_H \\
&\quad = \int_0^\infty \frac{d}{ds}\langle LT(t+s)x, T(t+s)y\rangle_H ds \\
&\quad = \langle LT(t+s)x, T(t+s)y\rangle_H\Big|_{s=0}^{\infty}.
\end{aligned}
\tag{2.1.22}
$$

Expanding the right-hand side of the preceding equality, noting $\lim_{s\to\infty} T(s)x = 0$ for all $x \in \mathscr{D}(A)$ and taking $t = 0$ in (2.1.22), we obtain the relation (2.1.20).

Conversely, assume that there exists a self-adjoint operator $P_I \geq 0$ such that (2.1.20) holds with $L = I$. Then for each $x \in H$ and $t \geq 0$, we define a function $\Lambda(x,t) = \langle P_I T(t)x, T(t)x\rangle_H$, $t \geq 0$, and also put $\Lambda(x) = \Lambda(x,0)$. Since P_I is nonnegative, this implies that $\Lambda(x,t) \geq 0$ for all $x \in H$, $t \geq 0$. Let $x \in \mathscr{D}(A)$, then $T(t)x \in \mathscr{D}(A)$, $t \geq 0$, and $\Lambda(x,t)$ is differentiable with derivative

$$
\frac{d\Lambda(x,t)}{dt} = \langle P_I AT(t)x, T(t)x\rangle_H + \langle P_I T(t)x, AT(t)x\rangle_H = -\|T(t)x\|_H^2.
$$

Hence, we have for each $t \geq 0$ that

$$
0 \leq \Lambda(x,t) = \Lambda(x) - \int_0^t \|T(s)x\|_H^2 ds,
$$

which further yields

$$
\Lambda(x) \geq \int_0^t \|T(s)x\|_H^2 ds \quad \text{for all} \quad t \geq 0, \quad x \in \mathscr{D}(A). \tag{2.1.23}
$$

For arbitrarily given $t \geq 0$, it is known that there exist constants $M \geq 1$, $\mu \geq 0$ such that for any $s \in [0,t]$,

$$
\|T(s)(x_n - x)\|_H \leq Me^{\mu t}\|x_n - x\|_H \qquad \text{for arbitrary} \qquad x_n,\ x \in H.
$$

This relation shows that if $x_n \to x$, then $T(s)x_n \to T(s)x$ uniformly on the compact interval $[0,t]$ as $n \to \infty$. Hence, by virtue of the Dominated Convergence Theorem we get that the inequality (2.1.23) is valid for all x in H since $\mathscr{D}(A)$ is dense in H and $\Lambda(\cdot)$ is continuous in H. This implies that

$$
\int_0^t \|T(s)x\|_H^2 ds \leq \Lambda(x) = \langle P_I x, x\rangle_H < \infty \qquad \text{for all} \qquad x \in H, \tag{2.1.24}
$$

which, letting $t \to \infty$, implies further the desired result by virtue of Theorem 2.1.5. Last, the relation (2.1.21) is immediate. The proof is thus complete. □

Remark 2.1.13 (i) According to the proofs of Theorem 2.1.12, this theorem can be also stated in another different way: the semigroup e^{tA}, $t \geq 0$, on H is exponentially stable if and only if for some self-adjoint, invertible positive operator $L \in \mathscr{L}(H)$, there exists a nonnegative, self-adjoint operator $P \in \mathscr{L}(H)$ such that

$$\langle Ax, Px\rangle_H + \langle Px, Ax\rangle_H = -\langle Lx, x\rangle_H \quad \text{for any} \quad x \in \mathscr{D}(A). \qquad (2.1.25)$$

(ii) Similarly, we can also obtain the following sufficient condition: the semigroup e^{tA}, $t \geq 0$, is exponentially stable if there exists a nonnegative, self-adjoint operator $P \in \mathscr{L}(H)$ such that

$$\langle Ax, Px\rangle_H + \langle Px, Ax\rangle_H \leq -\langle x, x\rangle_H \quad \text{for any} \quad x \in \mathscr{D}(A). \qquad (2.1.26)$$

Example 2.1.14 (Example 1.1.22 revisited) Suppose that $B = \beta I$, $\beta \in \mathbb{R}$, in (1.1.12).

First, note that from the definition of the resolvent sets $\rho(-A)$ and $\rho(\mathcal{A})$, we have that $\lambda \in \rho(\mathcal{A})$ if and only if $\lambda(\lambda - \beta) \in \rho(-A)$. Indeed, $\lambda \in \rho(\mathcal{A})$ if and only if the following equation system defining resolvent operators

$$\begin{pmatrix} \lambda I & -I \\ A & \lambda I - \beta I \end{pmatrix} \begin{pmatrix} y_1 \\ y_2 \end{pmatrix} = \begin{pmatrix} z_1 \\ z_2 \end{pmatrix}$$

has a unique solution in $\mathscr{D}(\mathcal{A})$ for any $(z_1, z_2) \in \mathcal{H}$, a situation that is possible if and only if $\lambda(\lambda - \beta) \in \rho(-A)$. From this, we have

$$\omega_s(\mathcal{A}) = \sup\{Re\,\lambda\colon\ \lambda \in \sigma(\mathcal{A})\} = \sup\{Re\,\lambda\colon\ \lambda(\lambda - \beta) \in \sigma(-A)\}.$$

In other words, $\omega_s(\mathcal{A})$ is the larger real part of the roots to the equation

$$\lambda^2 - \beta\lambda - \omega_s(-A) = 0 \qquad (2.1.27)$$

with $\lambda(\lambda - \beta) \in \sigma(-A)$. Since $-A$ is self-adjoint, $\sigma(-A)$ is a subset of $\mathbb{R}$, and so we have $\lambda(\lambda - \beta) \in \mathbb{R}$. Let $\lambda = a + ib$. Since $(a + ib)(a + ib - \beta) \in \mathbb{R}$, it follows that

$$b = 0 \qquad \text{or} \qquad b \neq 0,\ 2a - \beta = 0.$$

If $b = 0$, this means that $\omega_s(\mathcal{A})$ equals the larger real root of (2.1.27), which is

$$\omega_s(\mathcal{A}) = \frac{\beta}{2} + \sqrt{\frac{\beta^2}{4} + \omega_s(-A)} \quad \text{if} \quad \frac{\beta^2}{4} + \omega_s(-A) \geq 0.$$

If $b \neq 0$ and $2a - \beta = 0$, this means that (2.1.27) has in this case no real roots and $\omega_s(\mathcal{A}) = a = \beta/2$. In other words, we have the following relations of spectrum bounds between $\omega_s(-A)$ and $\omega_s(\mathcal{A})$

$$\omega_s(\mathcal{A}) = \begin{cases} \dfrac{\beta}{2} + \sqrt{\dfrac{\beta^2}{4} + \omega_s(-A)} & \text{if} \quad \dfrac{\beta^2}{4} + \omega_s(-A) \geq 0, \\ \dfrac{\beta}{2} & \text{otherwise.} \end{cases} \tag{2.1.28}$$

Moreover, we come to the conclusion as follows.

(i) If $\omega_s(-A) \geq 0$, we have $\omega_g(\mathcal{A}) \geq \omega_s(\mathcal{A}) \geq 0$ from (2.1.28), and system (1.1.13) is thus exponentially unstable for any $\beta \in \mathbb{R}$.
(ii) If $\omega_s(-A) < 0$ and $\beta \geq 0$, then (2.1.28) implies that $\omega_g(\mathcal{A}) \geq \omega_s(\mathcal{A}) \geq 0$, and system (1.1.13) is thus exponentially unstable.
(iii) If $\omega_s(-A) < 0$ and $\beta = -\alpha$, $\alpha > 0$, $B = -\alpha I$, we may show that the operator

$$\mathcal{P} = \begin{pmatrix} \dfrac{1}{\alpha} I + \dfrac{\alpha}{2} A^{-1} & \dfrac{1}{2} A^{-1} \\ \dfrac{1}{2} I & \dfrac{1}{\alpha} I \end{pmatrix} \tag{2.1.29}$$

is the unique self-adjoint, nonnegative solution of the Lyapunov equation

$$\left\langle \mathcal{A}\begin{pmatrix} y_1 \\ y_2 \end{pmatrix}, \mathcal{P}\begin{pmatrix} y_1 \\ y_2 \end{pmatrix} \right\rangle_{\mathcal{H}} + \left\langle \mathcal{P}\begin{pmatrix} y_1 \\ y_2 \end{pmatrix}, \mathcal{A}\begin{pmatrix} y_1 \\ y_2 \end{pmatrix} \right\rangle_{\mathcal{H}} = -\left\| \begin{pmatrix} y_1 \\ y_2 \end{pmatrix} \right\|_{\mathcal{H}}^2 \tag{2.1.30}$$

for any $y_1 \in \mathscr{D}(A)$, $y_2 \in \mathscr{D}(A^{1/2})$. In this case, the following estimates hold:

$$\gamma_- \left\| \begin{pmatrix} y_1 \\ y_2 \end{pmatrix} \right\|_{\mathcal{H}}^2 \leq \left\langle \mathcal{P}\begin{pmatrix} y_1 \\ y_2 \end{pmatrix}, \begin{pmatrix} y_1 \\ y_2 \end{pmatrix} \right\rangle_{\mathcal{H}} \leq \gamma_+ \left\| \begin{pmatrix} y_1 \\ y_2 \end{pmatrix} \right\|_{\mathcal{H}}^2, \qquad \begin{pmatrix} y_1 \\ y_2 \end{pmatrix} \in \mathcal{H}, \tag{2.1.31}$$

where

$$\gamma_- = \frac{1}{\alpha} \cdot \frac{\sqrt{1+\theta}}{1+\sqrt{1+\theta}} > 0, \qquad \gamma_+ = \frac{1}{\alpha}\left(1 + \frac{1+\sqrt{1+\theta}}{\theta}\right) > 0,$$

$$\theta = \frac{4|\omega_s(-A)|}{\alpha^2}.$$

Moreover, we have

$$\|e^{t\mathcal{A}}\| \leq \sqrt{\frac{\gamma_+}{\gamma_-}} e^{-\frac{1}{2\gamma_+} t}, \qquad t \geq 0.$$

That is, system (1.1.13) in this case is exponentially stable.

Indeed, it is straightforward to get symmetry and nonnegativity of $\mathcal{P}$ and (2.1.30) is easily verified by a direct calculation.

To show (2.1.31), note that for any $\begin{pmatrix} y_1 \\ y_2 \end{pmatrix} \in \mathcal{H}$,

$$\left\langle \mathcal{P}\begin{pmatrix} y_1 \\ y_2 \end{pmatrix}, \begin{pmatrix} y_1 \\ y_2 \end{pmatrix} \right\rangle_{\mathcal{H}} = \frac{1}{\alpha}\left(\|A^{1/2}y_1\|_H^2 + \frac{\alpha^2}{2}\|y_1\|_H^2 + \alpha\langle y_1, y_2\rangle_H + \|y_2\|_H^2 \right). \tag{2.1.32}$$

We first find the maximal $\gamma \geq 0$ such that for all $\begin{pmatrix} y_1 \\ y_2 \end{pmatrix} \in \mathcal{H}$,

$$\frac{1}{\alpha}\left(\|A^{1/2}y_1\|_H^2 + \frac{\alpha^2}{2}\|y_1\|_H^2 + \alpha\langle y_1, y_2\rangle_H + \|y_2\|_H^2 \right) \geq \gamma(\|A^{1/2}y_1\|_H^2 + \|y_2\|_H^2). \tag{2.1.33}$$

If $\gamma = 1/\alpha$, inequality (2.1.33) becomes

$$\frac{\alpha^2}{2}\|y_1\|_H^2 + \alpha\langle y_1, y_2\rangle_H \geq 0,$$

which clearly does not hold for all $\begin{pmatrix} y_1 \\ y_2 \end{pmatrix} \in \mathcal{H}$, e.g., let $y_2 = -\alpha y_1$, $y_1 \neq 0$. Therefore, $\gamma \in [0, 1/\alpha)$ and inequality (2.1.33) becomes

$$(1-\alpha\gamma)\|A^{1/2}y_1\|_H^2 + \frac{\alpha^2}{2}\|y_1\|_H^2 + \alpha\langle y_1, y_2\rangle_H + (1-\alpha\gamma)\|y_2\|_H^2 \geq 0.$$

On the other hand, for fixed y_1, we have

$$\begin{aligned} &\min_{y_2 \in H}\left\{\alpha\langle y_1, y_2\rangle_H + (1-\alpha\gamma)\|y_2\|_H^2\right\} \\ &= \min_{y_2 \in H}\left\{(1-\alpha\gamma)\left\| y_2 + \frac{\alpha y_1}{2(1-\alpha\gamma)} \right\|_H^2 - \frac{\alpha^2\|y_1\|_H^2}{4(1-\alpha\gamma)}\right\} \\ &= -\frac{\alpha^2}{4(1-\alpha\gamma)}\|y_1\|_H^2. \end{aligned}$$

Therefore, the required $\gamma \in [0, 1/\alpha)$ should be such that for $y_1 \in \mathscr{D}(A^{1/2})$,

$$(1-\alpha\gamma)\|A^{1/2}y_1\|_H^2 \geq \frac{\alpha^2}{2}\left(\frac{1}{2(1-\alpha\gamma)} - 1\right)\|y_1\|_H^2. \tag{2.1.34}$$

Since A is self-adjoint, it is known (see, e.g., Yosida [224]) that

$$|\omega_s(-A)| = \inf_{y_1 \neq 0} \frac{\|A^{1/2}y_1\|_H^2}{\|y_1\|_H^2}.$$

This implies, in addition to (2.1.34), that one is equivalently looking for the maximum value $\gamma \in [0, 1/\alpha)$ such that

$$|\omega_s(-A)| \geq \frac{\alpha^2}{4}\left(\frac{1}{(1-\alpha\gamma)^2} - \frac{2}{1-\alpha\gamma}\right),$$

or,

$$\theta \geq \frac{1}{(1-\alpha\gamma)^2} - \frac{2}{1-\alpha\gamma}.$$

This easily gives

$$\gamma_- = \frac{1}{\alpha} \cdot \frac{\sqrt{1+\theta}}{1+\sqrt{1+\theta}}.$$

In a similar way, the expression for γ_+ can be obtained by looking for a minimum value $\gamma > 0$ such that for all $\begin{pmatrix} y_1 \\ y_2 \end{pmatrix} \in \mathcal{H}$,

$$\frac{1}{\alpha}\Big(\|A^{1/2}y_1\|_H^2 + \frac{\alpha^2}{2}\|y_1\|_H^2 + \alpha\langle y_1, y_2\rangle_H + \|y_1\|_H^2 \Big) \leq \gamma\Big(\|A^{1/2}y_1\|_H^2 + \|y_1\|_H^2 \Big).$$

To prove the final part, we consider the mild solution y of the problem (1.1.13). Then from the Lyapunov equation (2.1.30), we have

$$\frac{d}{dt}\langle \mathcal{P}y(t), y(t)\rangle_{\mathcal{H}} = -\|y(t)\|_{\mathcal{H}}^2, \qquad t \geq 0,$$

which, in addition to (2.1.31), immediately implies

$$\frac{d}{dt}\langle \mathcal{P}y(t), y(t)\rangle_{\mathcal{H}} = -\|y(t)\|_{\mathcal{H}}^2 \leq -\frac{1}{\gamma_+}\langle \mathcal{P}y(t), y(t)\rangle_{\mathcal{H}}, \qquad t \geq 0.$$

A simple calculation further yields

$$\langle \mathcal{P}y(t), y(t)\rangle_{\mathcal{H}} \leq e^{-\frac{1}{\gamma_+}t}\langle \mathcal{P}y(0), y(0)\rangle_{\mathcal{H}} \leq \gamma_+ e^{-\frac{1}{\gamma_+}t}\|y(0)\|_{\mathcal{H}}^2, \qquad t \geq 0.$$

Hence, we have

$$\|y(t)\|_{\mathcal{H}}^2 \leq \frac{1}{\gamma_-}\langle \mathcal{P}y(t), y(t)\rangle_{\mathcal{H}} \leq \frac{\gamma_+}{\gamma_-} e^{-\frac{1}{\gamma_+}t}\|y(0)\|_{\mathcal{H}}^2, \qquad t \geq 0,$$

as desired.

2.2 Lyapunov Equations and Stochastic Stability

From this section on, we can consider stochastic stability of differential equations in a Hilbert space H. One of the main objectives in this section is to establish a stochastic version of Theorem 2.1.12. Based on this result, we shall establish sufficient conditions to capture the mean square and almost sure stability of linear stochastic systems.

2.2.1 Stochastic Stability in Mean Square

Let e^{tA}, $t \geq 0$, be a C_0-semigroup with its infinitesimal generator A on a Hilbert space H. Let $(\Omega, \mathscr{F}, \mathbb{P})$ be a complete probability space, equipped with a normal filtration $\{\mathscr{F}_t\}_{t\geq 0}$ with respect to which $\{W(t)\}_{t\geq 0}$, $t \geq 0$, is a Q-Wiener process on some Hilbert space K. Consider the following linear stochastic differential equation on H,

$$\begin{cases} dy(t) = Ay(t)dt + By(t)dW(t), & t \geq 0, \\ y(0) = y_0 \in H, \end{cases} \tag{2.2.1}$$

where $B \in \mathscr{L}(H, \mathscr{L}_2(K_Q, H))$. It is immediate from Theorem 1.4.2 that the equation (2.2.1) has a unique mild solution $y(t) \in C([0,T]; L^2(\Omega; H))$ for any $T \geq 0$.

Our arguments to establish a stochastic version of Theorem 2.1.12 basically involve a calculation of the term

$$\int_0^T \mathbb{E}\langle My(t), y(t)\rangle_H dt + \mathbb{E}\langle Gy(T), y(T)\rangle_H, \quad \forall T \geq 0, \tag{2.2.2}$$

where $M \geq 0$, $G \geq 0$ and both are self-adjoint in $\mathscr{L}(H)$. To this end, consider the following backward linear operator differential equation,

$$\begin{cases} \dfrac{d}{dt}\langle P(t)x, x\rangle_H + \langle Ax, P(t)x\rangle_H + \langle P(t)x, Ax\rangle_H + \langle [M + \Delta(P(t))]x, x\rangle_H = 0, \\ \qquad\qquad\qquad\qquad\qquad\qquad x \in \mathscr{D}(A), \quad 0 \leq t \leq T, \\ P(T) = G, \end{cases} \tag{2.2.3}$$

or its integral version (see Lemmas 2.2.2 and 2.2.3),

$$\begin{aligned} P(t)x = & \int_t^T e^{(s-t)A^*}[M + \Delta(P(s))]e^{(s-t)A}x ds \\ & + e^{(T-t)A^*}Ge^{(T-t)A}x, \quad \forall x \in H, \quad 0 \leq t \leq T, \end{aligned} \tag{2.2.4}$$

where $\Delta(\cdot) \in \mathscr{L}(H)$ is the unique operator defined by the sesquilinear form

$$\langle x, \Delta(P)y\rangle_H := Tr\big\{PB(x)Q^{1/2}(B(y)Q^{1/2})^*\big\}, \quad x, \ y \in H, \quad P \in \mathscr{L}(H). \tag{2.2.5}$$

Remark 2.2.1 To see the existence of operator $\Delta(P)\colon H \to H$, we note that for any $P \in \mathscr{L}(H)$ and fixed y, it is true that $Tr\big[PB(\cdot)Q^{1/2}(B(y)Q^{1/2})^*\big]$ is a continuous linear functional on H. By virtue of the Riesz Representation Theorem, there is a unique $y' \in H$ such that

$$Tr\big[PB(x)Q^{1/2}(B(y)Q^{1/2})^*\big] = \langle x, y'\rangle_H \quad \text{for all} \quad x \in H.$$

Define $\Delta(P)y = y'$. Then it is not difficult to check that $\Delta(P)$ is a bounded linear operator from H into itself. The uniqueness of $\Delta(P)$ in (2.2.5) is obvious.

Lemma 2.2.2 *There exists a unique solution $P(t) \in \mathscr{L}(H)$, $0 \le t \le T$, satisfying (2.2.4) in the strongly continuous family of linear, nonnegative, and self-adjoint operators on H.*

Proof Define a sequence of strongly continuous, self-adjoint, nonnegative operators $P_n(t)$, $n \in \mathbb{N}_0$, $0 \le t \le T$, by the equation system

$$\begin{cases} P_n(t)x = \int_t^T e^{(s-t)A^*}[M + \Delta(P_{n-1}(s))]e^{(s-t)A}xds \\ \qquad\qquad + e^{(T-t)A^*}Ge^{(T-t)A}x, \quad x \in H, \quad n \ge 1, \\ P_0(t) = 0, \quad 0 \le t \le T. \end{cases} \tag{2.2.6}$$

Then for any $n \ge 1$ and $0 \le t \le T$, we have the estimates

$$\|P_n(t)\| \le p_1(T) + a\int_t^T \|P_{n-1}(s)\|ds,$$

where $p_1(T) = \sup_{0 \le t \le T} \|P_1(t)\|$ and $a = a(T) > 0$. By induction, it is not difficult to show that

$$\|P_n(t)\| \le p_1(T)e^{a(T-t)} \quad \text{for any} \quad n \ge 1, \quad 0 \le t \le T.$$

That is, $P_n(t)$ is uniformly bounded in $t \in [0, T]$ and $n \ge 0$. It is easy to see by definition that $\Delta(P)$ is self-adjoint and $\Delta(P) \ge 0$ for each $P \ge 0$. Hence, $P_n(t)$ is self-adjoint, nonnegative, and increasing, i.e., $P_n(t) \le P_{n+1}(t)$ for each $t \ge 0$, and so there exists a strong limit $P(t)$ that is self-adjoint and nonnegative (cf. theorem 2, p. 304, [224]),

$$P(t)x = \lim_{n\to\infty} P_n(t)x, \quad t \ge 0, \quad x \in H, \tag{2.2.7}$$

which satisfies (2.2.4) and has the desired properties.

For uniqueness, suppose that there are operators $\tilde{P}(t)$ and $\bar{P}(t)$ satisfying (2.2.4), which clearly yields the following integral equation:

$$(\tilde{P}(t) - \bar{P}(t))x$$
$$= \int_t^T e^{(s-t)A^*}\Delta(\tilde{P}(s) - \bar{P}(s))e^{(s-t)A}xds, \quad x \in H, \quad 0 \le t \le T.$$

Then, we have in this case that

$$\|\tilde{P}(t) - \bar{P}(t)\| \le b\int_t^T \|\tilde{P}(s) - \bar{P}(s)\|ds, \quad \forall 0 \le t \le T,$$

for some $b = b(T) > 0$, which, together with the Gronwall inequality, immediately implies that $\|\tilde{P}(t) - \bar{P}(t)\| = 0$, i.e., $\tilde{P}(t) = \bar{P}(t)$ for all $0 \le t \le T$. The proof is now complete. □

Lemma 2.2.3 *The equations (2.2.3) and (2.2.4) are equivalent. Moreover, there exists a unique solution $P(t)$, $0 \le t \le T$, to (2.2.3) in the strongly continuous family of linear, self-adjoint, and nonnegative operators on H.*

Proof Obviously, it suffices to show by Lemma 2.2.2 the equivalence of (2.2.3) and (2.2.4). First suppose that $P(t)$, $t \ge 0$, satisfies (2.2.4). Then by differentiating $\langle P(t)x, x\rangle_H$, $x \in \mathscr{D}(A)$, we obtain (2.2.3) immediately.

Conversely, suppose $P(t)$, $t \ge 0$, satisfies (2.2.3). Let $x \in \mathscr{D}(A)$ and $0 < s \le t$, then $\langle P(t)e^{(t-s)A}x, e^{(t-s)A}x\rangle_H$ is differentiable in t and

$$\begin{aligned}
&\frac{d}{dt}\langle P(t)e^{(t-s)A}x, e^{(t-s)A}x\rangle_H \\
&= -\langle Ae^{(t-s)A}x, P(t)e^{(t-s)A}x\rangle_H - \langle P(t)e^{(t-s)A}x, Ae^{(t-s)A}x\rangle_H \\
&\quad + \langle P(t)e^{(t-s)A}x, Ae^{(t-s)A}x\rangle_H + \langle Ae^{(t-s)A}x, P(t)e^{(t-s)A}x\rangle_H \\
&\quad - \langle [M + \Delta(P(t))]e^{(t-s)A}x, e^{(t-s)A}x\rangle_H \\
&= -\langle [M + \Delta(P(t))]e^{(t-s)A}x, e^{(t-s)A}x\rangle_H.
\end{aligned} \tag{2.2.8}$$

By integrating (2.2.8) in t from s to T, we then obtain for any $0 \le s \le T$ that

$$\begin{aligned}
\langle P(s)x, x\rangle_H &= \langle Ge^{(T-s)A}x, e^{(T-s)A}x\rangle_H \\
&\quad + \int_s^T \langle [M + \Delta(P(u))]e^{(u-s)A}x, e^{(u-s)A}x\rangle_H du \quad \forall x \in \mathscr{D}(A).
\end{aligned} \tag{2.2.9}$$

Since $\mathscr{D}(A)$ is dense in H, (2.2.4) follows easily from (2.2.9). The proof is complete. □

Proposition 2.2.4 *Let $P(\cdot)$ be the unique solution of (2.2.3) and $y(t)$, $t \ge 0$, be the mild solution of equation (2.2.1), then the following relation holds: for any $0 \le t \le T$,*

$$\int_t^T \mathbb{E}\langle My(s), y(s)\rangle_H ds + \mathbb{E}\langle Gy(T), y(T)\rangle_H = \mathbb{E}\langle P(t)y(t), y(t)\rangle_H. \tag{2.2.10}$$

Proof We first introduce the following Yosida approximating system of (2.2.1):

$$\begin{cases} dy(t) = Ay(t)dt + nR(n, A)By(t)dW(t), & t \ge 0, \\ y(0) = nR(n, A)y_0 \in \mathscr{D}(A), \end{cases} \tag{2.2.11}$$

for $n \in \rho(A)$, the resolvent set of A, and $R(n, A)$ is the resolvent of A. Since $nAR(n, A) = n^2 R(n, A) - n$ is bounded, it is easy to verify that the conditions in Proposition 1.4.4 are satisfied so that there exists a unique solution $y_n(t)$, $t \geq 0$ of (2.2.11) in the strong sense. Then applying Itô's formula to $\langle P(t)y_n(t), y_n(t)\rangle_H$, we obtain for any $0 \leq t \leq T$ that

$$\int_t^T \mathbb{E}\langle My_n(s), y_n(s)\rangle_H ds + \mathbb{E}\langle Gy_n(T), y_n(T)\rangle_H = \mathbb{E}\langle P(t)y_n(t), y_n(t)\rangle_H, \tag{2.2.12}$$

which, by passing to the limit $n \to \infty$ and using Proposition 1.4.5, immediately yields (2.2.10). The proof is thus complete. □

Now we are in a position to establish the main results in this section.

Theorem 2.2.5 *Suppose $y(t, y_0)$, $t \geq 0$, is the unique mild solution of (2.2.1) with nonrandom initial datum $y_0 \in H$. Then the following statements are equivalent.*

(i) The solution $y(t, y_0)$, $t \geq 0$, is L^2-stable in mean square, i.e., it satisfies

$$\int_0^\infty \mathbb{E}\|y(t, y_0)\|_H^2 dt < \infty \quad \textit{for each} \quad y_0 \in H. \tag{2.2.13}$$

(ii) There exists a nonnegative, self-adjoint operator $P \in \mathscr{L}(H)$ such that

$$\langle Ax, Px\rangle_H + \langle Px, Ax\rangle_H + \langle \Delta(P)x, x\rangle_H = -\langle x, x\rangle_H \quad \textit{for any} \quad x \in \mathscr{D}(A), \tag{2.2.14}$$

where $\langle \Delta(P)y, y\rangle_H = Tr\{PB(y)Q^{1/2}(B(y)Q^{1/2})^*\}$ for each $y \in H$.

(iii) There exist positive numbers $M \geq 1$, $\mu > 0$ such that for all $t \geq 0$,

$$\mathbb{E}\|y(t, y_0)\|_H^2 \leq Me^{-\mu t}\|y_0\|_H^2.$$

Proof Let $T \geq 0$. Suppose that (i) holds and let $P_T(\cdot)$ be the unique solution of

$$\begin{cases} \dfrac{d}{dt}\langle P_T(t)x, x\rangle_H + \langle Ax, P_T(t)x\rangle_H + \langle P_T(t)x, Ax\rangle_H \\ \quad +\langle [I + \Delta(P_T(t))]x, x\rangle_H = 0, \qquad 0 \leq t \leq T, \quad x \in \mathscr{D}(A), \\ P_T(T) = 0. \end{cases} \tag{2.2.15}$$

Then, by applying Lemma 2.2.3, (2.2.4), and Proposition 2.2.4 to $M = I$, $G = 0$, we obtain for any $T \geq 0$ that

$$\langle P_T(0)y_0, y_0\rangle_H = \int_0^T \mathbb{E}\|y(s)\|_H^2 ds \leq \int_0^\infty \mathbb{E}\|y(s)\|_H^2 ds < \infty. \tag{2.2.16}$$

Define a mapping $B(\cdot,\cdot)\colon\ H \times H \to \mathbb{R}$ by

$$B(y_1, y_2) = \int_0^\infty \mathbb{E}\langle y(s, y_1), y(s, y_2)\rangle_H ds, \ \forall\ y_1,\ y_2 \in H.$$

Then it is easy to see that $B(\cdot,\cdot)$ is a bounded sesquilinear form, which implies (cf. corollary, p. 44 in [195]) the existence of a nonnegative, self-adjoint operator $\tilde{P}$ such that

$$B(y_1, y_2) = \langle \tilde{P} y_1, y_2\rangle_H,\ \forall\ y_1,\ y_2 \in H$$

and $P_T(0) \le \tilde{P}$ for any $T \ge 0$. Since $P_T(0)$ is monotonically increasing in T, there exists a limit $P = \lim_{T\to\infty} P_T(0) \ge 0$ that is nonnegative and self-adjoint. Applying (2.2.15) to $P_T(0)$ and letting $T \to \infty$, we can conclude that P satisfies (2.2.14). This shows that (i) implies (ii).

Now we suppose that (ii) holds. First, it is easy to see that (2.2.14) yields the relation

$$\langle Ax, Px\rangle_H + \langle Px, Ax\rangle_H \le -\langle x, x\rangle_H \quad \text{for any} \quad x \in \mathscr{D}(A),$$

which, by virtue of Remark 2.1.13, implies the exponential stability of C_0-semigroup e^{tA}, $t \ge 0$, i.e.,

$$\|e^{tA}\| \le C \cdot e^{-\gamma t} \quad \text{for some} \quad C \ge 1,\ \gamma > 0. \tag{2.2.17}$$

We may carry out an approximation system argument as in Proposition 2.2.4 to (2.2.1). By applying Itô's formula to $\Lambda(x) = \langle Px, x\rangle_H$, $x \in H$, and the Yosida approximating solution y_n of (2.2.1), and using Corollary 1.4.6, one can establish by virtue of (2.2.14) that

$$\langle P y_0, y_0\rangle_H = \mathbb{E}\langle P y(t, y_0), y(t, y_0)\rangle_H + \int_0^t \mathbb{E}\|y(s, y_0)\|_H^2 ds, \qquad t \ge 0. \tag{2.2.18}$$

Hence, it follows that

$$\int_0^\infty \mathbb{E}\|y(t, y_0)\|_H^2 dt \le \langle P y_0, y_0\rangle_H < \infty,$$

which implies the result (i).

On the other hand, from (2.2.18) we obtain

$$\frac{d}{dt}\mathbb{E}\langle P y(t), y(t)\rangle_H = -\mathbb{E}\|y(t)\|_H^2 \le -\|P\|^{-1}\mathbb{E}\langle P y(t), y(t)\rangle_H, \qquad t \ge 0, \tag{2.2.19}$$

which immediately yields

$$\mathbb{E}\langle P y(t), y(t)\rangle_H \le e^{-\|P\|^{-1} t}\langle P y_0, y_0\rangle_H, \qquad t \ge 0. \tag{2.2.20}$$

If $\|P\|^{-1} \neq 2\gamma$, by using (1.3.4), (2.2.17), (2.2.19), and (2.2.20), we get that for some number $C' > 0$,

$$\begin{aligned}
\mathbb{E}\|y(t, y_0)\|_H^2 &= \|e^{tA} y_0\|_H^2 \\
&\quad + \mathbb{E}\int_0^t Tr\{e^{(t-s)A} B(y(s)) Q^{1/2} (e^{(t-s)A} B(y(s)) Q^{1/2})^*\} ds \\
&\le C^2 e^{-2\gamma t} \|y_0\|_H^2 + C' \int_0^t e^{-2\gamma(t-s)} \mathbb{E}\|y(s)\|_H^2 ds \\
&= C^2 e^{-2\gamma t} \|y_0\|_H^2 - C' \int_0^t e^{-2\gamma(t-s)} \frac{d}{ds} \mathbb{E}\langle P y(s), y(s)\rangle_H ds \\
&\le C^2 e^{-2\gamma t} \|y_0\|_H^2 + C' e^{-2\gamma t} \langle P y_0, y_0\rangle_H \\
&\quad + \frac{C'}{2\gamma} \int_0^t e^{-2\gamma(t-s)} \mathbb{E}\langle P y(s), y(s)\rangle_H ds \\
&\le C^2 e^{-2\gamma t} \|y_0\|_H^2 \\
&\quad + \left(C' e^{-2\gamma t} + \frac{C'}{2\gamma(2\gamma - \|P\|^{-1})} e^{-\|P\|^{-1} t} \right) \langle P y_0, y_0\rangle_H .
\end{aligned}$$

If $\|P\|^{-1} = 2\gamma$, the last term in the preceding relation is replaced by $(C'/(2\gamma)) t e^{-2\gamma t} \langle P y_0, y_0\rangle_H$. This immediately implies (iii), i.e.,

$$\mathbb{E}\|y(t, y_0)\|_H^2 \le M e^{-\mu t} \|y_0\|_H^2 \quad \text{for some} \quad M \ge 1, \ \mu > 0.$$

Last, it is obvious that (iii) implies (i), so the proof is complete. □

Corollary 2.2.6 *The equation (2.2.14) has at most one solution. Thus, if such a solution P exists, it is unique and satisfies the relation*

$$\langle P y_0, y_0\rangle_H = \int_0^\infty \mathbb{E}\|y(t, y_0)\|_H^2 dt, \qquad y_0 \in H.$$

In particular, P is the strong limit of the operator family $P(t)$ in (2.2.7) as $t \to \infty$, which satisfies the equation

$$P = \int_0^\infty e^{tA^*} [I + \Delta(P)] e^{tA} dt.$$

Proof If there exists a solution P to (2.2.14), then the null solution of (2.2.1) is exponentially stable in mean square and the mild solution $y(t, y_0)$, $t \ge 0$, of (2.2.1) satisfies $\mathbb{E}\langle P y(t, y_0), y(t, y_0)\rangle_H \to 0$ as $t \to \infty$. Hence, by virtue of (2.2.18), it follows further that

$$\langle P y_0, y_0\rangle_H = \int_0^\infty \mathbb{E}\|y(t, y_0)\|_H^2 dt, \quad y_0 \in H. \tag{2.2.21}$$

From (2.2.21) and the fact that $\|P\| = \sup_{\|x\|_H=1} |\langle Px, x\rangle_H|$ (see theorem 9.2-2 in [112]), the uniqueness of P is verified. One can obtain the last equality from the construction of $P = \lim_{T\to\infty} P_T(0)$ and the equivalent integral form of equation (2.2.15). □

Example 2.2.7 Let H be a separable real Hilbert space and b_i, h_i, $i = 1, \ldots, n$, are some nonzero elements in H. Consider the following stochastic system

$$\begin{cases} dy(t) = Ay(t)dt + \sum_{i=1}^{n} b_i \langle h_i, y(t)\rangle_H dw_i(t), & t \geq 0, \\ y(0) = y_0 \in H, \end{cases} \tag{2.2.22}$$

where A is the generator of some C_0-semigroup e^{tA}, $t \geq 0$, on H and $w_i(t)$, $t \geq 0$, $i = 1, \ldots, n$, are a group of independent real-valued Brownian motions on $(\Omega, \mathscr{F}, \mathbb{P})$. We have here $K = \mathbb{R}^n$, $Q = I$, and $B = (B_1, \ldots, B_n)$ in (2.2.1) is given by $B_i = b_i \otimes h_i$ and

$$(By)k = \sum_{i=1}^{n} B_i(y)k_i = \sum_{i=1}^{n} (b_i \otimes h_i)(y)k_i, \quad k = (k_1, \ldots, k_n) \in \mathbb{R}^n, \quad y \in H. \tag{2.2.23}$$

Here for arbitrary x, $y \in H$, we denote by $x \otimes y$ the linear operator defined by $(x \otimes y)h = \langle y, h\rangle_H x$ for any $h \in H$. In this case, it is easy to see that

$$\Delta(P) = \sum_{i=1}^{n} \langle Pb_i, b_i\rangle_H h_i \otimes h_i, \qquad P \in \mathscr{L}(H). \tag{2.2.24}$$

Define an $n \times n$ matrix $M = (m_{ij})_{n\times n}$ by

$$m_{ij} = \int_0^\infty \langle e^{tA} b_i, h_j\rangle_H^2 dt, \quad i, j = 1, \ldots, n, \tag{2.2.25}$$

whenever the integrals in (2.2.25) exist, and let $\lambda_i \in \mathbb{C}$, $i = 1, \ldots, n$, be all the eigenvalues of M. We claim that if

(i) the semigroup e^{tA}, $t \geq 0$, is exponentially stable,
(ii) $|\lambda_i| < 1$ for all $i = 1, \ldots, n$,

then the system (2.2.22) is exponentially stable in mean square.

Indeed, condition (i) ensures the well-definedness of matrix M. Let $N = \int_0^\infty e^{tA^*} e^{tA} dt$ and $Y = (\langle b_1, Nb_1\rangle_H, \ldots, \langle b_n, Nb_n\rangle_H)^T$. Then (ii) implies that there exists a vector $X = (x_1, \ldots, x_n)^T = (I - M)^{-1} Y$, which is the unique solution to the equation

$$X - MX = Y.$$

In association with this solution, let P be the nonnegative, self-adjoint operator given by

$$P = \sum_{i=1}^{n} x_i \int_0^\infty e^{tA^*}(h_i \otimes h_i)e^{tA}dt + N. \tag{2.2.26}$$

Then it is easy to obtain

$$(\langle b_1, Pb_1\rangle_H, \dots, \langle b_n, Pb_n\rangle_H)^T = Y + MX = X. \tag{2.2.27}$$

Hence, (2.2.26) and (2.2.27) together imply

$$P = \sum_{i=1}^{n} \langle b_i, Pb_i\rangle_H \int_0^\infty e^{tA^*}(h_i \otimes h_i)e^{tA}dt + N.$$

That is, operator $P \in \mathscr{L}^+(H)$ satisfies the equation

$$P = \int_0^\infty e^{tA^*}\Big[I + \sum_{i=1}^{n} \langle b_i, Pb_i\rangle_H (h_i \otimes h_i)\Big]e^{tA}dt = \int_0^\infty e^{tA^*}[I + \Delta(P)]e^{tA}dt, \tag{2.2.28}$$

which, according to Lemma 2.2.3 and Corollary 2.2.6, immediately yields

$$\langle Ax, Px\rangle_H + \langle Px, Ax\rangle_H + \sum_{i=1}^{n} \langle B_i^* PB_i x, x\rangle_H = -\langle x, x\rangle_H \quad \text{for any} \quad x \in \mathscr{D}(A). \tag{2.2.29}$$

By virtue of Theorem 2.2.5, we have the desired stability of (2.2.22).

As a matter of fact, it is worth pointing out that by means of further arguments, one can actually show that (i) and (ii) are also necessary conditions for the stability of (2.2.22) (cf. Zabczyk [228]).

In application, it is not always possible to identify the operator P explicitly as we do in Example 2.2.7, whose stability can be completely determined by Theorem 2.2.5. On a variety of occasions, it is hoped that one can find effective sufficient conditions to capture the stability of equation (2.2.1). Motivated by Example 2.2.7, we assume the following condition for linear systems:

(H) semigroup e^{tA}, $t \geq 0$, is exponentially stable, i.e., there exist constants $M \geq 1$ and $\gamma > 0$ such that

$$\|e^{tA}\| \leq Me^{-\gamma t} \quad \text{for all} \quad t \geq 0, \tag{2.2.30}$$

and

$$\Big\| \int_0^\infty e^{tA^*}\Delta(I)e^{tA}dt \Big\| < 1. \tag{2.2.31}$$

Theorem 2.2.8 *Under the condition (H), system (2.2.1) has exponential stability in mean square. In other words, there exist positive constants $M \geq 1$, $\mu > 0$ such that*

$$\mathbb{E}\|y(t, y_0)\|_H^2 \leq M\|y_0\|_H^2 e^{-\mu t}, \qquad t \geq 0. \tag{2.2.32}$$

Proof By virtue of Theorem 2.2.5, it suffices to prove the existence of a nonnegative, self-adjoint operator $P \in \mathscr{L}(H)$ that is the solution of (2.2.14). This operator P can be constructed through a family of operators $\{P_n\}_{n\geq 1}$ defined by

$$0 \leq P_1 := \int_0^\infty e^{tA^*} e^{tA} dt \in \mathscr{L}(H),$$

and

$$P_{n+1} := P_1 + S(P_n), \qquad n \geq 1,$$

where

$$S(G) = \int_0^\infty e^{tA^*} \Delta(G) e^{tA} dt \qquad \text{for any} \qquad G \in \mathscr{L}(H).$$

Indeed, under the condition (H), it is clear that $S(G) \geq 0$ if $G \geq 0$, and P_n is thus nondecreasing as $n \to \infty$. On the other hand, we have

$$\|S(G)\| = \sup_{\|x\|_H = 1} \int_0^\infty \langle e^{tA^*} \Delta(G) e^{tA} x, x\rangle_H dt \leq \|G\| \|S(I)\|. \tag{2.2.33}$$

By virtue of (2.2.31), we have $\|S(I)\| < 1$, so (2.2.33) implies that mapping $S\colon \mathscr{L}(H) \to \mathscr{L}(H)$ is a contraction. This fact immediately implies the existence of the limit $P = \lim_{n\to\infty} P_n$, and

$$P = \int_0^\infty e^{tA^*} [I + \Delta(P)] e^{tA} dt, \tag{2.2.34}$$

or

$$\langle Ax, Px\rangle_H + \langle Px, Ax\rangle_H + \langle \Delta(P)x, x\rangle_H + \|x\|_H^2 = 0 \quad \text{for all} \quad x \in \mathscr{D}(A),$$

which, together with Theorem 2.2.5, immediately yields the desired result. The proof is thus complete. □

2.2.2 Almost Sure Stability

Consider the following linear stochastic evolution equation

$$\begin{cases} dy(t) = Ay(t)dt + By(t)dW(t), & t \geq 0, \\ y(0) = y_0 \in H, \end{cases} \tag{2.2.35}$$

where $A\colon \mathscr{D}(A) \subset H \to H$ generates a C_0-semigroup e^{tA}, $t \geq 0$, $B \in \mathscr{L}(H, \mathscr{L}_2)$, $\mathscr{L}_2 := \mathscr{L}_2(K_Q, H)$, and $W(t)$, $t \geq 0$, is a Q-Wiener process in Hilbert space K.

Lemma 2.2.9 *Assume that A generates an exponentially stable C_0-semigroup e^{tA}, $t \geq 0$, such that*

$$\|e^{tA}\| \leq Me^{-\lambda t}, \qquad t \geq 0,$$

for $M \geq 1$, $\lambda > 0$, and $M^2\|B\|^2 < 2\lambda$, where $\|B\| := \|B\|_{\mathscr{L}(H,\mathscr{L}_2)}$. Then for any self-adjoint, nonnegative operator $L \in \mathscr{L}(H)$, there exists a unique self-adjoint operator $P \in \mathscr{L}^+(H)$ satisfying the equation

$$\langle Ax, Px\rangle_H + \langle Px, Ax\rangle_H + \langle \Delta(P)x, x\rangle_H = -\langle Lx, x\rangle_H \quad \text{for any} \quad x \in \mathscr{D}(A), \tag{2.2.36}$$

where $\Delta(P)$ is given as in (2.2.5). In this case, the operator P satisfies

$$P = \int_0^\infty e^{tA^*}(\Delta(P) + L)e^{tA}dt, \tag{2.2.37}$$

and the system (2.2.35) has exponential stability in mean square. In other words, there exist positive constants $C \geq 1$, $\mu > 0$ such that

$$\mathbb{E}\|y(t, y_0)\|_H^2 \leq C\|y_0\|_H^2 e^{-\mu t}, \qquad t \geq 0. \tag{2.2.38}$$

Proof The proofs are similar to those of Theorem 2.2.8, but we sketch them here for the reader's convenience. Define a sequence of operators $\{P(j)\}_{j\geq 0}$ in H by

$$P(0) = 0, \qquad P(j+1) = \int_0^\infty e^{tA^*}(\Delta(P(j)) + L)e^{tA}dt, \quad j \in \mathbb{N}_+. \tag{2.2.39}$$

It is immediate to see that

$$\begin{aligned}
\|P(j+1) - P(j)\| &= \left\| \int_0^\infty e^{tA^*}\Delta(P(j) - P(j-1))e^{tA}dt \right\| \\
&\leq M^2\|B\|^2 \int_0^\infty e^{-2\lambda t}dt \|P(j) - P(j-1)\| \\
&= \frac{M^2\|B\|^2}{2\lambda}\|P(j) - P(j-1)\|.
\end{aligned}$$

As $M^2\|B\|^2/2\lambda < 1$, it is easy to see that $\{P(j)\}_{j=0}^\infty$ is a Cauchy sequence in $\mathscr{L}(H)$ so that, letting $j \to \infty$ in (2.2.39), we obtain the relation (2.2.37), or (2.2.36) by virtue of Lemma 2.2.3. The last claim is immediate by virtue of Theorem 2.2.5. □

Lemma 2.2.10 *Suppose that A generates a C_0-semigroup e^{tA}, $t \geq 0$, such that*

$$\|e^{tA}\| \leq Me^{-\lambda t}, \qquad t \geq 0,$$

for $M > 0$, $\lambda > 0$ and $M^2\|B\|^2 < 2\lambda$. Let A_n, $n \in \mathbb{N}$, be the Yosida approximation of A and $L \in \mathscr{L}(H)$ be a nonnegative, self-adjoint operator. For large $n \in \mathbb{N}$, let $P_n \geq 0$ denote the unique solution to the Lyapunov equation

$$A_n^* P_n + P_n A_n + \Delta(P_n) = -L. \tag{2.2.40}$$

Then as $n \to \infty$,

$$\langle x, P_n y\rangle_H \to \langle x, Py\rangle_H \quad \textit{for any} \quad x,\ y \in H,$$

where $P \in \mathscr{L}^+(H)$ is the unique solution to

$$A^* P + PA + \Delta(P) = -L. \tag{2.2.41}$$

Proof Recall that (see (A.13) in [53])

$$\|e^{tA_n}\| \leq Me^{-\frac{n\lambda t}{n-\lambda}}, \qquad n > \lambda, \qquad t \geq 0.$$

Thus the exponential stability of e^{tA}, $t \geq 0$, implies the same property of e^{tA_n}, $t \geq 0$, for large $n \in \mathbb{N}$. By virtue of Lemma 2.2.9, for any self-adjoint, nonnegative operator $L \in \mathscr{L}(H)$, there exists a unique solution $P_n \in \mathscr{L}(H)$ to (2.2.40) for each $n \in \mathbb{N}$ sufficiently large.

For each $n \in \mathbb{N}$, we construct a sequence

$$P_n(0) := 0 \quad \text{and} \quad P_n(j+1) := \int_0^\infty e^{tA_n^*}(\Delta(P_n(j))+L)e^{tA_n}dt, \quad j \in \mathbb{N}_+,$$

and in a similar way, let

$$P(0) := 0 \quad \text{and} \quad P(j+1) := \int_0^\infty e^{tA^*}(\Delta(P(j)) + L)e^{tA}dt, \quad j \in \mathbb{N}_+.$$

We show that for all $x,\ y \in H$ and $j \in \mathbb{N}$, the term $\langle x, (P_n(j)-P(j))y\rangle_H \to 0$ as $n \to \infty$. Indeed, we can get this by induction. For $j = 0$, the claim holds trivially. Now suppose that the claim holds for $j = k-1$, $k \geq 1$. Let $x,\ y \in H$, then for $j = k$, we have

$$
\begin{aligned}
&|\langle x, (P_n(k) - P(k))y\rangle_H| \\
&\quad = \left| \int_0^\infty \langle e^{tA_n}x, (\Delta(P_n(k-1)) + L)e^{tA_n}y\rangle_H dt \right. \\
&\qquad \left. - \int_0^\infty \langle e^{tA}x, (\Delta(P(k-1)) + L)e^{tA}y\rangle_H dt \right| \\
&\quad \le \int_0^\infty |\langle e^{tA_n}x, \Delta(P_n(k-1) - P(k-1))e^{tA_n}y\rangle_H| dt \qquad (2.2.42) \\
&\qquad + \left| \left\langle x, \int_0^\infty e^{tA_n^*}(\Delta(P(k-1)) + L)e^{tA_n}y dt \right. \right. \\
&\qquad \left. \left. - \int_0^\infty e^{tA^*}(\Delta(P(k-1)) + L)e^{tA}y dt \right\rangle_H \right| \\
&\quad =: I_1(n) + I_2(n).
\end{aligned}
$$

By the induction hypothesis, we have

$$
\langle x, (P_n(k-1) - P(k-1))y\rangle_H \to 0, \quad \forall x,\ y \in H, \quad \text{as } n \to \infty.
$$

This fact further implies $I_1(n) \to 0$ as $n \to \infty$ by the Dominated Convergence Theorem.

For $I_2(n)$, we have by Proposition 1.1.19 that

$$
\begin{aligned}
&\left| \left\langle x, \int_0^\infty \left(e^{tA_n^*}(\Delta(P(k-1)) + L)e^{tA_n}y - e^{tA^*}(\Delta(P(k-1)) + L)e^{tA}y \right) dt \right\rangle_H \right| \\
&\quad = \left| \int_0^\infty \langle e^{tA_n}x, (\Delta(P(k-1)) + L)(e^{tA_n} - e^{tA})y\rangle_H \right. \\
&\qquad \left. + \langle (e^{tA_n} - e^{tA})x, (\Delta(P(k-1)) + L)e^{tA}y\rangle_H dt \right| \\
&\quad \le \int_0^\infty \|e^{tA_n}\| \cdot \|x\|_H \|\Delta(P(k-1)) + L\| \cdot \|(e^{tA_n} - e^{tA})y\|_H dt \\
&\qquad + \int_0^\infty \|(e^{tA_n} - e^{tA})x\|_H \|\Delta(P(k-1)) + L\| \cdot \|e^{tA}\| \cdot \|y\|_H dt \\
&\quad \to 0 \qquad \text{as } n \to \infty,
\end{aligned}
$$

by virtue of the Dominated Convergence Theorem. Hence, the claim is proved by induction.

Last, by the proofs of Lemma 2.2.9 we have that $P(j) \to P$ and $P_n(j) \to P_n$, uniformly in n, as $j \to \infty$ in the norm topology of $\mathscr{L}(H)$. Therefore, we have

$$
\langle x, (P_n - P)y\rangle_H \to 0 \quad \text{as} \quad n \to \infty \quad \text{for any} \quad x,\ y \in H
$$

by using the relation

$$|\langle x, (P_n - P)y\rangle_H| \le |\langle x, (P_n - P_n(j))y\rangle_H| + |\langle x, (P_n(j) - P(j))y\rangle_H| + |\langle x, (P(j) - P)y\rangle_H|.$$

The proof is thus complete. □

Now consider the linear stochastic evolution equation on a real Hilbert space H,

$$\begin{cases} dy(t) = Ay(t)dt + \sum_{i=1}^{n} B_i y(t) dw_i(t), & t \ge 0, \\ y(0) = y_0 \in H, \end{cases} \tag{2.2.43}$$

where $B_i \in \mathscr{L}(H)$ and $w_i(t)$, $t \ge 0$, $i = 1, \ldots, n$, are a group of independent, standard, real-valued Brownian motions.

Proposition 2.2.11 *Let $B = (B_1, \ldots, B_n)$. Suppose that there exist $\beta = (\beta_1, \ldots, \beta_n) \in \mathbb{R}^n$ and $\lambda > 0$ such that the C_0-semigroup $e^{t(A+\beta\cdot B)}$ generated by $A + \beta \cdot B$ satisfies*

$$\|e^{t(A+\beta\cdot B)}\| \le e^{-\lambda t}, \qquad t \ge 0,$$

and

$$\frac{1}{4}\sum_{i=1}^{n} \beta_i^2 + \frac{1}{2}\sum_{i=1}^{n} \|B_i\|^2 < \lambda, \tag{2.2.44}$$

then the mild solution of (2.2.43) is almost surely exponentially stable with

$$\overline{\lim_{t\to\infty}} \frac{1}{t} \log \|y(t)\|_H \le -\Big(\lambda - \frac{1}{4}\sum_{i=1}^{n} \beta_i^2 - \frac{1}{2}\sum_{i=1}^{n} \|B_i\|^2\Big) \qquad a.s.$$

Proof Choose arbitrarily a number

$$\gamma \in \Big(0, \lambda - \frac{1}{4}\sum_{i=1}^{n} \beta_i^2 - \frac{1}{2}\sum_{i=1}^{n} \|B_i\|^2\Big).$$

Let us put $\|\beta\|^2 := \sum_{i=1}^{n} \beta_i^2$ and $\Lambda := A + \beta \cdot B + \frac{\|\beta\|^2}{4} + \gamma$, then Λ is the generator of a C_0-semigroup $T(t)$ that satisfies

$$\|T(t)\| \le e^{-(\lambda - \|\beta\|^2/4 - \gamma)t}, \qquad t \ge 0.$$

For each $m \in \mathbb{N}$, let $\Lambda_m := A_m + \beta \cdot B + \frac{\|\beta\|^2}{4} + \gamma$, where A_m is the Yosida approximation of A. Without loss of generality, it is assumed that for all $m \ge 1$, we have

$$\|T_m(t)\| \le e^{-(\lambda-\|\beta\|^2/4-\gamma)t}, \qquad t \ge 0. \tag{2.2.45}$$

Since semigroups $T(\cdot)$ and $T_m(\cdot)$, $m \ge 1$, are exponentially stable, by virtue of (2.2.44) and Lemma 2.2.9, one can find for $m \in \mathbb{N}$ and $\Delta(I) \in \mathscr{L}^+(H)$ a unique solution $Q_m \in \mathscr{L}(H)$ to the Lyapunov equations

$$\Lambda_m^* Q_m + Q_m \Lambda_m + \Delta(Q_m) = -\Delta(I). \tag{2.2.46}$$

In other words, for each $m \ge 1$, $P_m = Q_m + I \in \mathscr{L}(H)$ is the unique solution to the Lyapunov equation

$$\Lambda_m^*(P_m - I) + (P_m - I)\Lambda_m + \Delta(P_m - I) = -\Delta(I). \tag{2.2.47}$$

In an analogous manner, let $P \in \mathscr{L}(H)$ be the unique solution to

$$\Lambda^*(P - I) + (P - I)\Lambda + \Delta(P - I) = -\Delta(I).$$

Then, by virtue of Lemmas 2.2.9 and 2.2.10, it is easy to see that

$$P_m \ge I, \ m \in \mathbb{N}, \ P \ge I \quad \text{and} \quad \langle P_m x, y\rangle_H \to \langle Px, y\rangle_H, \quad \text{as } m \to \infty, \tag{2.2.48}$$

for all $x, y \in H$. By the definition of Λ_m, $m \ge 1$, and Proposition 1.1.18, it follows that

$$\langle(\Lambda_m + \Lambda_m^*)x, x\rangle_H \le 0 \quad \text{for any} \quad x \in H.$$

Hence, by a direct computation we find from relation (2.2.47) that

$$\begin{aligned}(A_m + \beta\cdot B)^* P_m + P_m(A_m + \beta\cdot B)&\\ + \Delta(P_m) + \frac{\|\beta\|^2}{2}P_m + 2\gamma P_m \le 0 \quad &\text{for each} \quad m \in \mathbb{N}.\end{aligned}$$

In other words, for any $m \in \mathbb{N}$ and $x \in H$, we have

$$\begin{aligned}&\langle P_m A_m x, x\rangle_H + \langle A_m x, P_m x\rangle_H + \langle\Delta(P_m)x, x\rangle_H + 2\gamma\langle P_m x, x\rangle_H\\ &\quad \le -\sum_{i=1}^n \beta_i\langle P_m x, B_i x\rangle_H - \sum_{i=1}^n \beta_i\langle B_i x, P_m x\rangle_H - \frac{\|\beta\|^2}{2}\langle P_m x, x\rangle_H,\end{aligned} \tag{2.2.49}$$

which further implies that for any $m \in \mathbb{N}$ and $x \in H$,

$$\begin{aligned}
&\langle P_m x, x\rangle_H \Big[\langle P_m x, A_m x\rangle_H + \langle A_m x, P_m x\rangle_H + \langle \Delta(P_m)x, x\rangle_H + 2\gamma\langle P_m x, x\rangle_H\Big] \\
&\quad \le -\sum_{i=1}^n \beta_i \langle P_m x, x\rangle_H \langle P_m x, B_i x\rangle_H \\
&\qquad - \sum_{i=1}^n \beta_i \langle P_m x, x\rangle_H \langle B_i x, P_m x\rangle_H - \frac{\|\beta\|^2}{2}\langle P_m x, x\rangle_H^2 \\
&\quad \le 2\sum_{i=1}^n \langle P_m x, B_i x\rangle_H^2.
\end{aligned} \tag{2.2.50}$$

On the other hand, let y_m denote the mild solution to the equation

$$\begin{cases} dy_m(t) = A_m y_m(t)dt + \sum_{i=1}^n B_i y_m(t) dw_i(t), \quad t \ge 0, \\ y_m(0) = y_0, \end{cases} \tag{2.2.51}$$

where $m \in \mathbb{N}$. Then there exists a sequence, still denote it by $y_m \in H$ (see [53]), such that

$$\sup_{t\in[0,T]} \|y_m(t) - y(t)\|_H \to 0 \quad \text{as} \quad m \to \infty \quad a.s. \tag{2.2.52}$$

If $y_0 = 0$, then $\mathbb{P}(y(t,0) = 0) = 1$ for $t \ge 0$, and the desired estimate holds trivially. Now suppose that $y_0 \ne 0$, then by uniqueness of the solutions and strict positivity of P_m, $\mathbb{P}(\langle P_m y_m(t), y_m(t)\rangle_H = 0) = 0$ for all $t \ge 0$. Letting $\theta_m(t) = \langle P_m y_m(t), y_m(t)\rangle_H$, $t \ge 0$, and applying Itô's formula to $\log\theta_m(t)$, we obtain by (2.2.50) that for $t \ge 0$,

$$\begin{aligned}
&\log\langle P_m y_m(t), y_m(t)\rangle_H \\
&\quad = \log\langle P_m y_0, y_0\rangle_H + \int_0^t \bigg\{\frac{1}{\theta_m(s)}\Big[\langle P_m y_m(s), A_m y_m(s)\rangle_H \\
&\qquad + \langle A_m y_m(s), P_m y_m(s)\rangle_H + \langle \Delta(P_m) y_m(s), y_m(s)\rangle_H\Big] \\
&\qquad - \frac{2}{\theta_m^2(s)} \sum_{i=1}^n \langle P_m y_m(s), B_i y_m(s)\rangle_H^2\bigg\} ds \qquad\qquad (2.2.53) \\
&\quad + \int_0^t \frac{2}{\theta_m(s)} \sum_{i=1}^n \langle P_m y_m(s), B_i y_m(s) dw_i(s)\rangle_H \\
&\quad \le \log\langle P_m y_0, y_0\rangle_H - 2\gamma t \\
&\qquad + 2\int_0^t \sum_{i=1}^n \left\langle \frac{P_m y_m(s)}{\langle P_m y_m(s), y_m(s)\rangle_H}, B_i y_m(s) dw_i(s)\right\rangle_H.
\end{aligned}$$

Letting $m \to \infty$ and using (2.2.48) and (2.2.52), we have from (2.2.53) that

$$\log\langle Py(t), y(t)\rangle_H \le \log\langle Py_0, y_0\rangle_H - 2\gamma t + 2\sum_{i=1}^{n}\int_0^t \left\langle \frac{Py(s)}{\langle Py(s), y(s)\rangle_H}, By(s)dw_i(s)\right\rangle_H. \tag{2.2.54}$$

On the other hand, it is easy to see that the stochastic integral

$$\int_0^t \left\langle \frac{2Py(s)}{\langle Py(s), y(s)\rangle_H}, B_i y(s)dw_i(s)\right\rangle_H, \quad t \ge 0, \quad i = 1, \dots, n,$$

is a real continuous local martingale, and since $\langle Px, x\rangle_H \ge \|x\|_H^2$ for any $x \in H$, it follows that

$$\sup_{x\neq 0,\, x\in H} \frac{\langle Px, B_i x\rangle_H^2}{\langle Px, x\rangle_H^2} \le \sup_{x\neq 0,\, x\in H} \frac{\|P\|^2\|B_i\|^2\|x\|_H^4}{\|x\|_H^4} < \infty.$$

Thus, one can employ Proposition 1.2.6 to get that for each $i \in \{1, \dots, n\}$,

$$\frac{1}{t}\int_0^t \left\langle \frac{2Py(s)}{\langle Py(s), y(s)\rangle_H}, B_i y(s)dw_i(s)\right\rangle_H \to 0 \quad \text{as} \quad t \to \infty \quad a.s.,$$

which, in addition to (2.2.54), immediately implies that

$$\overline{\lim_{t\to\infty}} \frac{1}{t}\log\|y(t)\|_H^2 \le \overline{\lim_{t\to\infty}} \frac{1}{t}\log\langle Py(t), y(t)\rangle_H \le -2\gamma \qquad a.s.$$

Last, letting

$$\gamma \to \lambda - \frac{1}{4}\sum_{i=1}^{n}\beta_i^2 - \frac{1}{2}\sum_{i=1}^{n}\|B_i\|^2,$$

then we obtain the desired result. □

Example 2.2.12 Consider the following linear stochastic differential equation in H,

$$dy(t) = Ay(t)dt + \sigma y(t)dw(t), \qquad t \ge 0, \qquad y(0) = y_0 \in H,$$

where $w(\cdot)$ is a standard real Brownian motion and A generates a C_0-semigroup e^{tA}, $t \ge 0$, satisfying $\|e^{tA}\| \le e^{\alpha t}$ for some $\alpha \in \mathbb{R}$ and $\sigma \in \mathbb{R}$. We claim that if $\alpha < \sigma^2/2$, the mild solution y is exponentially stable almost surely.

To see this, it suffices by Proposition 2.2.11 to find $\lambda > 0$ and $\beta \in \mathbb{R}$ such that

$$-\lambda > \alpha + \beta\sigma \qquad \text{and} \qquad \sigma^2 + \frac{\beta^2}{2} - 2\lambda < 0. \tag{2.2.55}$$

However, this is immediate since we can choose $\beta = -2\sigma$ and $\lambda = 2\sigma^2 - \alpha - \varepsilon$ with $\varepsilon > 0$ small enough by using the condition $\alpha < \sigma^2/2$ and, as a consequence, the mild solution y is pathwise exponentially stable with

$$\overline{\lim_{t\to\infty}} \frac{1}{t} \log \|y(t)\|_H \leq \alpha - \frac{\sigma^2}{2} \qquad a.s.$$

2.2.3 Unbounded Diffusion Operators

In (2.2.35), operator B is assumed to be bounded, precisely, $B \in \mathscr{L}(H, \mathscr{L}_2(K_Q, H))$. In this section, we shall relax this restriction and allow B to be unbounded in the sense that $B \in \mathscr{L}(V, \mathscr{L}_2(K_Q, H))$.

Recall that V is a Banach space that is densely and continuously embedded into a separable Hilbert space H. That is, we have the following relations:

$$V \hookrightarrow H \cong H^* \hookrightarrow V^*$$

with $\|\cdot\|_H^2 \leq \beta\|\cdot\|_V^2$ for some $\beta > 0$. For simplicity, we assume in this section that $W(t)$, $t \geq 0$, is a K-valued, Q-Wiener process with $Tr(Q) < \infty$ defined on some complete probability space $(\Omega, \mathscr{F}, \mathbb{P})$, equipped with a normal filtration $\{\mathscr{F}_t\}_{t\geq 0}$. Consider the following stochastic linear differential equation

$$\begin{cases} dy(t) = Ay(t) + By(t)dW(t), \\ y(0) = y_0 \in H, \end{cases} \tag{2.2.56}$$

where $A\colon V \to V^*$ is a bounded linear operator, $B \in \mathscr{L}(V, \mathscr{L}_2(K_Q, H))$, and there exist constants $\alpha > 0$, $\lambda \geq 0$ such that for any $x \in V$,

$$2\langle\langle x, Ax\rangle\rangle_{V,V^*} + \langle\langle x, \Delta(I)x\rangle\rangle_{V,V^*} \leq -\alpha\|x\|_V^2 + \lambda\|x\|_H^2, \qquad \forall x \in V. \tag{2.2.57}$$

Here, for arbitrary $P \in \mathscr{L}(H)$, $\Delta(P) \in \mathscr{L}(V, V^*)$ is the unique operator defined by

$$\langle\langle x, \Delta(P)y\rangle\rangle_{V,V^*} = Tr\big[P(Bx)Q^{1/2}((By)Q^{1/2})^*\big], \qquad x,\ y \in V. \tag{2.2.58}$$

According to theorem 3.1, p. 105 in [184], it is known that a unique strong solution to (2.2.56) exists in $L^2(\Omega; C([0,T]; H)) \cap L^2(\Omega \times [0,T]; V)$ and the associated energy equation (1.4.24) holds.

Remark 2.2.13 If $B \in \mathscr{L}(H, \mathscr{L}_2(K_Q, H))$, then we clearly have the relation

$$\langle\langle x, \Delta(I)x\rangle\rangle_{V,V^*} \leq \gamma\|x\|_H^2 \qquad \text{for some} \qquad \gamma > 0,$$

and in this case the condition (2.2.57) can reduce to a simple form as

$$2\langle\langle x, Ax\rangle\rangle_{V,V^*} \leq -\alpha\|x\|_V^2 + \lambda\|x\|_H^2, \qquad \forall x \in V. \tag{2.2.59}$$

Under condition (2.2.57) or (2.2.59), it may be shown (see, e.g., Chapter 3 in [203]) that A generates an analytic semigroup e^{tA}, $t \geq 0$, in both H and V^* such that $e^{tA}\colon\ V^* \to V$ for each $t > 0$. Hence, the condition (2.2.31) makes sense.

Theorem 2.2.14 *Under (2.2.57) and the conditions (2.2.30) and (2.2.31), the strong solution $y(t)$, $t \geq 0$, of (2.2.56) is exponentially stable in mean square, i.e., there exist constants $C \geq 1$, $M > 0$ and $\mu > 0$ such that*

$$\mathbb{E}\|y(t, y_0)\|_H^2 \leq C\|y_0\|_H^2 e^{-\mu t}, \qquad t \geq 0, \tag{2.2.60}$$

and

$$\mathbb{E}\Big(\sup_{0\leq t<\infty} \|y(t, y_0)\|_H^2\Big) \leq M\|y_0\|_H^2. \tag{2.2.61}$$

Proof By a similar argument to the proofs of Theorem 2.2.8, the relation (2.2.60) is easily proved. Moreover, it may be shown similarly to (2.2.18) that there exists a self-adjoint, nonnegative operator $P \in \mathscr{L}(H)$ such that

$$\mathbb{E}\langle Py(t), y(t)\rangle_H = \mathbb{E}\langle Py(s), y(s)\rangle_H - \int_s^t \mathbb{E}\|y(u)\|_H^2 du, \qquad \forall 0 \leq s \leq t. \tag{2.2.62}$$

To show (2.2.61), we first notice that the relations (1.4.24) and (2.2.57) yield

$$\begin{aligned}\mathbb{E}\|y(t)\|_H^2 \leq \mathbb{E}\|y(s)\|_H^2 &+ \lambda \int_s^t \mathbb{E}\|y(u)\|_H^2 du\\ &- \alpha \int_s^t \mathbb{E}\|y(u)\|_V^2 du, \qquad \forall 0 \leq s \leq t.\end{aligned} \tag{2.2.63}$$

Let us put $g(t) := \lambda\mathbb{E}\langle Py(t), y(t)\rangle_H + \mathbb{E}\|y(t)\|_H^2$, $t \geq 0$. Then it follows from (2.2.62) and (2.2.63) that

$$g(t) \leq g(s) - \alpha \int_s^t \mathbb{E}\|y(u)\|_V^2 du, \qquad \forall 0 \leq s \leq t. \tag{2.2.64}$$

On the other hand, we have by applying the energy equation (1.4.24) to $y(t, y_0)$, $t \geq 0$, that

$$\begin{aligned}\|y(t)\|_H^2 &= \|y_0\|_H^2 + 2\int_0^t \langle\langle y(s), Ay(s)\rangle\rangle_{V,V^*} ds\\ &\quad + \int_0^t \langle\langle y(s), \Delta(I)y(s)\rangle\rangle_{V,V^*} ds + 2\int_0^t \langle y(s), By(s)dW(s)\rangle_H\\ &\leq \|y_0\|_H^2 + \lambda\int_0^t \|y(s)\|_H^2 ds + 2\int_0^t \langle y(s), By(s)dW(s)\rangle_H, \quad t \geq 0.\end{aligned} \tag{2.2.65}$$

Hence, for any $T \geq 0$, we have

$$\mathbb{E}\sup_{0\leq t\leq T}\|y(t)\|_H^2 \leq \|y_0\|_H^2 + \lambda\int_0^T \mathbb{E}\|y(s)\|_H^2 ds + 2\mathbb{E}\sup_{0\leq t\leq T}\left|\int_0^t \langle y(s), By(s)dW(s)\rangle_H\right|. \tag{2.2.66}$$

However, by using Doob's inequality (1.2.9) and the relation $2ab \leq \frac{a^2}{6} + 6b^2$ for $a, b \geq 0$, we have that for $T \geq 0$,

$$\begin{aligned}
&2\mathbb{E}\sup_{0\leq t\leq T}\left|\int_0^t \langle y(s), By(s)dW(s)\rangle_H\right| \\
&\leq 6\mathbb{E}\left\{\int_0^T \|y(s)\|_H^2 \langle\langle y(s), \Delta(I)y(s)\rangle\rangle_{V,V^*}ds\right\}^{1/2} \\
&\leq 3\mathbb{E}\left\{2\sup_{0\leq t\leq T}\|y(t)\|_H\left[\int_0^T \langle\langle y(s), \Delta(I)y(s)\rangle\rangle_{V,V^*}ds\right]^{1/2}\right\} \\
&\leq \frac{1}{2}\mathbb{E}\left\{\sup_{0\leq t\leq T}\|y(t)\|_H^2\right\} + 18\int_0^T \mathbb{E}\langle\langle y(s), \Delta(I)y(s)\rangle\rangle_{V,V^*}ds.
\end{aligned} \tag{2.2.67}$$

Hence, by virtue of (2.2.64), we obtain for $T \geq 0$ that

$$\begin{aligned}
\int_0^T \mathbb{E}\langle\langle y(s), \Delta(I)y(s)\rangle\rangle_{V,V^*}ds &= \int_0^T \mathbb{E}\Big\{Tr[By(s)Q^{1/2}(By(s)Q^{1/2})^*]\Big\}ds \\
&\leq Tr(Q)\|B\|^2\int_0^T \mathbb{E}\|y(s)\|_V^2 ds \\
&\leq \alpha^{-1}Tr(Q)\|B\|^2 g(0) \\
&\leq \alpha^{-1}Tr(Q)\|B\|^2(\lambda\|P\|+1)\|y_0\|_H^2,
\end{aligned} \tag{2.2.68}$$

which, in addition to (2.2.60), (2.2.66), and (2.2.67), immediately yields that

$$\mathbb{E}\sup_{0\leq t\leq T}\|y(t)\|_H^2 \leq \Big(2+2\lambda C\mu^{-1}+36\alpha^{-1}Tr(Q)\|B\|^2(\lambda\|P\|+1)\Big)\|y_0\|_H^2 \quad \text{for any} \quad T \geq 0. \tag{2.2.69}$$

By letting $T \to \infty$ and using the well-known Fatou lemma, we can easily get the desired result. The proof is thus complete. □

Remark 2.2.15 There is a somewhat stronger condition that, together with the energy equation (1.4.24), yields (2.2.60). This condition may be stated as follows: there exists a constant $\nu > 0$ such that for arbitrary $y \in V$,

$$2\langle\langle y, Ay\rangle\rangle_{V,V^*} + \langle\langle y, \Delta(I)y\rangle\rangle_{V,V^*} \leq -\nu\|y\|_H^2. \tag{2.2.70}$$

At present, we will not go into further details about this because some general conditions that contain (2.2.70) as a special case will be introduced to deal with nonlinear stochastic systems in Theorem 3.3.1.

Now we are in a position to obtain the desired pathwise exponential stability of the strong solutions to (2.2.56). Let $E_1, E_2, \ldots$ be an infinite sequence of events from probability space $(\Omega, \mathscr{F}, \mathbb{P})$. Recall that the event that infinitely many of E_n occur is defined by $\bigcap_{n=1}^{\infty}\bigcup_{m=n}^{\infty} E_m$.

Lemma 2.2.16 (Borel–Cantelli Lemma) *Let* $E = \bigcap_n \bigcup_{m=n}^{\infty} E_m$ *be the event that infinitely many of* E_n *occur. Then*

(a) $\mathbb{P}(E) = 0$ *if* $\sum_{n=1}^{\infty} \mathbb{P}(E_n) < \infty$,
(b) $\mathbb{P}(E) = 1$ *if* $\sum_{n=1}^{\infty} \mathbb{P}(E_n) = \infty$, *and* $E_1, E_2, \ldots$ *are independent events.*

Theorem 2.2.17 *Under the same conditions as in Theorem 2.2.14, there exist a constant* $\nu > 0$ *and random variable* $T(\omega) \geq 0$ *such that for all* $t \geq T(\omega)$,

$$\|y(t, y_0)\|_H \leq \|y_0\|_H e^{-\nu t} \qquad a.s.$$

In other words, the null solution of (2.2.56) is exponentially stable in the almost sure sense.

Proof Without loss of generality, we assume that $\lambda > 0$ in (2.2.57). From the energy equation (1.4.24) and (2.2.57), it follows that for arbitrary $t \geq N > 0$, $N \in \mathbb{N}_+$,

$$\begin{aligned}
\|y(t)\|_H^2 &= \|y(N)\|_H^2 + 2\int_N^t \langle\langle y(s), Ay(s)\rangle\rangle_{V,V^*} ds \\
&\quad + \int_N^t \langle\langle y(s), \Delta(I)y(s)\rangle\rangle_{V,V^*} ds + 2\int_N^t \langle y(s), By(s)dW(s)\rangle_H \\
&\leq \|y(N)\|_H^2 + \lambda \int_N^t \|y(s)\|_H^2 ds + 2\int_N^t \langle y(s), By(s)dW(s)\rangle_H.
\end{aligned} \tag{2.2.71}$$

Hence, for any positive constant $\varepsilon_N > 0$, we have

$$\begin{aligned}
&\mathbb{P}\Big\{\sup_{N\leq t\leq N+1} \|y(t)\|_H \geq \varepsilon_N\Big\} \\
&\leq \mathbb{P}\{\|y(N)\|_H^2 \geq \varepsilon_N^2/3\} + \mathbb{P}\Big\{\int_N^{N+1} \|y(t)\|_H^2 dt \geq \varepsilon_N^2/3\lambda\Big\} \\
&\quad + \mathbb{P}\Big\{\sup_{N\leq t\leq N+1}\Big|\int_N^t \langle y(s), By(s)dW(s)\rangle_H\Big| \geq \varepsilon_N^2/6\Big\}.
\end{aligned} \tag{2.2.72}$$

On the other hand, by using (2.2.64), one can show, similarly to (2.2.68), that

$$\int_N^{N+1} \mathbb{E}\langle\langle y(s), \Delta(I)y(s)\rangle\rangle_{V,V^*} ds \le M\|y_0\|_H^2 e^{-\mu N} \quad \text{for some} \quad M > 0,$$

which, similarly to (2.2.60) and (2.2.67), implies that

$$\begin{aligned}
&\mathbb{P}\Big\{ \sup_{N\le t\le N+1} \Big| \int_N^t \langle y(s), By(s)dW(s)\rangle_H \Big| \ge \varepsilon_N^2/6 \Big\} \\
&\le 6\varepsilon_N^{-2}\Big\{\mathbb{E} \sup_{N\le t\le N+1} \Big| \int_N^t \langle y(s), By(s)dW(s)\rangle_H \Big|\Big\} \\
&\le \frac{3}{2}\varepsilon_N^{-2}\mathbb{E}\Big\{ \sup_{N\le t\le N+1} \|y(t)\|_H^2\Big\} + 54\varepsilon_N^{-2} \int_N^{N+1} \mathbb{E}\langle\langle y(s), \Delta(I)y(s)\rangle\rangle_{V,V^*} ds \\
&\le c_1\|y_0\|_H^2 e^{-\mu N}/\varepsilon_N^2, \qquad (2.2.73)
\end{aligned}$$

where $c_1 > 0$ is some constant. Therefore, it follows from (2.2.72), (2.2.73), and (2.2.60) that there exist positive constants $c_2, c_3, c_4 > 0$ such that

$$\begin{aligned}
\mathbb{P}\Big\{ \sup_{N\le t\le N+1} \|y(s)\|_H \ge \varepsilon_N\Big\} &\le c_2\|y_0\|_H^2 e^{-\mu N}\Big/\varepsilon_N^2 \\
&\quad + \Big\{c_3\|y_0\|_H^2 e^{-\mu N} + c_1\|y_0\|_H^2 e^{-\mu N}\Big\}\Big/\varepsilon_N^2 \\
&\le c_4 e^{-\mu N/2}
\end{aligned}$$

if we take $\varepsilon_N = \|y_0\|_H e^{-\mu N/4}$. Then Lemma 2.2.16 further implies the existence of a random time $N_1(\omega) > 0$ such that whenever $N \ge N_1(\omega)$,

$$\sup_{N\le t\le N+1} \|y(t)\|_H^2 \le \|y_0\|_H^2 e^{-\mu N/2} \qquad a.s.$$

and the desired result follows. The proof is now complete. □

2.3 Systems with Boundary Noise*

Suppose that $\mathcal{O}$ is a bounded open domain in $\mathbb{R}^n$ with smooth boundary $\partial\mathcal{O}$. Informally, consider a stochastic partial differential equation of mixed type in which noise sources appear on the boundary $\partial\mathcal{O}$,

$$\begin{cases}
\dfrac{\partial y(t,x)}{\partial t} = \Delta y(t,x), \quad x \in \mathcal{O}, \quad t \ge 0, \\
y(0,x) = y_0(x), \quad x \in \mathcal{O}, \\
\dfrac{\partial y(t,x)}{\partial n} + h(x)y(t,x) = g(x)f(y(t,x))\dot{w}(t), \quad x \in \partial\mathcal{O}, \quad t \ge 0,
\end{cases} \qquad (2.3.1)$$

* The material of this section is not used in other parts and the reader can skip this section or only visit it when the need arises.

where g, h, f are some appropriate functions, Δ is the standard Laplace operator, $\partial/\partial n$ is the outward normal derivative. Here $w(t)$ is a standard real Brownian motion and $\dot{w}(t)$, white noise process, is the informal derivative of $w(t)$ with respect to $t \geq 0$.

Let Σ be the mapping: $L^2(\partial\mathcal{O}) \to L^2(\mathcal{O})$ defined by $\Sigma g = u$ for $g \in L^2(\partial\mathcal{O})$, where u is the unique solution of the equation,

$$\begin{cases} \Delta u(x) = 0, & x \in \mathcal{O}, \\ \dfrac{\partial u(x)}{\partial n} + h(x)u(x) = g(x), & x \in \partial\mathcal{O}. \end{cases} \tag{2.3.2}$$

We take $H = L^2(\mathcal{O})$ and define $\mathscr{D}(\Delta)$ as the closure in $L^2(\mathcal{O})$ of the subspace of $C^2(\overline{\mathcal{O}})$, which consists of functions ϕ satisfying the boundary condition $\partial\phi/\partial n + h\phi = 0$. Let A be the restriction of Δ on $\mathscr{D}(\Delta)$. It turns out (see [95]) that under some regular conditions on g, h, and f, we can reformulate (2.3.1) to get an abstract stochastic evolution equation on H of the form

$$y(t) = e^{tA}y_0 + \int_0^t (-A)^\theta e^{(t-s)A} bf(y(s))dw(s), \qquad t \geq 0, \tag{2.3.3}$$

where $b = (-A)^{1-\theta}\Sigma g \in H$ and $\theta \in (1/4, 1/2)$. Any solution of (2.3.3) is also called the mild solution of (2.3.1). Here the remarkable feature in this formulation is that the operator $(-A)^\theta$ appears in the integral term of (2.3.3).

In this section, we sketch a theory similar to those in Section 2.2 on stability property of such systems as (2.3.3) and refer the reader to [95] for more details. To this end, let us employ a variational scheme as in Section 2.2.3. Let H be a real Hilbert space and suppose that

$$V = \mathscr{D}((-A)^\theta) \hookrightarrow H \cong H^* = (\mathscr{D}((-A)^\theta))^*, \qquad \theta \in (0, 1),$$

where A is some nonpositive, self-adjoint operator on H and $(\mathscr{D}((-A)^\theta))^*$ is the dual space of $\mathscr{D}((-A)^\theta)$. Moreover, there exists a constant $\alpha > 0$ such that

$$2\langle\langle x, Ax\rangle\rangle_{V,V^*} \leq -\alpha\|x\|_V^2, \qquad \forall x \in V.$$

It is known that A generates an exponentially stable, analytic semigroup $e^{tA}: V^* \to V$, $t \geq 0$, on both H and V^*. Consider the following stochastic evolution equation on H,

$$\begin{cases} y(t) = e^{tA}y_0 + \displaystyle\int_0^t (-A)^\theta e^{(t-s)A} bf(y(s))dw(s), & t \geq 0, \\ y(0) = y_0, \end{cases} \tag{2.3.4}$$

where $0 < \theta < 1/2$, $b \in H$, f is a real Lipschitz continuous function on H, $f(0) = 0$ and $w(t)$ is a real standard Brownian motion defined on probability space $(\Omega, \mathscr{F}, \{\mathscr{F}_t\}_{t\geq 0}, \mathbb{P})$. In [95], it was shown that when $y_0 \in (\mathscr{D}((-A)^\theta))^*$ and $0 \leq \theta < 1/2$, there exists a unique solution to (2.3.4) in $C([0,T]; L^2(\Omega; H))$.

To establish the stability property of (2.3.4), we need to show, in line with the spirit of Section 2.2.1, the existence of a solution of the operator integral equation

$$\begin{cases} \langle\!\langle P(t)y, y\rangle\!\rangle_\theta = \langle Ge^{(T-t)A}y, e^{(T-t)A}y\rangle_H + \displaystyle\int_t^T \Big[\langle Me^{(s-t)A}y, e^{(s-t)A}y\rangle_H \\ \qquad\qquad + f^2(e^{(s-t)A}y)\langle P(s)(-A)^\theta b, (-A)^\theta b\rangle_H\Big]ds, \\ \qquad\qquad\qquad\qquad\qquad y \in (\mathscr{D}((-A)^\theta))^*, \quad 0 \leq t \leq T, \\ P(T) = G, \end{cases} \tag{2.3.5}$$

where M, $G \in \mathscr{L}^+(H)$ and $\langle\!\langle \cdot, \cdot\rangle\!\rangle_\theta$ are the dual pairing between $\mathscr{D}((-A)^\theta)$ and $(\mathscr{D}((-A)^\theta))^*$, $\theta \in (0,1)$.

Proposition 2.3.1 *Suppose that $\theta \in (0, 1/2)$. Then there exists a unique solution $P(t) \in \mathscr{L}^+(H)$ to equation (2.3.5) such that $P(t) \in \mathscr{L}((\mathscr{D}((-A)^\theta))^*, \mathscr{D}((-A)^\theta))$, $t \in [0,T)$, and $P(t)$ is strongly continuous on $[0,T)$ with*

$$\|P(t)\|_\theta \leq \frac{C}{(T-t)^{2\theta}} \quad \textit{for some} \quad C = C(T) > 0,$$

where $\|\cdot\|_\theta$ is the operator norm in the space $\mathscr{L}((\mathscr{D}((-A)^\theta))^*, \mathscr{D}((-A)^\theta))$.

Proof Note that if $b \in \mathscr{D}((-A)^\theta)$, the proof is easily given as in Section 2.2.1.

We first show the existence and uniqueness of solutions for the following operator integral equation

$$\begin{cases} \langle Q(t)z, z\rangle_H = \langle Ge^{(T-t)A}(-A)^\theta z, e^{(T-t)A}(-A)^\theta z\rangle_H \\ \qquad + \displaystyle\int_t^T \Big[\langle Me^{(s-t)A}(-A)^\theta z, e^{(s-t)A}(-A)^\theta z\rangle_H \\ \qquad + \langle Q(s)b, b\rangle_H f^2(e^{(s-t)A}(-A)^\theta z)\Big]ds, \\ Q(T) = (-A)^\theta G(-A)^\theta, \quad z \in \mathscr{D}((-A)^\theta), \quad 0 \leq t \leq T, \end{cases} \tag{2.3.6}$$

where M, $G \in \mathscr{L}^+(H)$. To establish the uniqueness of solutions to (2.3.6), it suffices to consider the equation

$$\langle R(t)z, z\rangle_H = \int_t^T \langle R(s)b, b\rangle_H f^2(e^{(s-t)A}(-A)^\theta z)ds,$$
$$z \in \mathscr{D}((-A)^\theta), \qquad t \in [0,T].$$

This further implies that for all $t \in [0, T)$,

$$\|R(t)\| \leq C \int_t^T \frac{\|R(s)\|}{(s-t)^{2\theta}} ds \quad \text{for some} \quad C = C(T) > 0.$$

Thus by using Gronwall's inequality, we obtain $\|R(t)\| = 0$ almost surely with respect to $t \in [0, T]$.

In order to find a solution to (2.3.6), one can proceed by defining an operator $Q_0(t)$ for $t \in [0, T]$ as

$$\begin{aligned}\langle Q_0(t)z, z\rangle_H = &\langle Ge^{(T-t)A}(-A)^\theta z, e^{(T-t)A}(-A)^\theta z\rangle_H \\ &+ \int_t^T \langle Me^{(s-t)A}(-A)^\theta z, e^{(s-t)A}(-A)^\theta z\rangle_H ds, \qquad (2.3.7) \\ & \qquad\qquad z \in \mathscr{D}((-A)^\theta),\end{aligned}$$

and then iterating a sequence of nonnegative operators $\{Q_n(t)\}$, $n \in \mathbb{N}_+$, $t \in [0, T)$, by

$$\langle Q_n(t)z, z\rangle_H = \langle Q_0(t)z, z\rangle_H + \int_t^T \langle Q_{n-1}(s)b, b\rangle_H f^2(e^{(s-t)A}(-A)^\theta z) ds.$$

Then $Q_n(t)$ is well defined and strongly continuous on $[0, T)$. Moreover for each $t \in [0, T)$, $Q_n(t)$ is monotonically increasing in $n \in \mathbb{N}$. On the other hand, it is not difficult to get the following estimate

$$\|Q_n(t)\| \leq \frac{C_1}{(T-t)^{2\theta}} + C_1 \int_t^T \frac{\|Q_{n-1}(s)\|}{(s-t)^{2\theta}} ds, \qquad n \in \mathbb{N},$$

for some number $C_1 = C_1(T) > 0$. Thus, it follows that

$$\|Q_n(t)\| \leq \frac{C_2}{(T-t)^{2\theta}}, \qquad t \in [0, T), \qquad n \in \mathbb{N}, \qquad (2.3.8)$$

for some number $C_2 = C_2(T) > 0$ independent of n. Hence, for each $t \in [0, T)$ there exists a strong limit $Q(t)$ of $Q_n(t)$, and $Q(t)$ has the desired properties.

By virtue of the unique solution Q of (2.3.6), we define

$$P(t) = (-A)^{-\theta} Q(t)(-A)^{-\theta}, \qquad t \in [0, T].$$

Then it is easy to see that P is the unique solution of (2.3.5). Moreover, by virtue of (2.3.8), there is an inequality

$$\|P(t)\|_\theta \leq \frac{C_3}{(T-t)^{2\theta}}, \quad t \in [0, T) \quad \text{for some} \quad C_3 = C_3(T) > 0.$$

The proof is now complete. □

Corollary 2.3.2 *For $T > 0$, $y_0 \in (\mathscr{D}((-A)^\theta))^*$ and $0 < \theta < 1/2$, there exists $P_T(0) \in \mathscr{L}((\mathscr{D}((-A)^\theta))^*, \mathscr{D}((-A)^\theta)) \cap \mathscr{L}^+(H)$ such that*

$$\langle\langle P_T(0)y_0, y_0\rangle\rangle_\theta = \mathbb{E}\langle Gy(T,y_0), y(T,y_0)\rangle_H + \int_0^T \mathbb{E}\langle My(s,y_0), y(s,y_0)\rangle_H ds.$$

Proof If $b \in \mathscr{D}((-A)^\theta)$, the proof is similar to that in Proposition 2.2.4. The general case can be established by choosing a sequence $b_n \in \mathscr{D}((-A)^\theta)$ such that $b_n \to b$ as $n \to \infty$ and passing on a limit procedure. □

Now we can state our stability results for mild solutions of (2.3.4).

Theorem 2.3.3 *The following three statements are equivalent:*

(i) The null solution of (2.3.4) is L^2-stable in the sense that

$$\int_0^\infty \mathbb{E}\|y(t,y_0)\|_H^2 dt < \infty \quad \text{for any} \quad y_0 \in (\mathscr{D}((-A)^\theta))^*. \qquad (2.3.9)$$

(ii) There exists a solution $P \in \mathscr{L}((\mathscr{D}((-A)^\theta))^*, \mathscr{D}((-A)^\theta)) \cap \mathscr{L}^+(H)$ to the Lyapunov equation

$$\langle Py, Ay\rangle_H + \langle Ay, Py\rangle_H + \langle(-A)^\theta P(-A)^\theta b, b\rangle_H f^2(y) = -\|y\|_H^2, \quad y \in \mathscr{D}(A). \qquad (2.3.10)$$

(iii) For any $y_0 \in (\mathscr{D}((-A)^\theta))^*$ and $t > 0$,

$$\mathbb{E}\|y(t,y_0)\|_H^2 \le Ce^{-\alpha t}\|y_0\|^2_{(\mathscr{D}((-A)^\theta))^*}, \qquad (2.3.11)$$

for some $\alpha > 0$ and $C \ge 1$.

Proof We first assume (i) holds, then

$$\int_0^\infty \mathbb{E}\|y(t,(-A)^\theta z)\|_H^2 dt < \infty \quad \text{for any} \quad z \in H.$$

Thus from Corollary 2.3.2 with $G = 0$, $M = I$, we have

$$\langle(-A)^\theta P_T(0)(-A)^\theta z, z\rangle_H = \int_0^T \mathbb{E}\|y(t,(-A)^\theta z)\|_H^2 dt,$$

where $P_T(t)$ is the unique solution to (2.3.5). Hence, there exists $Q \in \mathscr{L}^+(H)$ such that

$$(-A)^\theta P_T(0)(-A)^\theta \uparrow Q \text{ as } T \to \infty \text{ in } \mathscr{L}(H).$$

Let $P = (-A)^{-\theta}Q(-A)^{-\theta}$, then

$$P_T(0) \to P \quad \text{in a strong sense as } T \to \infty,$$

in the space $\mathscr{L}((\mathscr{D}((-A)^\theta))^*, \mathscr{D}((-A)^\theta)) \cap \mathscr{L}^+(H)$, and it may be similarly concluded that P satisfies (2.3.10). Hence, (i) implies (ii). Conversely, suppose that

$$P \in \mathscr{L}((\mathscr{D}((-A)^\theta))^*, \mathscr{D}((-A)^\theta)) \cap \mathscr{L}^+(H)$$

is a solution of (2.3.10). Similarly to the proofs in Theorem 2.2.5, it can be shown that

$$\int_0^\infty \mathbb{E}\|y(t, y_0)\|_H^2 dt \le \langle\langle P y_0, y_0\rangle\rangle_\theta < \infty, \qquad y_0 \in (\mathscr{D}((-A)^\theta))^*.$$

Thus, (i) is true. Last, the equivalence of (i) and (iii) is straightforward. □

By some similar arguments to those in Theorem 2.2.5 and Corollary 2.2.6, it is possible to show further that the operator P in Theorem 2.3.3 is the unique solution of the equation

$$\langle\langle P y_0, y_0\rangle\rangle_\theta = \int_0^\infty \Big[\|e^{tA} y_0\|_H^2 + \langle(-A)^\theta P(-A)^\theta b, b\rangle_H f^2(e^{tA} y_0)\Big] dt,$$
$$y_0 \in (\mathscr{D}((-A)^\theta))^*. \quad (2.3.12)$$

By developing an analogous scheme to that in Section 2.2.1, one can obtain a sufficient condition for exponential stability as follows.

Proposition 2.3.4 *Suppose that $\theta \in (0, 1/2)$. If the relation*

$$\|b\|_H^2 \int_0^\infty f(e^{tA}(-A)^\theta y)^2 dt \le \rho\|y\|_H^2, \quad y \in H, \quad (2.3.13)$$

holds for some $0 \le \rho < 1$, then there exists a unique nonnegative solution to (2.3.12). In particular, if $f(y) = \langle y, \tilde{f}\rangle_H$ for some $\tilde{f} \in H$ and

$$e^{tA} y = \sum_{n=1}^\infty e^{-\lambda_n t}\langle y, e_n\rangle_H e_n, \qquad \lambda_n > 0,$$

for some orthonormal basis $\{e_n\}_{n\ge 1}$ in H, then the null solution of (2.3.4) is exponentially stable in mean square provided that

$$\|b\|_H^2 \sum_{n=1}^\infty \frac{f_n^2}{2\lambda_n^{1-2\theta}} \le \rho < 1, \quad f_n = \langle e_n, \tilde{f}\rangle_H. \quad (2.3.14)$$

Example 2.3.5 Consider the example in the beginning of this section with $h(\cdot) = 0$, $g(\cdot) = \alpha \in \mathbb{R}$. For simplicity, let $\mathcal{O} = (0, 1)$ and assume $f(y) = \langle y, \tilde{f}\rangle_H$ for some $\tilde{f} \in H$. Then (2.3.1) may be represented by

$$\begin{cases} \partial y(t,x)/\partial t = \partial^2 y(t,x)/\partial x^2, \quad x \in (0,1), \quad t \geq 0, \\ y(0,x) = y_0(x), \quad x \in (0,1), \\ \partial y(t,0)/\partial x = \alpha \langle y(t,0), \tilde{f} \rangle_H \dot{w}(t), \quad \partial y(t,1)/\partial x = 0, \quad t \geq 0. \end{cases} \tag{2.3.15}$$

Let $A = d^2/dx^2$, $\mathscr{D}(A) = \{y \in H^2(0,1) \colon y'(0) = y'(1) = 0\}$ and $y_0 \in \bigcap_{\varepsilon>0} \mathscr{D}((-A)^{-1/4+\varepsilon})$; it is known (see [95]) that (2.3.15) may be formulated in terms of the semigroup model (2.3.3) with $\theta \in (1/4, 1/2)$. Let us assume further that $\|\tilde{f}\|_H = 1$, then (2.3.14) implies that if

$$\alpha^2 \sum_{n=1}^{\infty} \frac{1}{n^2\pi^2} < 1,$$

that is, if $|\alpha| < \sqrt{6}$, then the mild solution of (2.3.15) is exponentially stable in mean square.

2.4 Exponentially Stable Stationary Solutions

Consider the following linear stochastic differential equation with additive noise on the Hilbert space H,

$$\begin{cases} dy(t) = Ay(t)dt + BdW(t), \qquad t \geq 0, \\ y(0) = y_0 \in L^2(\Omega, H), \end{cases} \tag{2.4.1}$$

where A generates a C_0-semigroup e^{tA}, $t \geq 0$, $B \in \mathscr{L}_2(K_Q, H)$ and W is a Q-Wiener process on Hilbert space K. It is clear that zero is not generally a solution for this equation. However, we shall show the existence of a nontrivial stationary solution that is exponentially stable.

Definition 2.4.1 A mild solution $y = \{y(t); t \geq 0\}$ of (2.4.1) is called *(strongly) stationary* if the finite-dimensional distribution of the solution is invariant under time shift, that is,

$$\mathbb{P}\big\{y(s+t_k) \in \Gamma_k,\ k = 1, \ldots, n\big\} = \mathbb{P}\big\{y(t_k) \in \Gamma_k,\ k = 1, \ldots, n\big\}$$

for all $s \geq 0$, $t_k \geq 0$ and sets $\Gamma_k \in \mathscr{B}(H)$, $k = 1, \ldots, n$, or equivalently, if for any $h_1, \ldots, h_n \in H$,

$$\mathbb{E}\left[\exp\left(i \sum_{k=1}^{n} \langle y(t_k+s), h_k \rangle_H\right)\right] = \mathbb{E}\left[\exp\left(i \sum_{k=1}^{n} \langle y(t_k), h_k \rangle_H\right)\right]. \tag{2.4.2}$$

We say that (2.4.1) has a stationary solution y if there exists some initial $y_0 \in L^2(\Omega, H)$ such that $y(t, y_0)$, $t \geq 0$, is a stationary solution of (2.4.1) with $y(0) = y_0 \in L^2(\Omega, H)$.

Definition 2.4.2 A stationary solution is said to be *unique* if any two stationary solutions of (2.4.1) have the same finite-dimensional distribution.

The following theorem establishes conditions under which there exists a unique stationary solution to stochastic system (2.4.1).

Theorem 2.4.3 *Suppose that A generates a C_0-semigroup e^{tA}, $t \geq 0$, which is exponentially stable, i.e., there are constants $M \geq 1$ and $\mu > 0$ such that*

$$\|e^{tA}\| \leq Me^{-\mu t} \quad \textit{for all} \quad t \geq 0, \tag{2.4.3}$$

then there exists a unique stationary solution of (2.4.1). Moreover, this stationary solution is a zero mean Gaussian process with its covariance operator K given by

$$K(t,s) = \int_0^\infty e^{(t-s+u)A} BQ^{1/2}(e^{uA}BQ^{1/2})^* du, \qquad t,\ s \geq 0. \tag{2.4.4}$$

Proof Let $\{\widetilde{W}(t)\}_{t\geq 0}$ be an independent copy of $\{W(t)\}_{t\geq 0}$ defined on $(\Omega, \mathscr{F}, \mathbb{P})$. We first extend W to obtain a two-sided Q-Wiener process $\overline{W}$ on the whole real axis $\mathbb{R}$ by

$$\overline{W}(t) = \begin{cases} W(t), & t \geq 0, \\ \widetilde{W}(-t), & t < 0, \end{cases} \tag{2.4.5}$$

with the filtration $\overline{\mathscr{F}}_t := \bigcap_{s>t} \widetilde{\mathscr{F}}_s^0$, where

$$\widetilde{\mathscr{F}}_s^0 := \sigma\Big(\{\overline{W}(r_2) - \overline{W}(r_1) : \ -\infty < r_1 \leq r_2 \leq s\}, N\Big)$$

and $N := \{E \in \mathscr{F} : \ \mathbb{P}(E) = 0\}$. It is not difficult to see that $\overline{W}(t)$, $t \in \mathbb{R}$, is a Q-Wiener process with respect to $\{\overline{\mathscr{F}}_t\}_{t\in\mathbb{R}}$. By virtue of condition (2.4.3), it makes sense to define a process

$$U(t) = \int_{-\infty}^t e^{(t-s)A} Bd\overline{W}(s), \qquad t \geq 0. \tag{2.4.6}$$

It is immediate that $\mathbb{E}U(t) = 0$ for any $t \geq 0$. Moreover, let $0 \leq t_1 < \cdots < t_n$; we have for any $h_1, \ldots, h_n \in H$ that

$$\mathbb{E}\exp\left(i\sum_{k=1}^{n}\langle h_k, U(t_k)\rangle_H\right)$$
$$= \exp\left\{-\frac{1}{2}\left[\int_0^\infty \sum_{j\geq i=1}^{n}\langle e^{(t_j-t_i+s)A}BQ^{1/2}(e^{sA}BQ^{1/2})^* h_i, h_j\rangle_H ds\right]\right\}. \tag{2.4.7}$$

It is easy to see from (2.4.7) that $\mathbb{E}\exp(i\sum_{k=1}^{n}\langle h_k, U(t_k)\rangle_H)$ is invariant when t_k is replaced by $t_k + s$, $k = 1,\ldots,n$, for any $s \geq 0$. Thus, process U is stationary. Moreover, it is a Gaussian process, a fact that follows directly from the definition of stochastic integral, with the covariance operator $K(t_j, t_i)$, $j \geq i$, given by

$$K(t_j, t_i) = \int_0^\infty e^{(t_j-t_i+s)A}BQ^{1/2}(e^{sA}BQ^{1/2})^* ds.$$

Thus, the covariançe operator $K(t_j, t_i)$ is a map of $t_j - t_i$ only.

Next, we show that $U(t)$, $t \geq 0$, in (2.4.6) is a mild solution of (2.4.1). Indeed, by using the well-known stochastic Fubini's theorem we have for any $t \geq 0$ and $h \in \mathscr{D}(A^*)$ that

$$\int_0^t \langle A^*h, U(s)\rangle_H ds + \langle h, U(0)\rangle_H$$
$$= \int_0^t \left\langle A^*h, \int_{-\infty}^{s} e^{(s-u)A}Bd\overline{W}(u)\right\rangle_H ds + \left\langle h, \int_{-\infty}^{0} e^{-uA}Bd\overline{W}(u)\right\rangle_H$$
$$= \int_{-\infty}^{0}\left\langle \int_0^t \frac{d}{ds}B^*e^{(s-u)A^*}h ds, d\overline{W}(u)\right\rangle_H$$
$$+ \int_0^t \left\langle \int_u^t \frac{d}{ds}B^*e^{(s-u)A^*}h ds, d\overline{W}(u)\right\rangle_H + \left\langle h, \int_{-\infty}^{0} e^{-uA}Bd\overline{W}(u)\right\rangle_H$$
$$= \int_{-\infty}^{t}\langle B^*e^{(t-u)A^*}h, d\overline{W}(u)\rangle_H - \int_{-\infty}^{0}\langle B^*e^{-uA^*}h, d\overline{W}(u)\rangle_H$$
$$- \int_0^t \langle B^*h, d\overline{W}(u)\rangle_H + \left\langle h, \int_{-\infty}^{0} e^{-uA}Bd\overline{W}(u)\right\rangle_H$$
$$= \langle h, U(t)\rangle_H - \left\langle h, \int_0^t Bd\overline{W}(s)\right\rangle_H.$$

Hence U is a weak solution that, according to theorem 5.4 in Da Prato and Zabczyk [53], is also a mild solution of (2.4.1).

Last, we show the uniqueness of stationary solutions. Suppose that there are two stationary solutions $y(t, y_0)$ and $z(t, z_0)$ to (2.4.1). Note that for any initial $y_0 \in L^2(\Omega, H)$, the mild solution $y(t, y_0)$ to (2.4.1) is represented by

$$y(t) = e^{tA} y_0 + \int_0^t e^{(t-s)A} B dW(s), \qquad t \geq 0.$$

Thus, we have

$$\mathbb{E}\|y(t, y_0) - z(t, z_0)\|_H^2 \leq \|e^{tA}\|^2 \mathbb{E}\|y_0 - z_0\|_H^2 \to 0 \quad \text{as} \quad t \to \infty,$$

since, by assumption, $\|e^{tA}\| \to 0$ as $t \to \infty$. As $y(t, y_0) - z(t, z_0)$ is stationary, $y(t, y_0)$ and $z(t, z_0)$ must have the same distribution, and the proof is thus complete. □

Let $y_0^* \in L^2(\Omega, H)$ be an initial datum such that the corresponding mild solution $y(\cdot, y_0^*)$ of (2.4.1) is stationary. The following corollary shows that this unique stationary solution of equation (2.4.1) is exponentially stable in mean square.

Corollary 2.4.4 *Suppose that (2.4.3) holds. Then the unique stationary solution $y(\cdot, y_0^*)$ of (2.4.1) has mean square exponential stability in the sense that for any mild solution $y(\cdot, y_0)$ of (2.4.1), there exist constants $M > 0$, $\mu > 0$ such that*

$$\mathbb{E}\|y(t, y_0) - y(t, y_0^*)\|_H^2 \leq M^2 e^{-2\mu t} \mathbb{E}\|y_0 - y_0^*\|_H^2, \qquad t \geq 0.$$

Proof We know from (2.4.6) that

$$y(t, y_0^*) = \int_{-\infty}^t e^{(t-s)A} B d\overline{W}(s), \qquad t \geq 0,$$

is the unique stationary solution with initial $y_0^* \in L^2(\Omega, H)$. Let $y(t, y_0)$ be an arbitrary mild solution of (2.4.1) with initial $y_0 \in L^2(\Omega, H)$. Then for any $t \geq 0$, we have

$$\begin{aligned} y(t, y_0) - y(t, y_0^*) &= e^{tA} y_0 + \int_0^t e^{(t-s)A} B d\overline{W}(s) - \int_{-\infty}^t e^{(t-s)A} B d\overline{W}(s) \\ &= e^{tA} y_0 - \int_{-\infty}^0 e^{(t-s)A} B d\overline{W}(s) = e^{tA}(y_0 - y_0^*), \end{aligned} \tag{2.4.8}$$

which immediately implies the existence of constants $M > 0$, $\mu > 0$ such that

$$\mathbb{E}\|y(t, y_0) - y(t, y_0^*)\|_H^2 \leq M^2 e^{-2\mu t} \mathbb{E}\|y_0 - y_0^*\|_H^2, \qquad t \geq 0.$$

The proof is now complete. □

2.5 Some Examples

In this section, we shall study several examples to illustrate the theory established in the previous sections.

Example 2.5.1 Consider a one-dimensional rod of length π whose ends are maintained at 0° and whose sides are insulated. Assume that there is an exothermic reaction taking place inside the rod with heat being produced proportionally to the temperature. The temperature, denoted by $y(t,x)$ at time t and location x, in the rod may be modeled as a solution of the following equation

$$\begin{cases} \dfrac{\partial y(t,x)}{\partial t} = \dfrac{\partial^2 y(t,x)}{\partial x^2} + ry(t,x), \quad t > 0, \quad 0 < x < \pi, \\ y(t,0) = y(t,\pi) = 0, \quad t \geq 0, \\ y(0,x) = y_0(x), \quad 0 \leq x \leq \pi, \end{cases} \tag{2.5.1}$$

where $r \in \mathbb{R}$ depends on the rate of reaction. If we assume $r = r_0 > 0$, a constant, then we can solve by the standard partial differential equation theory the equation in an explicit form:

$$y(t,x) = \sum_{n=1}^{\infty} a_n e^{-(n^2 - r_0)t} \sin nx, \qquad t \geq 0.$$

Hence, we obtain exponential stability if $n^2 > r_0$ for all $n \in \mathbb{N}_+$, or equivalently, $r_0 < 1$. This is exactly the condition (2.2.30) in (H).

Suppose now that r is random, and assume it is informally modeled as $r = r_0 + r_1 \dot{w}(t)$, so that (2.5.1) becomes

$$y(t,x) = y_0(x) + \int_0^t \left(\frac{\partial^2}{\partial x^2} + r_0 \right) y(s,x) ds + r_1 \int_0^t y(s,x) dw(s), \qquad t \geq 0, \tag{2.5.2}$$

where $w(t)$ is a standard one-dimensional Brownian motion. We can formulate this into a form of (2.2.1) by letting $K = \mathbb{R}$, $H = L^2(0,\pi)$ and

$$A = \frac{\partial^2}{\partial x^2} + r_0.$$

Now, by a simple computation $\Delta(P)$ defined in (2.2.5) can be shown to be $r_1^2 P$ so that (2.2.31) in (H) becomes $r_1^2 < 2(1 - r_0)$. Hence, if the unperturbed system (2.5.1) is stable, i.e., r_0 is sufficiently less than one, then the perturbations (i.e., r_1) can be reasonably large, and according to Theorem 2.2.8, we have

$$\mathbb{E}\|y(t,y_0)\|_H^2 \leq M\|y_0\|_H^2 e^{-\mu t}, \qquad t \geq 0,$$

for some constants $M \geq 1$, $\mu > 0$.

In order to obtain pathwise almost sure stability, one can apply the theory in Chapter 3, i.e., Theorem 3.3.9, to (2.5.2). However, we can employ the theory of strong solutions on this occasion. Precisely, we set for any $y,\ z \in V$,

$$V = H_0^1(0,\pi), \qquad \langle\langle y, Az\rangle\rangle_{V,V^*} = \int_0^{\pi}\Big(-\frac{\partial y(x)}{\partial x}\frac{\partial z(x)}{\partial x} + r_0 y(x)z(x)\Big)dx.$$

Then by the well-known Poincaré inequality, we have $\langle\langle y, Ay\rangle\rangle_{V,V^*} \le -\|y\|_V^2 + r_0\|y\|_H^2$, so $\alpha = 2$ and $\lambda = 2r_0$ in (2.2.59). Then by Theorems 2.2.14 and 2.2.17, it follows that there exist a constant $\nu > 0$ and a random variable $T(\omega) \ge 0$ such that for all $t \ge T(\omega)$,

$$\|y(t, y_0)\|_H \le \|y_0\|_H e^{-\nu t} \qquad a.s.$$

Example 2.5.2 Let us consider the following stochastic partial differential equation

$$y(t,x) = y_0(x) + \int_0^t \frac{\partial^2}{\partial x^2}y(s,x)ds + \int_0^t \gamma(x)\frac{\partial y(s,x)}{\partial x}dw(s), \qquad t \ge 0, \tag{2.5.3}$$

where $\gamma(x) \in L^{\infty}([0,\pi];\ \mathbb{R})$, i.e., we are observing heat diffusion in a rod relative to an origin moving with velocity $\gamma(\cdot)\dot{w}(t)$. Let $K,\ V$, and H be formulated as before, then

$$\langle\langle y, Ay\rangle\rangle_{V,V^*} = -\int_0^{\pi}\frac{\partial y}{\partial x}\frac{\partial y}{\partial x}dx, \qquad \forall y \in V,$$

$$\langle\langle y, \Delta(I)y\rangle\rangle_{V,V^*} = \int_0^{\pi}\gamma(x)^2\frac{\partial y}{\partial x}\frac{\partial y}{\partial x}dx, \qquad \forall y \in V.$$

Hence, condition (2.2.57) becomes

$$2\langle\langle y, Ay\rangle\rangle_{V,V^*} + \langle\langle y, \Delta(I)y\rangle\rangle_{V,V^*} \le -2\|y\|_V^2 + \|\gamma(\cdot)\|_{\infty}^2\|y\|_V^2,$$

which immediately shows that for $\|\gamma(\cdot)\|_{\infty}^2 < 2$, the null solution of (2.5.3) is exponentially stable in both mean square and almost sure senses.

Next, let us investigate exponential stability of some stochastic second-order partial differential equations arising in random vibration models of mechanically flexible systems.

Example 2.5.3 Consider a model about the lateral displacement of a damped stretched string subjected to random loading,

$$\begin{cases} \partial_t\Big[\Big(\dfrac{\partial y(t)}{\partial t}\Big) + \alpha y(t)\Big] - \dfrac{\partial^2 y(t)}{\partial x^2}dt \\ \qquad\qquad + \beta\dfrac{\partial y(t)}{\partial x}dw(t) = 0,\ t > 0,\ x \in (0,\pi), \\ y(0,t) = y(\pi,t) = 0, \quad t \ge 0, \end{cases} \tag{2.5.4}$$

where ∂_t denotes the partial differentiation with respect to t; $w(t)$, $t \geq 0$, is a standard one-dimensional Brownian motion; and $\alpha > 0$, $\beta \in \mathbb{R}$ are real constants. Let

$$H = L^2(0,\pi), \qquad A = -\frac{\partial^2}{\partial x^2},$$
$$\mathscr{D}(A) = \big\{y \in H:\ y'_x,\, y''_{xx} \in H \ \text{and}\ y(0) = y(\pi) = 0\big\}.$$

Then by the standard property of the Laplace operator, it is known that the spectral bound $\omega_s(-A) = -1 < 0$. Suppose that

$$\mathcal{A} = \begin{pmatrix} 0 & I \\ -A & -\alpha I \end{pmatrix} \quad \text{with domain} \quad \mathscr{D}(\mathcal{A}) = \mathscr{D}(A) \times \mathscr{D}(A^{1/2}).$$

Then, by virtue of Example 2.1.14 we have

$$\|e^{t\mathcal{A}}\| \leq \sqrt{\frac{\gamma_+}{\gamma_-}} \cdot e^{-\frac{t}{2\gamma_+}}, \quad t \geq 0,$$

where

$$\gamma_+ = \frac{4 + \alpha^2 + \alpha\sqrt{4+\alpha^2}}{4\alpha} > 0, \qquad \gamma_- = \frac{\sqrt{\alpha^2+4}}{\alpha^2 + \alpha\sqrt{\alpha^2+4}} > 0.$$

On the other hand, we have

$$\left\| \int_0^\infty e^{t\mathcal{A}^*}\Delta(I)e^{t\mathcal{A}}dt \right\| \leq \int_0^\infty \|e^{t\mathcal{A}}\|^2 \|\Delta(I)\| dt \leq \frac{\beta^2\gamma_+^2}{\gamma_-}. \tag{2.5.5}$$

Hence, by Theorems 2.2.14 and 2.2.17 we have the exponential stability of the null solution in both mean square and almost sure senses if

$$\beta < \sqrt{\gamma_-}/\gamma_+. \tag{2.5.6}$$

Example 2.5.4 Now let us consider a model from the vibration of a panel in supersonic flow subjected to random end loads,

$$\begin{cases} \partial_t\Big[\Big(\dfrac{\partial y(t)}{\partial t}\Big) + \alpha y(t)\Big] + \dfrac{\partial^4 y(t)}{\partial x^4}dt + \theta\dfrac{\partial^2 y(t)}{\partial x^2}dt \\ \qquad\qquad + \beta\dfrac{\partial^2 y(t)}{\partial x^2}dw(t) = 0, \quad t > 0, \quad x \in (0,1), \\ y(0,t) = y(1,t) = 0, \quad t \geq 0, \end{cases} \tag{2.5.7}$$

where $w(t)$, $t \geq 0$, is a standard one-dimensional Brownian motion and α, θ, and β are real constants with $\alpha > 0$. Let

$$H = L^2(0,1), \qquad A = \frac{\partial^4}{\partial x^4} + \theta\frac{\partial^2}{\partial x^2},$$

and

$$\mathscr{D}(A) = \Big\{ y \in H: \; y'_x, y''_{xx}, y'''_{xxx}, y''''_{xxxx} \in H,$$
$$y(0) = y(1) = 0 \text{ and } y''_{xx}(0) = y''_{xx}(1) = 0 \Big\}.$$

Note that A is self-adjoint and positive, and we know by standard functional analysis theory $\omega_s(-A) = \pi^2(\theta^2 - \pi^2)$. Thus we obtain by Example 2.1.14 that $\mathcal{A} = \begin{pmatrix} 0 & I \\ -A & -\alpha I \end{pmatrix}$ is stable if $\theta < \pi$, with

$$\|e^{t\mathcal{A}}\| \le \sqrt{\frac{\gamma_+}{\gamma_-}} \cdot e^{-t/2\gamma_+} \quad \text{for all} \quad t \ge 0$$

where

$$\gamma_+ = \frac{4\pi^2(\pi^2 - \theta^2) + \alpha^2 + \alpha\sqrt{\alpha^2 + 4\pi^2(\pi^2 - \theta^2)}}{4\alpha\pi^2(\pi^2 - \theta^2)} > 0,$$

and

$$\gamma_- = \frac{\sqrt{\alpha^2 + 4\pi^2(\pi^2 - \theta^2)}}{\alpha^2 + \alpha\sqrt{\alpha^2 + 4\pi^2(\pi^2 - \theta^2)}} > 0.$$

As in the previous example, it is easy to show that

$$\left\| \int_0^\infty e^{t\mathcal{A}^*} \Delta(I) e^{t\mathcal{A}} dt \right\| \le \frac{\beta^2 \gamma_+^2}{\gamma_-}$$

and the sufficient condition for the mean square and almost sure exponential stability of the null solution is

$$\beta < \sqrt{\gamma_-}/\gamma_+.$$

Last, we would like to mention that the theory established in this chapter can be also applied to stochastic delay differential equations, a subject that will be systematically developed in Chapter 4.

2.6 Notes and Comments

There exist quite a few stability concepts for stochastic systems, and the one in which we are especially interested within this book is exponential stability. In the deterministic case, its relation to the existence of a nonnegative, self-adjoint operator solution of a Lyapunov equation, i.e., Theorem 2.1.12, has been established in Datko [55]. The generalization from L^2-stability to L^p-stability, $p \ge 1$, i.e., Theorem 2.1.5, was presented in Pazy [187], but the statements here are taken from Prichard and Zabczyk [191]. Also, Zabczyk [225] relaxes (2.1.3) by replacing $\|T(t)x\|_X^p$ into $m(\|T(t)x\|)$, where $m(\cdot)$ is a strictly increasing convex function with $m(0) = 0$. The arguments in the proof of Theorem 2.1.5 cannot go through in the case of $p \in (0, 1)$ because p-integrable functions, $p \in (0, 1)$, do not form a Banach space. However, by a different approach, it is possible to show (see Ichikawa [94]) that the result in Theorem 2.1.5 remains valid for $0 < p < 1$. The results in Example 2.1.14 were announced in Prichard and Zabczyk [191], while the presentation here is a version of that from Zabczyk [229]. The properties of Lyapunov functions in infinite-dimensional spaces, i.e., Theorems 2.1.5 and 2.1.11, are mainly taken from Datko [56].

There exists an extensive literature attempting to establish a stability theory for stochastic differential equations in infinite dimensions. Theorem 2.2.5 is well known in finite-dimensional spaces (cf. Wonham [218]). If a C_0-semigroup $T(t)$, $t \geq 0$, is generated by some bounded generator on Banach spaces, this result in the deterministic case can be found in Daletskii and Krein [54]. Theorem 2.2.5 is a generalization of Datko's result for semigroups in [55] to stochastic systems. The main proofs of Theorem 2.2.5 employed here are taken from Ichikawa [89, 90] and Haussmann [87]. System (2.2.22) is frequently said to be of Lurie type, which is one of a couple of cases to have a necessary and sufficient condition for its exponential stability (cf. [228]).

In the early stages (1940s to 1960s) of the study of stochastic stability, investigators were primarily concerned with moment stability and stability in probability. The mathematical theory for the study of almost sure (sample path) stability was not fully developed then. Although moment stability criteria for a stochastic system may be easier to determine, it is the very sample paths that are frequently observed in the real world, and the stability properties of the sample paths can be most closely related to their deterministic counterpart. After 1960s, work along these lines gradually appeared in the literature. Since then, pathwise stability with probability one of stochastic systems has attracted increasing attention from researchers. In association with the material of Section 2.2, some systematic statements about the study of almost sure stability in finite-dimensional spaces can be found in, for instance, Khas'minskii [103] and Mao [165]. The material in Sections 2.2.2 and 2.2.3 is based on Bierkens [19] and Haussmann [87]. See also Ichikawa [89, 95] and Skorokhod [202] for some related topics on stability of linear stochastic evolution equations.

When zero is not the equilibrium solution of a stochastic system such as (2.4.1), it is natural to consider the existence and uniqueness of nontrivial equilibrium solutions and the associated stability problems. For a class of semilinear stochastic evolution equations, Caraballo et al. [27] investigated exponentially stable stationary solutions by treating them as certain random dynamical systems and considering their exponentially attracting fixed points. Examples 2.5.3 and 2.5.4 are taken from Plaut and Infante [189] and Curtain [46]. In connection with this, the reader is also referred to Curtain and Pritchard [48], Curtain and Zwart [49], and Pritchard and Zabczyk [191] for a more detailed investigation of deterministic systems.

3

Stability of Nonlinear Stochastic Differential Equations

In this chapter, we shall study stability of nonlinear stochastic differential equations (SDEs) on Hilbert spaces. To this end, we will employ semigroup and variational methods in a systematic way to deal with semilinear and nonlinear, nonautonomous systems, respectively.

3.1 An Extension of Linear Stability Criteria

Consider the following autonomous semilinear differential equation on the Hilbert space H,

$$\begin{cases} dy(t) = [Ay(t) + F(y(t))]dt, & t \geq 0, \\ y(0) = y_0 \in H, \end{cases} \tag{3.1.1}$$

where $F: H \to H$ is a nonlinear, Lipschitz continuous function with $F(0) = 0$. The question we are concerned about is whether the following two statements are equivalent in some appropriate sense:

(a) $\|y(t, y_0)\|_H^2 \leq Me^{-\mu t}\|y_0\|_H^2$, $\forall\, y_0 \in H$, for some $M \geq 1$ and $\mu > 0$;

(b) $\displaystyle\int_0^\infty \|y(t, y_0)\|_H^2 dt \leq C\|y_0\|_H^2$, $\forall\, y_0 \in H$, for some $C > 0$.

If F is linear, it is in Theorem 2.2.5 that we show the validity of this equivalence. In this section, we intend to establish a nonlinear stochastic version of this result.

It is worth mentioning that this question is also important even from Lyapunov's direct method viewpoint. This is due to the fact that there are many occasions when one can easily find a Lyapunov-like function $\Lambda(\cdot)$ and obtain the exponential stability of $\Lambda(y(t, y_0))$, whereas Λ is not strictly positive definite, i.e., $\Lambda(\cdot) \not\geq c\|\cdot\|_H^2$ for any constant $c > 0$, so it is not straightforward to get (a). To be more specific, recall that if the C_0-semigroup e^{tA}, $t \geq 0$, is exponentially stable, then, according to the theory in Section 2.1.2, there exists a nonnegative, self-adjoint operator $P \in \mathscr{L}(H)$ such that

$$\langle Px, Ax\rangle_H + \langle Ax, Px\rangle_H = -\|x\|_H^2, \qquad \forall x \in \mathscr{D}(A), \tag{3.1.2}$$

where P satisfies the equality

$$\langle Px, x\rangle_H = \int_0^\infty \|e^{tA}x\|_H^2 dt, \qquad \forall x \in H. \tag{3.1.3}$$

With the aim of establishing stability of (3.1.1), it seems plausible to consider a Lyapunov function of the form $\Lambda(x) = \langle Px, x\rangle_H$, $x \in H$, and manage to show, of course, under some additional conditions on F, a relation of the form

$$\langle Py(t, y_0), y(t, y_0)\rangle_H \leq Me^{-\mu t}\|y_0\|_H^2, \quad \forall t \geq 0, \tag{3.1.4}$$

for some $M > 0$ and $\mu > 0$. Unfortunately, since operator P is only nonnegative in general, it is not immediate to conclude from (3.1.4) that

$$\|y(t, y_0)\|_H^2 \leq C \cdot e^{-\nu t}\|y_0\|_H^2, \quad \forall t \geq 0, \quad \text{for some} \quad C \geq 1 \quad \text{and} \quad \nu > 0,$$

as shown in Remark 3.1.1. However, it frequently happens that one can establish the following L^2-stability

$$\int_s^\infty \|y(t, y(s))\|_H^2 dt \leq C\|y(s)\|_H^2, \qquad \text{for some } C > 0 \text{ and any } s \geq 0,$$

which may further imply exponential stability of this system.

Remark 3.1.1 For the form $\Lambda(x) = \langle Px, x\rangle_H$, $x \in H$, where P is given in (3.1.2) or (3.1.3), it is generally untrue that $\Lambda(x)$ is equivalent to $\|x\|_H^2$, $x \in H$, i.e., there does not exist constant $c > 0$ such that

$$c\|x\|_H^2 \leq \Lambda(x) \qquad \text{for all} \qquad x \in H. \tag{3.1.5}$$

Indeed, suppose that the contrary is true. If A in (3.1.3) is an unbounded operator that generates an exponentially stable, analytic semigroup e^{tA}, $t \geq 0$, it was shown by Pazy [186] that (3.1.5) is true if and only if there exist $t_0 > 0$ and a constant $C > 0$ such that $\|e^{t_0 A}x\|_H \geq C\|x\|_H$, $x \in H$. Therefore, if $x \in \mathscr{D}(A)$, we have

$$\|Ae^{t_0 A}x\|_H = \|e^{t_0 A}Ax\|_H \geq C\|Ax\|_H.$$

However, since e^{tA}, $t \geq 0$, is analytic, $Ae^{t_0 A}$ is a bounded operator, and so

$$\|Ax\|_H \leq \frac{1}{C}\|Ae^{t_0 A}\|\|x\|_H \quad \text{for every} \quad x \in \mathscr{D}(A).$$

Since $\mathscr{D}(A)$ is dense in H, this means that A is bounded, which is a contradiction.

In the remainder of this section, we shall establish a version of the equivalence between (a) and (b) for a large class of nonlinear stochastic evolution equations, a case that could be regarded as a nonlinear generalization of Theorem 2.2.5. To state this, let us consider a stochastic version of the system (3.1.1), initializing from any time $s \geq 0$ in the following form:

$$\begin{cases} dy(t) = (Ay(t) + F(y(t)))dt + B(y(t))dW(t), \quad t \geq s \geq 0, \\ y(s) \in L^p_{\mathscr{F}_s}(\Omega, \mathscr{F}, \mathbb{P}; H), \end{cases} \tag{3.1.6}$$

where $W(t)$, $t \geq 0$, is a Q-Wiener process on $(\Omega, \mathscr{F}, \{\mathscr{F}_t\}_{t\geq 0}, \mathbb{P})$ in Hilbert space K. Here A generates a C_0-semigroup e^{tA}, $t \geq 0$, on H and the coefficients $F(\cdot)$ and $B(\cdot)$ are two nonlinear measurable mappings satisfying a Lipschitz type of condition:

$$\|F(y) - F(z)\|_H + \|B(y) - B(z)\|_{\mathscr{L}_2(K_Q, H)} \leq \alpha \|y - z\|_H, \quad \alpha > 0, \quad \forall y,\ z \in H, \tag{3.1.7}$$

and $F(0) = 0$, $B(0) = 0$. For $p \geq 1$ and $t \geq 0$, let $L^p_{\mathscr{F}_t}(\Omega, \mathscr{F}, \mathbb{P}; H)$, $t \geq 0$, denote the space of all H-valued mappings ξ defined on $(\Omega, \mathscr{F}, \mathbb{P})$ such that ξ is $\mathscr{F}_t$-measurable and satisfies $\mathbb{E}\|\xi\|^p_H < \infty$. By a standard Picard iteration procedure or a probabilistic fixed-point theorem argument, it is possible to establish an existence and uniqueness theorem of mild solutions to (3.1.6).

Theorem 3.1.2 *Let $p \geq 1$ and assume that condition (3.1.7) holds. Then there exists a unique, $\{\mathscr{F}_t\}$-adapted mild solution $y(t) \in L^p(\Omega, \mathscr{F}, \mathbb{P}; H))$, $t \geq s$, to equation (3.1.6).*

We begin our stability analysis by stating a simple lemma.

Lemma 3.1.3 *Let $0 < r < 1$, $L > 0$ and n be a nonnegative integer. Then $nL \leq t \leq (n+1)L$ implies $e^{-at} \leq r^n \leq (1/r)e^{-at}$ where $a = -(\log r)/L > 0$.*

Proof Note that $r^n = e^{n \log r}$ and $\log r < 0$, then the result is easily obtained. □

Theorem 3.1.4 *Let $p \geq 1$ and $y(t) \in H$, $t \geq s$, be the mild solution of equation (3.1.6). Then the following statements are equivalent:*

(i) $\displaystyle\int_s^\infty \mathbb{E}\|y(t, y(s))\|^p_H dt \leq C\mathbb{E}\|y(s)\|^p_H$ *for some $C > 0$;*

(ii) $\mathbb{E}\|y(t, y(s))\|^p_H \leq M \cdot e^{-\lambda(t-s)}\mathbb{E}\|y(s)\|^p_H$ *for some $M \geq 1$, $\lambda > 0$ and all $t \geq s \geq 0$.*

Proof Obviously, it suffices to verify that (i) implies (ii). By the standard theory of SDE, it is possible to show that there exist constants $M \geq 1$ and $\mu > 0$ such that

$$\mathbb{E}\|y(u, y(v))\|_H^p \leq Me^{\mu(u-v)}\mathbb{E}\|y(v)\|_H^p \quad \text{for all} \quad u \geq v \geq 0. \tag{3.1.8}$$

For $u \geq v \geq 0$, we have by the Markov property of the solution $y(\cdot)$ to (3.1.6) and (3.1.8) that

$$\begin{aligned}
\mathbb{E}\|y(u, y(v))\|_H^p \int_v^u e^{-\mu(u-\tau)}d\tau &= \int_v^u e^{-\mu(u-\tau)}\mathbb{E}\|y(u, y(v))\|_H^p d\tau \\
&= \int_v^u e^{-\mu(u-\tau)}\mathbb{E}\|y(u, y(\tau, y(v)))\|_H^p d\tau \\
&\leq M \int_v^u e^{-\mu(u-\tau)}e^{\mu(u-\tau)}\mathbb{E}\|y(\tau, y(v))\|_H^p d\tau \\
&= M \int_v^u \mathbb{E}\|y(\tau, y(v))\|_H^p d\tau,
\end{aligned} \tag{3.1.9}$$

which, together with condition (i), immediately implies for some constant $C > 0$ that

$$\mathbb{E}\|y(u, y(v))\|_H^p \int_v^u e^{-\mu(u-\tau)}d\tau \leq C \cdot \mathbb{E}\|y(v)\|_H^p, \qquad \forall u \geq v \geq 0. \tag{3.1.10}$$

Let $L_1 > 0$ be an arbitrary but fixed number, then we have from (3.1.10) that for any $u > v \geq 0$ with $u - v > L_1$,

$$\mathbb{E}\|y(u, y(v))\|_H^p \leq \frac{C\mu}{1 - e^{-\mu L_1}}\mathbb{E}\|y(v)\|_H^p.$$

This relation, together with (3.1.8), implies the existence of a constant

$$\alpha = \max\left\{\frac{C\mu}{1 - e^{-\mu L_1}}, Me^{L_1 M}\right\} > 0$$

such that

$$\mathbb{E}\|y(u, y(v))\|_H^p \leq \alpha \cdot \mathbb{E}\|y(v)\|_H^p \quad \text{for any} \quad u \geq v \geq 0. \tag{3.1.11}$$

Now let $u > v$; then it follows by the Markov property of y and condition (i) that

$$\begin{aligned}
(u - v)\mathbb{E}\|y(u, y(v))\|_H^p &= \int_v^u \mathbb{E}\|y(u - \tau, y(\tau, y(v)))\|_H^p d\tau \\
&\leq \alpha \int_v^u \mathbb{E}\|y(\tau, y(v))\|_H^p d\tau \\
&\leq \alpha C \cdot \mathbb{E}\|y(v)\|_H^p,
\end{aligned} \tag{3.1.12}$$

which immediately yields

$$\mathbb{E}\|y(u, y(v))\|_H^p \le \frac{\alpha C}{u-v}\mathbb{E}\|y(v)\|_H^p, \qquad \forall u > v.$$

Hence, for each $0 < r < 1$, we can choose a number $L_2 = L_2(r) > 0$ such that

$$\mathbb{E}\|y(u, y(v))\|_H^p \le r \cdot \mathbb{E}\|y(v)\|_H^p, \quad \text{whenever} \quad u - v \ge L_2. \tag{3.1.13}$$

For $t - s \ge L_2$, there is an integer $n \ge 1$ such that $nL_2 \le t - s \le (n+1)L_2$. Let $t_0 = s$, $t_1 = s + L_2$, $\ldots$, $t_n = s + nL_2$, then by using the Markov property of y, relation (3.1.13) n times, and (3.1.11), we obtain

$$\begin{aligned}
\mathbb{E}\|y(t, y(s))\|_H^p &= \mathbb{E}\|y(t, y(t_n, y(s)))\|_H^p \le \alpha\mathbb{E}\|y(t_n, y(s)))\|_H^p,\\
\mathbb{E}\|y(t_n, y(s))\|_H^p &= \mathbb{E}\|y(t_n, y(t_{n-1}, y(s)))\|_H^p \le r\mathbb{E}\|y(t_{n-1}, y(s))\|_H^p,\\
&\cdots\\
\mathbb{E}\|y(t_1, y(s))\|_H^p &= \mathbb{E}\|y(t_1, y(s, y(s)))\|_H^p \le r\mathbb{E}\|y(s)\|_H^p,
\end{aligned}$$

which immediately imply that

$$\mathbb{E}\|y(t, y(s))\|_H^p \le \alpha r^n \mathbb{E}\|y(s)\|_H^p.$$

Now Lemma 3.1.3 further implies that

$$\mathbb{E}\|y(t, y(s))\|_H^p \le \beta \cdot e^{-\lambda(t-s)}\mathbb{E}\|y(s)\|_H^p \quad \text{for any} \quad t - s \ge L_2,$$

where $\beta = \alpha/r$ and $\lambda = -(\log r)/L_2 > 0$. Combining this with (3.1.11), we conclude that

$$\mathbb{E}\|y(t, y(s))\|_H^p \le M \cdot e^{-\lambda(t-s)}\mathbb{E}\|y(s)\|_H^p, \qquad \forall t \ge s \ge 0,$$

where $M = \max\{\beta, \alpha e^{\lambda L_2}\} > 0$. The proof is now complete. □

Proposition 3.1.5 *Suppose that there exists a nonnegative Itô functional $\Lambda(\cdot)$ on H such that for some $C_1 > 0$ and $p \ge 1$,*

$$\Lambda(y) \le C_1\|y\|_H^p \quad \textit{for any} \quad y \in H, \tag{3.1.14}$$

and for some $C_2 > 0$,

$$\begin{aligned}
\mathcal{L}\Lambda(y) &:= \langle \Lambda'(y), Ay + F(y)\rangle_H + \frac{1}{2}Tr\big[\Lambda''(y)B(y)Q^{1/2}(B(y)Q^{1/2})^*\big]\\
&\le -C_2\|y\|_H^p,
\end{aligned} \tag{3.1.15}$$

for any $y \in \mathscr{D}(A)$. Then the mild solution $y(\cdot)$ of equation (3.1.1) satisfies

$$\mathbb{E}\|y(t, y_0)\|_H^p \le M\|y_0\|_H^p e^{-\mu t} \quad \forall t \ge 0, \quad \textit{for some} \quad M \ge 1, \ \ \mu > 0.$$

For the proof, we need to consider the following Yosida approximation systems of (3.1.6):

$$\begin{cases} dy_n(t) = Ay_n(t)dt + R(n)F(y_n(t))dt + R(n)B(y_n(t))dW(t), & t \geq s \geq 0, \\ y_n(s) = R(n)y(s) \in \mathscr{D}(A), \end{cases} \tag{3.1.16}$$

where $n \in \rho(A)$, and $R(n) = nR(n,A)$, $R(n,A)$ is the resolvent of A.

Proof Applying Itô's formula to the function $\Lambda(y)$, $y \in H$, and solution $y_n(t) \in \mathscr{D}(A)$, $t \geq s \geq 0$, of (3.1.16), we have that for all $t \geq s \geq 0$,

$$\begin{aligned} &\Lambda(y_n(t)) - \Lambda(y_n(s)) \\ &\quad = \int_s^t \langle \Lambda'(y_n(u)), Ay_n(u) + R(n)F(y_n(u))\rangle_H du \\ &\quad\quad + \int_s^t \langle \Lambda'(y_n(u)), R(n)B(y_n(u))dW(u)\rangle_H \\ &\quad\quad + \frac{1}{2}\int_s^t Tr\Big(\Lambda''(y_n(u))R(n)B(y_n(u))Q^{1/2}(R(n)B(y_n(u))Q^{1/2})^*\Big)du. \end{aligned}$$

Therefore, by virtue of (3.1.15) we can obtain

$$\begin{aligned} \mathbb{E}\Lambda(y_n(t)) \leq \mathbb{E}\Lambda(y_n(s)) &- C_2\int_s^t \mathbb{E}\|y_n(u)\|_H^p du \\ &+ \int_s^t \mathbb{E}\Big\{\langle \Lambda'(y_n(u)), (R(n) - I)F(y_n(u))\rangle_H \\ &+ \frac{1}{2}Tr\Big[\Lambda''(y_n(u))R(n)B(y_n(u))Q^{1/2}(R(n)B(y_n(u))Q^{1/2})^* \\ &- \Lambda''(y_n(u))B(y_n(u))Q^{1/2}(B(y_n(u)Q^{1/2})^*\Big]\Big\}du. \end{aligned} \tag{3.1.17}$$

By virtue of Corollary 1.4.6, there exists a subsequence of $\{y_n\}_{n\geq 1}$, still denote it by $\{y_n\}_{n\geq 1}$, such that $y_n(t) \to y(t)$ almost surely as $n \to \infty$, uniformly on any compact set of $[s, \infty)$. Consequently, by letting $n \to \infty$ in (3.1.17), we obtain, for arbitrary $t \geq s \geq 0$, that

$$\mathbb{E}\Lambda(y(t)) \leq \mathbb{E}\Lambda(y(s)) - C_2\int_s^t \mathbb{E}\|y(u)\|_H^p du.$$

Let $t \to \infty$, then this inequality implies, in addition to (3.1.14), that

$$\int_s^\infty \mathbb{E}\|y(u, y(s))\|_H^p du \leq \frac{C_1}{C_2}\cdot \mathbb{E}\|y(s)\|_H^p \quad \text{for any} \quad s \geq 0.$$

In particular, by using Theorem 3.1.4 and letting $s = 0$ one can immediately obtain

$$\mathbb{E}\|y(t, y_0)\|_H^p \le M\|y_0\|_H^p e^{-\mu t}, \quad y_0 \in H, \quad t \ge 0,$$

for some $M \ge 1$, $\mu > 0$. The proof is now complete. □

Example 3.1.6 Consider the following stochastic partial differential equation

$$\begin{cases} y(x,t) = y_0(x) + \displaystyle\int_0^t \frac{\partial^2 y(x,s)}{\partial x^2} ds \\ \qquad + \displaystyle\int_0^t b(x) f(y(\cdot,s)) dw(s), \quad t \ge 0, \quad 0 < x < 1, \\ y(0,t) = y(1,t) = 0, \quad t \ge 0; \quad y(x,0) = y_0(x), \quad 0 \le x \le 1, \end{cases} \tag{3.1.18}$$

where $w(t)$, $t \ge 0$, is a real standard Brownian motion, $b(\cdot) \in L^2(0,1)$ with $\|b\|_{L^2(0,1)} = 1$ and f is a real Lipschitz continuous function on $H = L^2(0,1)$ satisfying $|f(y)| \le c\|y\|_H$, $y \in L^2(0,1)$, $c > 0$. Here we take $A = d^2/dx^2$ with $\mathscr{D}(A) = H_0^1(0,1) \cap H^2(0,1)$. Then

$$\langle y, Ay\rangle_H \le -\pi^2\|y\|_H^2 \quad \text{for any} \quad y \in \mathscr{D}(A).$$

Let $\Lambda(y) = \|y\|_H^2$, $y \in H$, then

$$\mathcal{L}\|y\|_H^2 \le 2\langle y, Ay\rangle_H + \|b\|_H^2 |f(y)|^2 \le -(2\pi^2 - c^2)\|y\|_H^2, \qquad \forall y \in \mathscr{D}(A).$$

Hence, if $c^2 < 2\pi^2$, the null solution of (3.1.18) is exponentially stable in mean square according to Proposition 3.1.5.

On the other hand, consider an operator $P \in \mathscr{L}(H)$ given by

$$P = \sum_{n=1}^{\infty} \left(\frac{1}{2n^2\pi^2}\right) e_n \otimes e_n, \tag{3.1.19}$$

where $e_n(x) = \sqrt{2}\sin n\pi x$, $x \in (0,1)$, and for each $g, h \in H$, $g \otimes h \in \mathscr{L}(H)$ is defined by $(g \otimes h)y = g\langle h, y\rangle_H \in H$, $y \in H$. Then P is a self-adjoint, nonnegative operator that is not coercive since

$$\langle Pe_n, e_n\rangle_H = \frac{1}{2n^2\pi^2} \to 0 \qquad \text{as} \quad n \to \infty.$$

However, we have from (3.1.19) and $Ae_n = -n^2\pi^2 e_n$, $n \in \mathbb{N}_+$, that

$$\langle Py, Ay\rangle_H + \langle Ay, Py\rangle_H = -\|y\|_H^2, \qquad \forall y \in \mathscr{D}(A).$$

On this occasion, let $\Lambda(y) = \langle Py, y\rangle_H$, $y \in H$, then a direct computation yields

$$\begin{aligned} \mathcal{L}\Lambda(y) &= \langle Py, Ay\rangle_H + \langle Ay, Py\rangle_H + \langle Pb, b\rangle_H |f(y)|^2 \\ &\le -\big[1 - \langle Pb, b\rangle_H c^2\big]\|y\|_H^2, \quad \forall y \in \mathscr{D}(A). \end{aligned} \tag{3.1.20}$$

Hence, in view of Proposition 3.1.5 we obtain the region of exponential stability in mean square:

$$c^2 < 1/\langle Pb, b\rangle_H.$$

This is larger than $\{c^2 < 2\pi^2\}$ if $\langle Pb, b\rangle_H < 1/2\pi^2$. For example, if $b = e_k$ for some $k \geq 2$, then

$$\langle Pb, b\rangle_H = \frac{1}{2k^2\pi^2} < \frac{1}{2\pi^2}.$$

Hence, system (3.1.18) is exponentially stable in the mean square sense if $c^2 < 2k^2\pi^2$. In other words, $\Lambda(y) = \langle Py, y\rangle_H$ plays at present the role of a Lyapunov function although it is not equivalent to $\|y\|_H^2$, $y \in H$.

3.2 Comparison Approach

In this section, we shall show that the mean square stability of a class of nonlinear stochastic evolution equations is secured by the same stability of some associated linear stochastic evolution equations provided noise terms in the former are dominated by those of the latter.

Consider the following stochastic differential equations in H,

$$\begin{cases} dy(t) = Ay(t)dt + B_i(y(t))dW_i(t), \quad t \geq 0, \quad i = 1, 2, \\ y(0) = y_0 \in H, \end{cases} \tag{3.2.1}$$

where A generates a strongly continuous semigroup e^{tA}, $t \geq 0$, on H, $B_i : H \to \mathscr{L}_2(K^i_{Q_i}, H)$ is Lipschitz continuous with $B_i(0) = 0$ and $W_i(t)$ is a Wiener process in some real separable Hilbert space K^i with incremental covariance operator Q_i, $Tr(Q_i) < \infty$, $i = 1, 2$. Let $y_i(t) = y_i(t, y_0)$, $t \geq 0$, $i = 1, 2$, be the unique mild solutions of (3.2.1) with initial $y_0 \in H$.

Proposition 3.2.1 *Let G and M be two nonnegative self-adjoint operators in $\mathscr{L}(H)$. Suppose that B_1 is linear, i.e., $B_1 \in \mathscr{L}(H, \mathscr{L}_2(K^1_{Q_1}, H))$ and*

$$B_2(y)Q_2^{1/2}(B_2(y)Q_2^{1/2})^* \leq B_1(y)Q_1^{1/2}(B_1(y)Q_1^{1/2})^* \quad \text{for any} \quad y \in H. \tag{3.2.2}$$

Then condition (3.2.2) implies that for any $T \geq 0$,

$$\begin{aligned} &\mathbb{E}\langle Gy_2(T), y_2(T)\rangle_H + \int_0^T \mathbb{E}\langle My_2(t), y_2(t)\rangle_H dt \\ &\leq \mathbb{E}\langle Gy_1(T), y_1(T)\rangle_H + \int_0^T \mathbb{E}\langle My_1(t), y_1(t)\rangle_H dt. \end{aligned} \tag{3.2.3}$$

Proof By analogy with Lemma 2.2.2, there exists a unique, strongly continuous, self-adjoint operator solution $P_1(t) \geq 0$ to the equation:

$$\begin{cases} \dfrac{d}{dt}\langle P(t)y, y\rangle_H + \langle Ay, P(t)y\rangle_H + \langle P(t)y, Ay\rangle_H + \langle My, y\rangle_H \\ \qquad\qquad + Tr[P(t)B_1(y)Q_1^{1/2}(B_1(y)Q_1^{1/2})^*] = 0, \qquad y \in \mathscr{D}(A), \\ P(T) = G, \quad 0 \leq t \leq T. \end{cases} \tag{3.2.4}$$

Moreover, by virtue of Proposition 2.2.4 we have

$$\langle P_1(0)y_0, y_0\rangle_H = \mathbb{E}\langle Gy_1(T), y_1(T)\rangle_H + \int_0^T \mathbb{E}\langle My_1(t), y_1(t)\rangle_H dt.$$

On the other hand, condition (3.2.2) implies (cf. lemma 2.1, [92]) that for any nonnegative self-adjoint operator $P \in \mathscr{L}(H)$,

$$\begin{aligned} &Tr[PB_2(y)Q_2^{1/2}(B_2(y)Q_2^{1/2})^*] \\ &\qquad \leq Tr[PB_1(y)Q_1^{1/2}(B_1(y)Q_1^{1/2})^*] \quad \text{for any} \quad y \in H, \end{aligned} \tag{3.2.5}$$

which, together with (3.2.4), further yields that for $t \in [0, T]$, $y \in \mathscr{D}(A)$,

$$\begin{aligned} &\frac{d}{dt}\langle P_1(t)y, y\rangle_H + \langle Ay, P(t)y\rangle_H + \langle P(t)y, Ay\rangle_H + \langle My, y\rangle_H \\ &\qquad + Tr[P_1(t)B_2(y)Q_2^{1/2}(B_2(y)Q_2^{1/2})^*] \leq 0. \end{aligned} \tag{3.2.6}$$

Hence, by a similar argument to the proof of Proposition 2.2.4, i.e., using an approximation procedure and applying Itô's formula to $\langle P_1(t)y, y\rangle_H$, one can obtain

$$\begin{aligned} &\mathbb{E}\langle Gy_2(T), y_2(T)\rangle_H + \int_0^T \mathbb{E}\langle My_2(t), y_2(t)\rangle_H dt \\ &\quad \leq \langle P_1(0)y_0, y_0\rangle_H = \mathbb{E}\langle Gy_1(T), y_1(T)\rangle_H + \int_0^T \mathbb{E}\langle My_1(t), y_1(t)\rangle_H dt. \end{aligned}$$

The proof is thus complete. □

Corollary 3.2.2 *Under the same conditions as in Proposition 3.2.1, we have*

$$\int_0^T \mathbb{E}\|y_2(t, y_0)\|_H^2 dt \leq \int_0^T \mathbb{E}\|y_1(t, y_0)\|_H^2 dt \quad \textit{for all} \quad T \geq 0,$$

and

$$\mathbb{E}\|y_2(t, y_0)\|_H^2 \leq \mathbb{E}\|y_1(t, y_0)\|_H^2 \quad \textit{for all} \quad t \geq 0.$$

Proof These inequalities follow immediately by taking $G = 0$, $M = I$ and $G = I$, $M = 0$ in (3.2.3), respectively. □

Corollary 3.2.3 *Suppose that B_1 is linear and the null solution of the associated equation in (3.2.1) is exponentially stable in mean square. Then condition (3.2.2) implies that the null solution of the other equation in (3.2.1) is exponentially stable in mean square. That is, for any $y_0 \in H$,*

$$\mathbb{E}\|y_2(t, y_0)\|_H^2 \le M \cdot e^{-\mu t}\|y_0\|_H^2 \quad \textit{for some} \quad M \ge 1 \quad \textit{and} \quad \mu > 0.$$

Proof This is immediate by virtue of Corollary 3.2.2. □

In order to illustrate the theory established in the preceding, let us consider a class of stochastic differential equations in which $K^i = \mathbb{R}$, $W_i(t) = w(t)$, a standard one-dimensional Brownian motion $i = 1, 2$, $B_1(y) = \alpha b\langle a, y\rangle_H$, $\alpha > 0$, $b, a \in H$, and $B_2(y) = bg(\langle a, y\rangle_H)$ where $g(\cdot)\colon \mathbb{R} \to \mathbb{R}$, $g(0) = 0$, is some continuous function. In this case, (3.2.1) turns out to be two stochastic differential equations:

$$dy(t) = Ay(t)dt + \alpha b\langle a, y(t)\rangle_H dw(t), \qquad t \ge 0, \qquad y(0) = y_0, \tag{3.2.7}$$

$$dy(t) = Ay(t)dt + bg(\langle a, y(t)\rangle_H)dw(t), \qquad t \ge 0, \qquad y(0) = y_0. \tag{3.2.8}$$

Now, let Σ_α be the set of all real Lipschitz continuous functions $g(\cdot)$ satisfying $|g(x)| \le \alpha|x|$ for any $x \in \mathbb{R}$. If $g \in \Sigma_\alpha$, condition (3.2.2) is satisfied since for any $x, y \in H$,

$$\begin{aligned}\langle B_1(y)B_1^*(y)x, x\rangle_H &= \alpha^2\langle b, x\rangle_H^2\langle a, y\rangle_H^2 \ge \langle b, x\rangle_H^2|g(\langle a, y\rangle_H)|^2\\ &= \langle B_2(y)B_2^*(y)x, x\rangle_H.\end{aligned} \tag{3.2.9}$$

Hence, the nonlinear stochastic system (3.2.8) is exponentially stable in mean square for any $g \in \Sigma_\alpha$ if linear system (3.2.7) is exponentially stable in mean square. Therefore, all we need to do is to find conditions to ensure exponential stability of (3.2.7). Since (3.2.7) is a linear equation, we have already established some stability results such as those in Example 2.2.7.

Proposition 3.2.4 *Suppose that A generates an exponentially stable, strongly continuous semigroup e^{tA}, $t \ge 0$, and*

$$\langle e^{tA}b, a\rangle_H^2 \le Ce^{-2\mu t}, \quad C > 0,\ \mu > 0 \quad \textit{with} \quad \alpha^2 C < 2\mu.$$

Then system (3.2.8) is exponentially stable in mean square for any $g \in \Sigma_\alpha$. If, in particular, $\langle e^{tA}b, a\rangle_H = 0$, $t \ge 0$, then it is exponentially stable in mean square for any Lipschitz continuous function $g(\cdot)$ with $g(0) = 0$.

Proof It is known from Example 2.2.7 that the null solution of (3.2.7) is exponentially stable in mean square if

$$\alpha^2 \int_0^\infty \langle e^{tA}b, a\rangle_H^2 dt < 1.$$

This just shows the first part. The second part also follows since we can take an arbitrarily large $\mu > 0$. □

Example 3.2.5 Consider the stochastic heat equation

$$\begin{cases} y(x,t) = y_0(x) + \displaystyle\int_0^t \frac{\partial^2 y(x,s)}{\partial x^2} ds \\ \qquad + \displaystyle\int_0^t b(x) g\Big(\int_0^1 a(x) y(x,s) dx\Big) dw(s), \quad 0 < x < 1,\ t \geq 0, \\ y(0,t) = \ y(1,t) = 0, \quad t \geq 0; \ \ y(x,0) = y_0(x), \quad 0 \leq x \leq 1, \end{cases} \tag{3.2.10}$$

where $w(t)$ is a one-dimensional standard Brownian motion.

On this occasion, we take $H = L^2(0,1)$ and $A = d^2/dx^2$ with domain

$$\mathscr{D}(A) = \{y \in H :\ y,\ y' \text{ are absolutely continuous},\ y'' \in H,\ y(0) = y(1) = 0\}.$$

If we take $b(x) = \sin \pi x$ and $a(\cdot) = 1$, then

$$\int_0^\infty \langle e^{tA}b, a\rangle_H^2 dt = 2/\pi^4.$$

Thus by Proposition 3.2.4, we have the desired exponential stability in mean square of (3.2.10) for any $g \in \Sigma_\alpha$ with $\alpha^2 < \pi^4/2$. If we take $b(x) = \cos \pi x$ and $a(x) = \sin \pi x$, then

$$\langle e^{tA}b, a\rangle_H = 0, \qquad t \geq 0.$$

Hence, we have the same stability for any Lipschitz continuous function $g(\cdot)$ with $g(0) = 0$.

3.3 Nonautonomous Stochastic Systems

In this section, we intend to consider exponential decay of nonautonomous, nonlinear stochastic differential equations in both the pth moment and almost sure senses.

3.3.1 A Coercive Decay Condition

Consider the following nonlinear stochastic differential equation with nonrandom initial datum $y_0 \in H$,

$$\begin{cases} y(t) = y_0 + \int_0^t A(s, y(s))ds + \int_0^t B(s, y(s))dW(s), & t \geq 0, \\ y(0) = y_0 \in H, \end{cases} \tag{3.3.1}$$

where $A(\cdot,\cdot): \mathbb{R}_+ \times V \to V^*$, $B(\cdot,\cdot): \mathbb{R}_+ \times V \to \mathscr{L}_2(K_Q, H)$, are two measurable families of nonlinear mappings and satisfy all the conditions (a)–(d) in Section 1.4.2 so as that a unique global strong solution exists for this equation. We want to establish an exponential decay criterion based on a version of the coercive condition (a) there. In particular, we don't assume that $A(t,0) = 0$ and $B(t,0) = 0$, $t \in \mathbb{R}_+$ in (3.3.1) on this occasion.

(A) There exist constants $p > 1$, $\alpha > 0$, $\lambda \in \mathbb{R}$, and a continuous function $\gamma: \mathbb{R}_+ \to \mathbb{R}_+$ such that

$$2\langle\langle y, A(t,y)\rangle\rangle_{V,V^*} + \|B(t,y)\|^2_{\mathscr{L}_2(K_Q,H)} \leq -\alpha\|y\|_V^p + \lambda\|y\|_H^2 + \gamma(t), \quad y \in V, \quad t \geq 0,$$

where γ satisfies

$$\int_0^\infty \gamma(t)e^{\mu t}dt < \infty \quad \text{for some} \quad \mu > 0. \tag{3.3.2}$$

(B) There exists a constant $M > 0$ such that

$$\|B(t,y) - B(t,z)\|^2_{\mathscr{L}_2(K_Q,H)} \leq M\|y - z\|_V^2, \qquad \forall\, y, z \in V.$$

Theorem 3.3.1 *Assume that $y(t, y_0)$, $t \geq 0$, is a global strong solution of equation (3.3.1) and condition (A) holds. If (i) $\alpha > \lambda\beta$, $p = 2$ or (ii) $\lambda < 0$, $p \neq 2$, then there exist numbers $\theta > 0$, $C = C(y_0) > 0$ such that*

$$\mathbb{E}\|y(t, y_0)\|_H^2 \leq C(y_0) \cdot e^{-\theta t}, \qquad \forall t \geq 0, \tag{3.3.3}$$

where $\beta > 0$ is the constant given in the relation $\|\cdot\|_H^2 \leq \beta\|\cdot\|_V^2$.

Proof We only show (i) since (ii) can be similarly proved. Firstly, from the relation $\|\cdot\|_H^2 \leq \beta\|\cdot\|_V^2$ and condition (A) it is easy to deduce that for any $t \geq 0$,

$$2\langle\langle x, A(t,x)\rangle\rangle_{V,V^*} + \|B(t,x)\|^2_{\mathscr{L}_2(K_Q,H)} \leq -\nu\|x\|_H^2 + \gamma(t), \quad x \in V, \tag{3.3.4}$$

where $\nu = (\alpha - \lambda\beta)/\beta > 0$. Applying Itô's formula to the function $\Lambda(t,x) = e^{\theta t}\|x\|_H^2$, $\theta = \min\{\nu, \mu\}$, $x \in H$, and strong solution $y(t, y_0) \in V$, $t \geq 0$, we get that for all $t \geq 0$,

$$
\begin{aligned}
e^{\theta t}\|y(t)\|_H^2 - \|y_0\|_H^2 = \theta \int_0^t e^{\theta s}\|y(s)\|_H^2 ds \\
+ 2\int_0^t e^{\theta s}\langle\langle y(s), A(s, y(s))\rangle\rangle_{V,V^*} ds \\
+ 2\int_0^t e^{\theta s}\langle y(s), B(s, y(s))dW(s)\rangle_H \\
+ \int_0^t e^{\theta s} Tr(B(s, y(s))Q^{1/2}(B(s, y(s))Q^{1/2})^*) ds.
\end{aligned}
\tag{3.3.5}
$$

Since $M(t) := \int_0^t e^{\theta s}\langle y(s), B(s, y(s))dW(s)\rangle_H$ is a continuous local martingale, there exists a sequence of stopping times $\{\tau_n\}$, $n \geq 1$, such that $\tau_n \uparrow \infty$ as $n \to \infty$, and for each $n \in \mathbb{N}_+$,

$$
\int_0^{t\wedge\tau_n} e^{\theta s}\langle y(s), B(s, y(s))dW(s)\rangle_H, \qquad t \in \mathbb{R}_+,
$$

is a continuous martingale. This yields that for any $n \geq 1$,

$$
\mathbb{E}\left(\int_0^{t\wedge\tau_n} e^{\theta s}\langle y(s), B(s, y(s))dW(s)\rangle_H\right) = 0, \quad t \in \mathbb{R}_+.
$$

This fact, together with (3.3.4) and (3.3.5), further implies that for all $t \geq 0$ and $n \in \mathbb{N}_+$,

$$
\mathbb{E}e^{\theta(t\wedge\tau_n)}\|y(t\wedge\tau_n)\|_H^2 \leq \|y_0\|_H^2 + \mathbb{E}\int_0^{t\wedge\tau_n} \gamma(s)e^{\theta s} ds. \tag{3.3.6}
$$

Then letting $n \to \infty$ and applying Fatou's lemma to (3.3.6) immediately yield that

$$
\mathbb{E}\|y(t)\|_H^2 \leq \left(\|y_0\|_H^2 + \int_0^\infty \gamma(s)e^{\mu s} ds\right)e^{-\theta t}, \qquad t \geq 0.
$$

In other words, the strong solution y is exponentially decayable in mean square and the proof is complete. □

Remark 3.3.2 In general, a similar nonautonomous version of equivalence to Theorem 3.1.4 between L^p-stability and pth moment exponential stability of (3.3.1) does not exist. To illustrate this, let us consider a simple one-dimensional stochastic differential equation

$$
\begin{cases}
dy(t) = -\dfrac{\alpha}{1+t}y(t)dt + (1+t)^{-\alpha}dw(t), & t \geq s \geq 0, \\
y(s) \in L^2_{\mathscr{F}_s}(\Omega, \mathscr{F}, \mathbb{P}),
\end{cases}
\tag{3.3.7}
$$

where $\alpha > 1$ is some constant and $w(t)$, $t \geq 0$, is a standard real Brownian motion.

It is easy to obtain the explicit solution of this equation

$$y(t, y(s)) = (1+t)^{-\alpha}(y(s) + w(t)), \qquad t \geq s \geq 0.$$

By a direct computation, we obtain the following equality

$$\lim_{t\to\infty} \frac{\log \mathbb{E}|y(t, y(s))|^2}{t} = 0.$$

That is, the solution of equation (3.3.7) is not exponentially decayable in mean square. However, we have

$$\begin{aligned}\int_s^\infty \mathbb{E}|y(t, y(s))|^2 dt &\leq 2\int_s^\infty \frac{\mathbb{E}y^2(s) + \mathbb{E}w^2(t)}{(1+t)^{2\alpha}} dt\\ &\leq \frac{2\mathbb{E}(y^2(s))}{2\alpha - 1} + \frac{1}{\alpha - 1} < \infty, \qquad s \geq 0.\end{aligned}$$

The following result shows that the conditions in Theorem 3.3.1, together with (B), assure not only the exponential decay of equation (3.3.1) in mean square, but also its almost sure pathwise exponential decay.

Theorem 3.3.3 *In addition to (B), there exists, under the same conditions as in Theorem 3.3.1, a random variable $T(\omega) \geq 0$ such that*

$$\|y(t, y_0)\|_H \leq C(y_0) \cdot e^{-\gamma t}, \qquad \forall t \geq T(\omega) \qquad a.s. \tag{3.3.8}$$

for some constant $\gamma > 0$ and number $C = C(y_0) > 0$.

Proof Once again, we only show the case (i). For any large integer $N > 0$, applying Itô's formula to the strong solution $y(t, y_0)$, $t \geq 0$, of (3.3.1) immediately yields that for any $t \geq N$,

$$\begin{aligned}\|y(t, y_0)\|_H^2 - \|y(N, y_0)\|_H^2 &= 2\int_N^t \langle\langle y(s), A(s, y(s))\rangle\rangle_{V,V^*} ds\\ &\quad + \int_N^t \|B(s, y(s))\|^2_{\mathscr{L}_2(K_Q, H)} ds\\ &\quad + 2\int_N^t \langle y(s), B(s, y(s))dW(s)\rangle_H,\end{aligned} \tag{3.3.9}$$

which, together with condition (A), further implies that

$$\begin{aligned}\|y(t, y_0)\|_H^2 \leq \|y(N, y_0)\|_H^2 + (\lambda \vee 0)\int_N^t \|y(s)\|_H^2 ds + \int_N^t \gamma(s) ds\\ + \left|2\int_N^t \langle y(s), B(s, y(s))dW(s)\rangle_H\right|,\end{aligned} \tag{3.3.10}$$

where $\lambda \vee 0 = \max\{\lambda, 0\}$. Let $I_N = [N, N+1]$. Then we can obtain from (3.3.10) that

$$\begin{aligned}\mathbb{E}\Big[\sup_{t\in I_N}\|y(t,y_0)\|_H^2\Big] \le \mathbb{E}\|y(N,y_0)\|_H^2 \\ + (\lambda\vee 0)\int_N^{N+1}\mathbb{E}\|y(s,y_0)\|_H^2 ds + \int_N^{N+1}\gamma(s)ds \\ + \mathbb{E}\Big\{\sup_{t\in I_N}\Big|2\int_N^t\langle y(s), B(s,y(s))dW(s)\rangle_H\Big|\Big\}.\end{aligned} \tag{3.3.11}$$

By using the Burkholder–Davis–Gundy lemma, we can estimate the last term in (3.3.11) (in what follows, C_1, C_2, ... denote some positive constants)

$$\begin{aligned}&\mathbb{E}\Big\{\sup_{t\in I_N}\Big|2\int_N^t\langle y(s), B(s,y(s))dW(s)\rangle_H\Big|\Big\} \\ &\le C_1\mathbb{E}\Big[\int_N^{N+1}\|y(s)\|_H^2\|B(s,y(s))\|^2_{\mathscr{L}_2(K_Q,H)}ds\Big]^{1/2} \\ &\le C_1\mathbb{E}\Big[\sup_{t\in I_N}\|y(t)\|_H\Big(\int_N^{N+1}\|B(s,y(s))\|^2_{\mathscr{L}_2(K_Q,H)}ds\Big)^{1/2}\Big] \\ &\le \frac{1}{2}\mathbb{E}\Big[\sup_{t\in I_N}\|y(t)\|_H^2\Big] + C_2\int_N^{N+1}\mathbb{E}\|B(s,y(s))\|^2_{\mathscr{L}_2(K_Q,H)}ds,\end{aligned} \tag{3.3.12}$$

which, in addition to (3.3.11) and the well-known Markov inequality, immediately implies that for any $\varepsilon_N > 0$,

$$\begin{aligned}\mathbb{P}\Big\{\sup_{t\in I_N}\|y(t)\|_H^2 \ge \varepsilon_N^2\Big\} \le \varepsilon_N^{-2}\mathbb{E}\Big[\sup_{t\in I_N}\|y(t)\|_H^2\Big] \\ \le C_3\varepsilon_N^{-2}\Big[\mathbb{E}\|y(N,y_0)\|_H^2 + \int_N^{N+1}\mathbb{E}\|y(t,y_0)\|_H^2 dt \\ + \int_N^{N+1}\gamma(t)dt + \int_N^{N+1}\mathbb{E}\|B(t,y(t))\|^2_{\mathscr{L}_2(K_Q,H)}dt\Big].\end{aligned} \tag{3.3.13}$$

On the other hand, it is easy to see the existence of a positive constant $C_4 > 0$ such that

$$\int_N^{N+1}\gamma(s)ds \le e^{-\theta N}\int_N^{N+1}\gamma(t)e^{\mu t}dt < C_4\cdot e^{-\theta N}, \tag{3.3.14}$$

which, together with (3.3.3), implies the existence of a constant $C_5 > 0$ such that

$$\mathbb{E}\|y(N,y_0)\|_H^2 + \int_N^{N+1} \mathbb{E}\|y(s,y_0)\|_H^2 ds + \int_N^{N+1} \gamma(s)ds \le C_5 \cdot e^{-\theta N}. \tag{3.3.15}$$

For the last term on the right-hand side of (3.3.13), we may claim the following result: there exists a positive constant $c > 0$ such that for $N > 0$,

$$\int_N^{N+1} \mathbb{E}\|B(t,y(t))\|^2_{\mathscr{L}_2(K_Q,H)} dt \le c \cdot e^{-\theta N/2}. \tag{3.3.16}$$

Let $\varepsilon_N^2 = e^{-\theta N/4}$. Then (3.3.13) through (3.3.16) imply that

$$\begin{aligned}\mathbb{P}\Big\{\sup_{t\in I_N} \|y(t,y_0)\|_H^2 \ge \varepsilon_N^2\Big\} &\le C_6 \cdot \varepsilon_N^{-2} e^{-\theta N/2}\\ &= C_6 \cdot e^{-\theta N/4} \quad \text{for some} \quad C_6 > 0,\end{aligned} \tag{3.3.17}$$

and finally a Borel–Cantelli's lemma type of argument completes the proof.

Lastly, let us prove the claim (3.3.16). Indeed, as in Theorem 3.3.1 we obtain by using (A) and Itô's formula that there exists an increasing sequence of stopping times $\{\tau_n\}$, $n \ge 1$, such that

$$\begin{aligned}&\mathbb{E}e^{\frac{\theta}{2}(t\wedge\tau_n)}\|y(t\wedge\tau_n)\|_H^2\\ &\le \|y_0\|_H^2 + \frac{\theta}{2}\mathbb{E}\int_0^{t\wedge\tau_n} e^{\frac{\theta}{2}s}\|y(s)\|_H^2 ds - \alpha\mathbb{E}\int_0^{t\wedge\tau_n} e^{\frac{\theta}{2}s}\|y(s)\|_V^2 ds\\ &\quad + (\lambda\vee 0)\mathbb{E}\int_0^{t\wedge\tau_n} e^{\frac{\theta}{2}s}\|y(s)\|_H^2 ds + \mathbb{E}\int_0^{t\wedge\tau_n}\gamma(s)e^{\frac{\theta}{2}s}ds.\end{aligned} \tag{3.3.18}$$

By virtue of Theorem 3.3.1, this further implies the existence of a number $C_7 = C_7(y_0) > 0$ such that

$$\begin{aligned}&\mathbb{E}\int_0^{t\wedge\tau_n} e^{\frac{\theta}{2}s}\|y(s)\|_V^2 ds\\ &\le \frac{1}{\alpha}\Big[\|y_0\|_H^2 + \Big(\frac{\theta}{2}+\lambda\vee 0\Big)\int_0^{t\wedge\tau_n} e^{\frac{\theta}{2}s}\mathbb{E}\|y(s)\|_H^2 ds + \int_0^{t\wedge\tau_n}\gamma(s)e^{\frac{\theta}{2}s}ds\Big]\\ &\le \frac{1}{\alpha}\Big[\|y_0\|_H^2 + \Big(\frac{\theta}{2}+\lambda\vee 0\Big)C(y_0)\int_0^{\infty} e^{-\frac{\theta}{2}s}ds + \int_0^{\infty}\gamma(s)e^{\mu s}ds\Big]\\ &\le C_7.\end{aligned} \tag{3.3.19}$$

Letting $n \to \infty$ in (3.3.19), we easily conclude that for any $0 \le s \le t$,

$$\int_s^t \mathbb{E}\|y(u)\|_V^2 du \le e^{-\frac{\theta}{2}s}\int_0^t e^{\frac{\theta}{2}u}\mathbb{E}\|y(u)\|_V^2 du \le C_7 \cdot e^{-\frac{\theta}{2}s}. \tag{3.3.20}$$

By taking into account the conditions (A), (B), and (3.3.20), one can obtain

$$\begin{aligned}\int_N^{N+1} & \mathbb{E}\|B(u,y(u))\|^2_{\mathscr{L}_2(K_Q,H)}du \\ &\le 2\int_N^{N+1} \mathbb{E}\|B(u,y(u)) - B(u,0)\|^2_{\mathscr{L}_2(K_Q,\,H)}du \\ &\quad + 2\int_N^{N+1} \mathbb{E}\|B(u,0)\|^2_{\mathscr{L}_2(K_Q,\,H)}du \\ &\le C_8\Big(\int_N^{N+1} \mathbb{E}\|y(u)\|_V^2 du\Big) + 2\int_N^{N+1} \gamma(u)du \\ &\le c\cdot e^{-\theta N/2} \qquad \text{for some} \quad c > 0.\end{aligned}$$

Hence, the proof is now complete. □

Remark 3.3.4 In general, it is impossible to relax the rate of $\gamma(\cdot)$ in (3.3.2) to obtain the desired exponential decay. To see this, let us consider an example by assuming that $y(t)$, $t \ge 0$, satisfies a real-valued stochastic differential equation

$$\begin{cases} dy(t) = -py(t)dt + (1+t)^{-q}dw(t), & t \ge 0, \\ y(0) = 0, \end{cases} \tag{3.3.21}$$

where p, q are two positive constants and $w(t)$, $t \ge 0$, is a standard one-dimensional Brownian motion.

Clearly, in this case the associated coercive condition (A) becomes

$$2vA(t,v) + |B(t,v)|^2 = -2pv^2 + (1+t)^{-2q}, \quad \forall v \in \mathbb{R}.$$

However, the solution y at present does not decay exponentially. Indeed, it is easy to obtain the explicit solution

$$y(t) = e^{-pt}\int_0^t e^{ps}\cdot(1+s)^{-q}dw(s) =: e^{-pt}M(t), \qquad t \ge 0,$$

which implies the mean square Lyapunov exponent $\lim_{t\to\infty} \frac{\log \mathbb{E}|y(t)|^2}{t} = 0.$ In a similar manner, by virtue of the well-known iterated logarithmic law of martingales (cf. Revuz and Yor [197]), we have

$$\overline{\lim_{t\to\infty}} \frac{M(t)}{\sqrt{2[M(t)]\log\log[M(t)]}} = 1 \qquad a.s.,$$

where $[M(t)]$ is the quadratic variation of $M(t)$, and

$$\lim_{t\to\infty}\frac{1}{t}\log[M(t)] = \lim_{t\to\infty}\frac{1}{t}\log\Big(\int_0^t e^{2ps}(1+s)^{-2q}ds\Big) = 2p.$$

Hence, it is possible to get by a direct calculation that the almost sure Lyapunov exponent

$$\overline{\lim_{t\to\infty}} \frac{\log|y(t)|}{t} = 0 \qquad a.s.$$

That is, in spite of the fact that the ordinary differential equation

$$\begin{cases} dy(t) = -py(t)dt, & t \geq 0, \\ y(0) = y_0 \in \mathbb{R}, \end{cases}$$

has its obvious exponential stability, a polynomial decay type of coefficient in the additive noise term is not enough to guarantee the exponential decay property of its stochastically perturbed system. In fact, we shall see in Section 3.8 that this stochastic system has a slower decay, i.e., polynomial decay.

As a direct consequence of Theorems 3.3.1 and 3.3.3, we may impose the following fractional power type of coercive condition.

(A*) There exist constants $p > 1, \alpha > 0, \lambda \in \mathbb{R}, 0 < \sigma < 1$ and nonnegative continuous functions $\gamma(t), \zeta(t), t \in \mathbb{R}_+$, such that

$$\begin{aligned} &2\langle\langle u, A(t,u)\rangle\rangle_{V,V^*} + \|B(t,u)\|^2_{\mathscr{L}_2(K_Q,H)} \\ &\leq -\alpha\|u\|_V^p + \lambda\|u\|_H^2 + \zeta(t)\|u\|_H^{2\sigma} + \gamma(t), \quad u \in V, \end{aligned} \tag{3.3.22}$$

where $\gamma(t), \zeta(t)$ satisfy the property that there exist constants $\delta > 0$, $\mu > 0$ such that $\gamma(t)e^{\mu t}$ and $\zeta(t)e^{\delta t}$ are integrable on $[0, \infty)$.

Corollary 3.3.5 *Suppose that the condition (A*) holds. If (i) $\alpha > \lambda\beta$, $p = 2$ or (ii) $\lambda < 0$, $p \neq 2$, then there exist numbers $\theta > 0$, $C = C(y_0) > 0$ such that*

$$\mathbb{E}\|y(t, y_0)\|_H^2 \leq C(y_0) \cdot e^{-\theta t}, \qquad \forall t \geq 0. \tag{3.3.23}$$

Further, if condition (B) holds, then there exist positive numbers $C = C(y_0) > 0$, $\gamma > 0$ and a random variable $T(\omega) \geq 0$ such that

$$\|y(t, y_0)\|_H \leq C(y_0) \cdot e^{-\gamma t}, \qquad \forall t \geq T(\omega) \qquad a.s.$$

Proof By virtue of Young's inequality

$$a \cdot b \leq \frac{a^p}{p} + \frac{b^q}{q} \quad \text{for any} \quad a \geq 0,\ b \geq 0,\ p,\ q > 1 \ \text{ with } \ \frac{1}{p} + \frac{1}{q} = 1,$$

we have for arbitrary $\varepsilon > 0$ that the third term on the right-hand side of (3.3.22) satisfies

$$\zeta(t)\|u\|_H^{2\sigma} \leq \sigma\varepsilon^{1/\sigma}\|u\|_H^2 + (1-\sigma)\varepsilon^{\frac{1}{1-\sigma}}\zeta(t)^{\frac{1}{1-\sigma}} \qquad u \in V,$$

which, together with (3.3.22), implies that

$$2\langle\langle u, A(t,u)\rangle\rangle_{V,V^*} + \|B(t,u)\|^2_{\mathscr{L}_2(K_Q,H)} \le -\alpha\|u\|_V^p + \lambda\|u\|_H^2 + \sigma\varepsilon^{1/\sigma}\|u\|_H^2$$
$$+\gamma(t) + (1-\sigma)\varepsilon^{\frac{1}{1-\sigma}}\zeta(t)^{\frac{1}{1-\sigma}}, \quad u \in V. \tag{3.3.24}$$

In the case (i), by virtue of Theorems 3.3.1 and 3.3.3 it is easy to deduce that if $\alpha - \lambda\beta > \beta\sigma\varepsilon^{1/\sigma}$, the solution is exponentially decayable in both mean square and almost sure senses. Since $\varepsilon > 0$ is an arbitrary number, the proof is thus complete by choosing $\varepsilon > 0$ small enough. The case (ii) can be shown in a similar manner. □

Example 3.3.6 Let $p > 2$ and $\mathcal{O}$ be an open, bounded subset in $\mathbb{R}^n$, $n \ge 1$, with regular boundary. Consider the Sobolev space $V = W_0^{1,p}(\mathcal{O})$, $H = L^2(\mathcal{O})$ with their usual norms, and the monotone operator $A\colon V \to V^*$ which is defined as for any $u, v \in V$,

$$\langle\langle v, A(u)\rangle\rangle_{V,V^*} = -\sum_{i=1}^n \int_{\mathcal{O}} \left|\frac{\partial u(x)}{\partial x_i}\right|^{p-2} \frac{\partial u(x)}{\partial x_i}\frac{\partial v(x)}{\partial x_i}dx - \int_{\mathcal{O}} a(x)u(x)v(x)dx,$$

where $a \in L^\infty(\mathcal{O}; \mathbb{R})$ satisfies $a(x) \ge \tilde{a} > 0$, $x \in \mathcal{O}$, for some constant $\tilde{a}$. We also set $B(t, u(x)) = g(u(x))$, $u \in V$, in (3.3.1) where $g\colon \mathbb{R} \to \mathbb{R}$ is some Lipschitz continuous function satisfying

$$|g(x) - g(y)| \le L|x-y|, \qquad \forall x,\ y \in \mathbb{R},$$

for some constant $L > 0$ such that $L^2 < 2\tilde{a}$ and $g(0) = 0$. Let $w(t)$, $t \ge 0$, be a standard real Brownian motion. In this case, we can claim that there exists a unique strong solution to equation (3.3.1). In particular, (A) holds with $\gamma(\cdot) = 0$, $\lambda = -(2\tilde{a} - L^2) < 0$, $p > 2$, $\alpha = 2$. Consequently, by using Theorems 3.3.1 and 3.3.3, we may obtain the exponential decay in both mean square and almost sure sense of the equation represented by

$$\begin{cases} dy(t,x) = \left[-\displaystyle\sum_{i=1}^n \frac{\partial}{\partial x_i}\left(\left|\frac{\partial y(t,x)}{\partial x_i}\right|^{p-2}\frac{\partial y(t,x)}{\partial x_i}\right) - a(x)y(t,x)\right]dt \\ \qquad\qquad + g(y(t,x))dw(t), \quad t > 0, \ x \in \mathcal{O}; \\ y(0;x) = y_0(x) \in H \ \text{ in } \mathcal{O}; \\ y(t;x) = 0 \ \text{almost surely in } (0,\infty) \times \partial\mathcal{O}. \end{cases} \tag{3.3.25}$$

3.3.2 Semigroup Methods

The material presented in Section 3.3.1 is, in essence, a Lyapunov function scheme where the Itô functional is taken as the quadratic function $\|\cdot\|_H^2$. A similar theory for general Lyapunov functions is possible, and the program is developed in this section through a semigroup method.

Consider the following semilinear stochastic evolution equation: for any $0 \le t < \infty$,

$$\begin{cases} y(t) = e^{tA}y_0 + \displaystyle\int_0^t e^{(t-s)A}F(s, y(s))ds + \int_0^t e^{(t-s)A}B(s, y(s))dW(s), \\ y(0) = y_0 \in H, \end{cases} \tag{3.3.26}$$

where e^{tA}, $t \ge 0$, is some C_0-semigroup of bounded linear operators on H with its infinitesimal generator A. Here $W(t)$, $t \ge 0$, is some given Q-Wiener process in Hilbert space K and F, B are two nonlinear measurable mappings from $\mathbb{R} \times H$ into H and $\mathscr{L}_2(K_Q, H)$, respectively, which satisfy the following Lipschitz and linear growth conditions

$$\|F(t,y) - F(t,z)\|_H + \|B(t,y) - B(t,z)\|_{\mathscr{L}_2(K_Q,H)} \le L\|y - z\|_H, \quad t \in [0,\infty),$$

$$\|F(t,y)\|_H + \|B(t,y)\|_{\mathscr{L}_2(K_Q,H)} \le L(1 + \|y\|_H), \quad t \in [0,\infty), \tag{3.3.27}$$

for some constant $L > 0$ and arbitrary $y, z \in H$. Then by virtue of Theorem 1.4.2, there is a unique global mild solution $y(t)$, $t \ge 0$, to (3.3.26) for arbitrary $y_0 \in H$.

Let $\Lambda: H \to \mathbb{R}_+$ be a nonnegative Itô functional in the sense of that introduced in Section 1.3. For any $x \in \mathscr{D}(A)$, $t \ge 0$, let

$$(\mathcal{L}\Lambda)(t,x) := \langle \Lambda'(x), Ax + F(t,x)\rangle_H + \frac{1}{2}Tr\big[\Lambda''(x)B(t,x)Q^{1/2}(B(t,x)Q^{1/2})^*\big], \tag{3.3.28}$$

whenever the right-hand side of (3.3.28) is meaningful. We are concerned with the exponential decay property of the solution $y(t, y_0)$, $t \ge 0$, to (3.3.26) by implementing a Lyapunov function program.

Theorem 3.3.7 *Let $\Lambda: H \to \mathbb{R}_+$ be an Itô functional such that*

(i) there exist constants $p \ge 1$ and $c > 0$ such that

$$|\Lambda(x)| + \|x\|_H\|\Lambda'(x)\|_H + \|x\|_H^2\|\Lambda''(x)\| \le c\|x\|_H^p, \quad \forall x \in H; \tag{3.3.29}$$

(ii) *there exist a constant* $\alpha > 0$ *and nonnegative function* $\gamma(t)$, $t \in \mathbb{R}_+$, *such that*

$$(\mathcal{L}\Lambda)(t,x) \leq -\alpha\Lambda(x) + \gamma(t), \quad x \in \mathscr{D}(A), \quad t \geq 0, \tag{3.3.30}$$

where γ *satisfies*

$$\int_0^\infty \gamma(t)e^{\theta t}dt < \infty \quad \text{for some} \quad \theta > 0.$$

Then there exists a number $C = C(y_0) > 0$ *such that*

$$\mathbb{E}\Lambda(y(t,y_0)) \leq C(y_0) \cdot e^{-(\alpha\wedge\theta)t}, \qquad \forall t \geq 0, \tag{3.3.31}$$

where $\alpha \wedge \theta = \min\{\alpha, \theta\}$. *In addition, if* $M\|x\|_H^p \leq \Lambda(x)$, $x \in H$, *for some constant* $M > 0$, *then*

$$\mathbb{E}\|y(t,y_0)\|_H^p \leq \frac{C(y_0)}{M}e^{-(\alpha\wedge\theta)t}, \qquad \forall t \geq 0. \tag{3.3.32}$$

That is, the solution of system (3.3.26) has exponential decay in the pth moment sense.

Proof To prove this theorem, we introduce the following Yosida approximating systems of (3.3.26) and use Proposition 1.4.5,

$$\begin{cases} dy(t) = Ay(t)dt + R(n)F(t,y(t))dt + R(n)B(t,y(t))dW(t), & t \geq 0, \\ y(0) = R(n)y_0 \in \mathscr{D}(A), \end{cases} \tag{3.3.33}$$

where $n \in \rho(A)$, the resolvent set of A and $R(n) = nR(n,A)$. We apply Itô's formula to the function $v(t,x) = e^{\mu t}\Lambda(x)$, $\mu = \alpha \wedge \theta$, and solution $y_n(t)$ of (3.3.33) for each $n \in \rho(A)$ to get that for any $t \geq 0$,

$$\begin{aligned} &e^{\mu t}\Lambda(y_n(t)) - \Lambda(y_n(0)) \\ &= \mu\int_0^t e^{\mu s}\Lambda(y_n(s))ds + \int_0^t e^{\mu s}\langle\Lambda'(y_n(s)), Ay_n(s) + R(n)F(s,y_n(s))\rangle_H ds \\ &\quad + \int_0^t e^{\mu s}\langle\Lambda'(y_n(s)), R(n)B(s,y_n(s))dW(s)\rangle_H \\ &\quad + \frac{1}{2}\int_0^t e^{\mu s}Tr\Big(\Lambda''(y_n(s))R(n)B(s,y_n(s))Q^{1/2}[R(n)B(s,y_n(s))Q^{1/2}]^*\Big)ds. \end{aligned} \tag{3.3.34}$$

By virtue of (3.3.27) and (3.3.29), we see that the process

$$\int_0^t e^{\mu s}\langle\Lambda'(y_n(s)), R(n)B(s,y_n(s))dW(s)\rangle_H$$

is a continuous local martingale for each $n \in \rho(A)$. Therefore, by carrying out a similar scheme of stopping times as in the proofs of Theorem 3.3.1 and taking expectation on both sides of (3.3.34), we can obtain that for $t \geq 0$,

$$\begin{aligned} e^{\mu t}\mathbb{E}\Lambda(y_n(t)) \leq{}& \Lambda(y_n(0)) + (\mu - \alpha)\int_0^t e^{\mu s}\mathbb{E}\Lambda(y_n(s))ds + \int_0^t \gamma(s)e^{\mu s}ds \\ &+ \int_0^t e^{\mu s}\mathbb{E}\Big\{\langle \Lambda'(y_n(s)), (R(n) - I)F(s, y_n(s))\rangle_H \\ &+ \frac{1}{2}Tr\Big[\Lambda''(y_n(s))R(n)B(s, y_n(s))Q^{1/2}\big(R(n)B(s, y_n(s))Q^{1/2}\big)^* \\ &- \Lambda''(y_n(s))B(s, y_n(s))Q^{1/2}(B(s, y_n(s))Q^{1/2})^*\Big]\Big\}ds. \end{aligned} \tag{3.3.35}$$

Since $\|R(n)\| \leq M$ for some $M > 0$ and all $n \in \mathbb{N}$ and $R(n) \to I$ strongly as $n \to \infty$, we may use Proposition 1.4.5 and let $n \to \infty$ in (3.3.35) to obtain, for any $t \geq 0$,

$$e^{\mu t}\mathbb{E}\Lambda(y(t)) \leq \Lambda(y_0) + \int_0^\infty \gamma(s)e^{\theta s}ds. \tag{3.3.36}$$

Then, we can find a number $C = C(y_0) > 0$ such that

$$\mathbb{E}\Lambda(y(t)) \leq C(y_0) \cdot e^{-(\alpha\wedge\theta)t}, \qquad t \geq 0.$$

Last, the conclusion (3.3.32) is straightforward. The proof is thus complete. □

As a direct application of Theorem 3.3.7, we may consider stability of the following autonomous version of stochastic differential equation (3.3.26),

$$\begin{cases} dy(t) = [Ay(t) + F(y(t))]dt + B(y(t))dW(t), & t \geq 0, \\ y(0) = y_0 \in H, \end{cases} \tag{3.3.37}$$

where both $F(\cdot)$, $B(\cdot)$ satisfy the condition (3.3.27) and $F(0) = 0$, $B(0) = 0$.

Corollary 3.3.8 *Let $\Lambda\colon H \to \mathbb{R}$ satisfy all the conditions in Theorem 3.3.7 and*

$$(\mathcal{L}\Lambda)(x) \leq -\alpha\Lambda(x), \qquad x \in \mathscr{D}(A),$$

for some constant $\alpha > 0$. Then the mild solution $y(t, y_0)$, $t \geq 0$, of (3.3.37) satisfies

$$\mathbb{E}\Lambda(y(t, y_0)) \leq C\|y_0\|_H^p \cdot e^{-\alpha t}, \qquad \forall t \geq 0,$$

for some constant $C > 0$. In addition, if $M\|x\|_H^p \leq \Lambda(x)$ for some constant $M > 0$ and all $x \in H$, then

$$\mathbb{E}\|y(t, y_0)\|_H^p \leq \frac{C}{M}\|y_0\|_H^p e^{-\alpha t}, \qquad \forall t \geq 0.$$

Now we state that under the same conditions as in Theorem 3.3.7, the solution y of (3.3.26) has exponential decay in the almost sure sense as well.

Theorem 3.3.9 *Assume that condition (3.3.27) holds and* $F(t,0) = 0$, $B(t,0) = 0$ *for* $t \geq 0$. *Suppose that the Itô functional* $\Lambda\colon H \to \mathbb{R}_+$ *satisfies (i)–(ii) in Theorem 3.3.7 and* $\Lambda(x) \geq M\|x\|_H^p$, $x \in H$, *for some constant* $M > 0$, *then*

(a) $\mathbb{E}\sup_{0\leq t<\infty} \Lambda(y(t,y_0)) \leq c_1(y_0)$ *for some* $c_1(y_0) > 0$*;*
(b) there exists a nonnegative random variable $T(\omega) \geq 0$ *and numbers* $c_2 = c_2(y_0) > 0$, $\mu > 0$ *such that for all* $t \geq T(\omega)$,

$$\|y(t,y_0)\|_H \leq c_2(y_0)\cdot e^{-\mu t} \qquad a.s.$$

Proof First, we apply Itô's formula to the Itô functional Λ and solution process $y_n(t)$ determined by (3.3.33). Then by a similar limit argument to that in Theorem 3.3.7, we will easily obtain that for any $T \geq t \geq t_0 \geq 0$,

$$\Lambda(y(t)) \leq \Lambda(y(t_0)) + \int_{t_0}^t \gamma(s)ds + \int_{t_0}^t \langle \Lambda'(y(s)), B(s,y(s))dW(s)\rangle_H. \tag{3.3.38}$$

To show (a), we first have by using the conditions in this theorem and Doob's inequality that for $t_0 \leq t \leq T$,

$$\begin{aligned}
&\mathbb{E}\left\{\sup_{t_0\leq t\leq T}\left|\int_{t_0}^t \langle \Lambda'(y(s)), B(s,y(s))dW(s)\rangle_H\right|\right\}\\
&\leq 3L\mathbb{E}\left\{\int_{t_0}^T \|\Lambda'(y(s))\|_H^2\cdot\|y(s)\|_H^2 ds\right\}^{1/2}\\
&\leq \frac{3Lc}{M}\mathbb{E}\left\{\int_{t_0}^T \Lambda(y(s))^2 ds\right\}^{1/2}\\
&\leq \frac{3Lc}{M}\mathbb{E}\left\{\left(\frac{M}{3Lc}\sup_{t_0\leq t\leq T}\Lambda(y(t))\right)^{1/2}\left(\frac{3Lc}{M}\int_{t_0}^T \Lambda(y(s))ds\right)^{1/2}\right\}\\
&\leq \frac{1}{2}\mathbb{E}\sup_{t_0\leq t\leq T}\Lambda(y(t)) + \frac{(3Lc)^2}{2M^2}\int_{t_0}^T \mathbb{E}\Lambda(y(t))dt.
\end{aligned} \tag{3.3.39}$$

Then by using (3.3.38) and Theorem 3.3.7, we obtain that

$$\begin{aligned}
\mathbb{E}\sup_{t_0\leq t\leq T}\Lambda(y(t)) &\leq 2\mathbb{E}\Lambda(y(t_0)) + \frac{(3Lc)^2}{M^2}\int_0^\infty C(y_0)e^{-(\alpha\wedge\theta)t}dt\\
&\quad + 2\int_0^\infty \gamma(s)ds \leq c_1(y_0)
\end{aligned} \tag{3.3.40}$$

for some number $c_1 = c_1(y_0) > 0$. Let $t_0 = 0$ and $T \to \infty$ in (3.3.40), then the desired (a) is proved.

To show (b), we employ the well-known Markov's inequality and apply Theorem 3.3.7, (a), and Hölder's inequality to (3.3.39) to get that for any $\varepsilon_N > 0$, $N \in \mathbb{N}_+$,

$$\begin{aligned}
&\mathbb{P}\left\{\sup_{N\le t\le N+1}\left|\int_N^t \langle \Lambda'(y(s)), B(s, y(s))dW(s)\rangle_H\right| \ge \varepsilon_N/2\right\}\\
&\le \frac{2}{\varepsilon_N}\mathbb{E}\left\{\sup_{N\le t\le N+1}\left|\int_N^t \langle \Lambda'(y(s)), B(s, y(s))dW(s)\rangle_H\right|\right\}\\
&\le \frac{2}{\varepsilon_N}\frac{3Lc}{M}\mathbb{E}\left\{\int_N^{N+1} \Lambda(y(s))^2 ds\right\}^{1/2}\\
&\le \frac{C_1(y_0)}{\varepsilon_N}\left(\mathbb{E}\sup_{N\le s\le N+1}(\Lambda(y(s)))\right)^{1/2}\left(\mathbb{E}\int_N^{N+1}\Lambda(y(s))ds\right)^{1/2}\\
&\le \frac{C_1(y_0)c_1(y_0)^{1/2}C(y_0)^{1/2}}{\varepsilon_N}e^{-(\alpha\wedge\theta)N/2},
\end{aligned}$$

for some $C_1(y_0) > 0$. Hence, a similar argument to Theorem 2.2.17, in addition to (3.3.38), yields

$$\mathbb{P}\left\{\sup_{N\le t\le N+1}\Lambda(y(t)) \ge \varepsilon_N\right\} \le (C_2(y_0)/\varepsilon_N)e^{-(\alpha\wedge\theta)N/2}$$

for some $C_2(y_0) > 0$. If we take $\varepsilon_N = e^{-(\alpha\wedge\theta)N/4}$ and carry out a Borel–Cantelli lemma type of argument, it can be concluded that there exists a random variable $T(\omega) \ge 0$ such that whenever $t \ge T(\omega)$,

$$\|y(t, y_0)\|_H \le c_2(y_0)\cdot e^{-(\alpha\wedge\theta)t/4} \qquad a.s.$$

for some $c_2(y_0) > 0$. The proof is now complete. □

Example 3.3.10 Let us consider the following semilinear stochastic partial differential equation:

$$\begin{cases} dy(t,x) = \dfrac{\partial^2}{\partial x^2}y(t,x)dt + e^{-\mu t}\alpha\big(y(t,x)\big)dw(t), & t > 0, \quad x \in (0,1),\\ y(0,x) = y_0(x), \quad 0 \le x \le 1; \quad y(t,0) = y(t,1) = 0, & t \ge 0, \end{cases} \tag{3.3.41}$$

where $w(t)$, $t \ge 0$, is a real standard Brownian motion. Here $\alpha(\cdot)\colon \mathbb{R} \to \mathbb{R}$ is some bounded, Lipschitz continuous function and μ is a positive number. We can formulate this example by taking $H = L^2(0,1)$, and $A = \partial^2/\partial x^2$ with

$$\mathscr{D}(A) = \Big\{ y \in H: \, y, \, y' \; \text{ are absolutely continuous with} \\ y', \, y'' \in H \text{ almost surely}, \; y(0) = y(1) = 0 \Big\},$$

and $F(t,u) = 0$, $B(t,u) = e^{-\mu t}\alpha(u)$, $u \in H$, in (3.3.26).

Obviously, the coefficient B satisfies condition (3.3.27). On the other hand, let $\Lambda(u) = \|u\|_H^2$, $u \in H$, then it is easy to see that for arbitrary $u \in \mathscr{D}(A)$,

$$2\langle u, Au\rangle_H + \|B(t,u)\|^2 \le -2\pi\|u\|_H^2 + Me^{-2\mu t}, \tag{3.3.42}$$

where M is some positive constant. Since all the hypotheses in Theorems 3.3.7 and 3.3.9 are fulfilled, we may deduce that the mild solution of this equation has exponential decay in mean square, that is, there exist positive numbers $\gamma > 0$, $C = C(y_0) > 0$ such that

$$\mathbb{E}\|y(t, y_0)\|_H^2 \le C(y_0)e^{-\gamma t}, \qquad \forall t \ge 0,$$

and meanwhile it has also exponential decay in the almost sure sense.

3.4 Stability in Probability and Sample Path

From time to time, it is useful to consider stochastic stability in a weak sense of systems without exponential decay. In this case, it is frequently possible to relax those conditions given in the previous sections. In this section, we shall employ the variational method to consider some stability of this type for the strong solution of the following stochastic differential equation,

$$\begin{cases} dy(t) = A(y(t))dt + B(y(t))dW(t), & t \ge 0, \\ y(0) = y_0 \in H, \end{cases} \tag{3.4.1}$$

where $A(0) = 0$ and $B(0) = 0$, and A and B satisfy all the conditions (a)–(d) in Section 1.4.2.

Theorem 3.4.1 *Suppose that there exists an Itô type of functional* $\Lambda(\cdot)$ *on* $\{x: \|x\|_H < r\}$ *for some* $r > 0$ *such that*

(i) $\Lambda(0) = 0$, $\Lambda(x) > 0$ *for* $0 < \|x\|_H < r$ *and*

$$b(\delta) := \inf_{\|x\|_H = \delta} \Lambda(x) > 0 \qquad \textit{for any} \qquad 0 < \delta < r;$$

(ii) $(\mathcal{L}\Lambda)(x) \le 0$ *for any* $x \in V$ *with* $\|x\|_H < r$ *where* $\mathcal{L}\Lambda$ *is given as in (1.4.26).*

Then the null solution of (3.4.1) is strongly stable in probability.

Proof Since $\Lambda(0) = 0$ and Λ is continuous, for any $0 < \varepsilon_1, \varepsilon_2 < r$ there exists a number $\delta = \delta(\varepsilon_1, \varepsilon_2) > 0$ such that

$$\frac{\Lambda(x)}{b(\varepsilon_2)} < \varepsilon_1 \quad \text{if} \quad \|x\|_H < \delta < r. \tag{3.4.2}$$

For any initial y_0 with $0 < \|y_0\|_H < \delta < r$, we define a stopping time

$$\tau = \tau(\varepsilon_2) = \inf\{t > 0: \|y(t, y_0)\|_H \geq \varepsilon_2\}.$$

By analogy with Theorem 3.3.9, it is not difficult to obtain for any $t > 0$ that

$$\Lambda(y(t \wedge \tau, y_0)) \leq \Lambda(y_0) + \int_0^{t\wedge\tau} \langle \Lambda'(y(s, y_0)), B(y(s, y_0))dW(s)\rangle_H. \tag{3.4.3}$$

Since the last term in (3.4.3) is a martingale, we thus have for any $t > 0$ that

$$\mathbb{P}\{\tau < t\} \inf_{\|x\|_H = \varepsilon_2} \Lambda(x) \leq \mathbb{E}\mathbf{1}_{\{\tau<t\}}\Lambda(y(t\wedge\tau, y_0)) \leq \mathbb{E}\Lambda(y(t\wedge\tau, y_0)) \leq \Lambda(y_0).$$

This implies, in addition to (3.4.2), that

$$\mathbb{P}\{\tau < t\} \leq \frac{\Lambda(y_0)}{b(\varepsilon_2)} < \varepsilon_1 \qquad \forall t > 0.$$

From this result, we obtain by letting $t \uparrow \infty$ and using the Fatou lemma that

$$\mathbb{P}\{\tau < \infty\} \leq \frac{\Lambda(y_0)}{b(\varepsilon_2)} < \varepsilon_1.$$

Therefore, from the definition of τ we immediately get

$$\mathbb{P}\left\{\sup_{t\geq 0} \|y(t, y_0)\|_H > \varepsilon_2\right\} \leq \mathbb{P}\{\tau < \infty\} < \varepsilon_1 \quad \text{if} \quad \|y_0\|_H < \delta < r.$$

The proof is now complete. □

Corollary 3.4.2 *Suppose that all the conditions in Theorem 3.4.1 hold for $r = \infty$. Assume also $\Lambda(x) \to \infty$ as $\|x\|_H \to \infty$. Then the strong solution $y(t; y_0)$ of (3.4.1) is bounded almost surely for any initial $y_0 \in H$, i.e.,*

$$\mathbb{P}\left\{\sup_{t\geq 0} \|y(t, y_0)\|_H < \infty\right\} = 1.$$

Proof Let $b(\cdot)$ and $\tau(\cdot)$ be defined as in Theorem 3.4.1. The claim of this corollary is easily seen by the following relations

$$\begin{aligned}\mathbb{P}\left\{\sup_{t\geq 0}\|y(t,y_0)\|_H=\infty\right\}&=\mathbb{P}\left\{\bigcap_{n=1}^{\infty}\left(\sup_{t\geq 0}\|y(t,y_0)\|_H>n\right)\right\}\\&\leq\mathbb{P}\{\tau(n)<\infty\text{ for any }n\in\mathbb{N}\}\\&\leq\frac{\Lambda(y_0)}{b(n)}\to 0\quad\text{as}\quad n\to\infty.\end{aligned}\tag{3.4.4}$$

□

Now we establish a result on asymptotic stability in sample path. Let Σ denote the family of all functions $\gamma:\mathbb{R}_+\to\mathbb{R}_+$, which are continuous, strictly increasing, and $\gamma(0)=0$. Let Σ_∞ be the subfamily of functions $\gamma\in\Sigma$ satisfying $\gamma(x)\to\infty$ as $x\to\infty$.

Theorem 3.4.3 *Assume that there exists an Itô type of functional Λ on H, α_1, $\alpha_2\in\Sigma_\infty$, and $\alpha_3\in\Sigma$ such that*

(i) $\alpha_1(\|x\|_H)\leq\Lambda(x)\leq\alpha_2(\|x\|_H)$ for all $x\in H$,
(ii) $(\mathcal{L}\Lambda)(x)\leq-\alpha_3(\|x\|_H)$ for all $x\in V$.

Then the null solution of (3.4.1) is global asymptotically stable in sample path, i.e.,

$$\mathbb{P}\left\{\lim_{t\to\infty}\|y(t,y_0)\|_H=0\right\}=1,\qquad\forall\, y_0\neq 0.$$

Proof We first consider the case $y_0\in V$, $y_0\neq 0$. Note that the stability in probability of the null solution is immediate according to Theorem 3.4.1.

We want to show that $\lim_{t\to\infty}\alpha_3(\|y(t)\|_H)=0$ almost surely. To do this, let

$$\Xi_1=\left\{\omega:\ \underline{\lim}_{t\to\infty}\,\alpha_3(\|y(t)\|_H)>0\right\},$$

$$\Xi_2=\left\{\omega:\ \underline{\lim}_{t\to\infty}\,\alpha_3(\|y(t)\|_H)=0\text{ and }\overline{\lim}_{t\to\infty}\,\alpha_3(\|y(t)\|_H)>0\right\}.$$

Then we show that $\mathbb{P}(\Xi_1)=\mathbb{P}(\Xi_2)=0$. By analogy with the proofs in Theorem 3.3.1, we may obtain by introducing a proper stopping time and using Itô's formula that for any $t\geq 0$,

$$\mathbb{E}\Lambda(y(t))\leq\Lambda(y_0)-\mathbb{E}\int_0^t\alpha_3(\|y(s)\|_H)ds.$$

Let $t\to\infty$, then it follows that

$$\mathbb{E}\int_0^\infty\alpha_3(\|y(s)\|_H)ds\leq\Lambda(y_0)<\infty.$$

This implies

$$\int_0^\infty \alpha_3(\|y(s)\|_H)ds < \infty \qquad a.s. \tag{3.4.5}$$

Hence,

$$\varliminf_{t\to\infty} \alpha_3(\|y(t)\|_H) = 0 \qquad a.s.,$$

that is, $\mathbb{P}(\Xi_1) = 0$.

To show $\mathbb{P}(\Xi_2) = 0$, first note that if the relation $\mathbb{P}(\Xi_2) = 0$ is not true, then there exist $\varepsilon_1 > 0$ and $\delta > 0$ such that

$$\mathbb{P}\Big\{\alpha_3(\|y(\cdot)\|_H) \text{ crosses from below } \varepsilon_1 \text{ to above } 2\varepsilon_1 \text{ and repeats infinitely many times}\Big\} \geq \delta. \tag{3.4.6}$$

On the other hand, note that we have the coercivity condition,

$$2\langle\langle u, A(u)\rangle\rangle_{V,V^*} \leq -\alpha\|u\|_V^p + \lambda\|u\|_H^2, \quad u \in V, \quad p > 1. \tag{3.4.7}$$

By using Itô's formula and (3.4.7), we obtain, for $t \geq 0$,

$$\begin{aligned}\|y(t) - y_0\|_H^2 \leq& -\alpha\int_0^t \|y(s)\|_V^p ds + \lambda\int_0^t \|y(s)\|_H^2 ds \\ &+ 2\int_0^t \|A(y(s))\|_{V^*}\|y_0\|_V ds \\ &+ \int_0^t \|B(y(s))\|^2_{\mathscr{L}_2(K_Q,H)} ds \\ &+ 2\int_0^t \langle y(s) - y_0, B(y(s))dW(s)\rangle_H.\end{aligned} \tag{3.4.8}$$

Note that $\|A(u)\|_{V^*} \leq c\|u\|_V^{p-1}$, $c > 0$, for any $u \in V$ and recall also the elementary inequality

$$a^r b^{1-r} \leq ra + (1-r)b, \qquad 0 \leq r \leq 1, \quad a,\ b > 0.$$

It then follows that

$$\begin{aligned}2&\int_0^t \|A(y(s))\|_{V^*}\|y_0\|_V ds \\ &\leq \int_0^t \left[\frac{\alpha p}{2(p-1)}\|y(s)\|_V^p\right]^{\frac{p-1}{p}} \left[\left(\frac{2(p-1)}{\alpha p}\right)^{p-1}(2c\|y_0\|_V)^p\right]^{\frac{1}{p}} ds \\ &\leq \frac{\alpha}{2}\int_0^t \|y(s)\|_V^p ds + \sigma\|y_0\|_V^p t,\end{aligned} \tag{3.4.9}$$

where $\sigma = (2c)^p/p(2(p-1)/\alpha p)^{p-1} > 0$. For any number $R \geq \|y_0\|_H$, we define a stopping time τ_R by

$$\tau_R = \inf\{t \geq 0\colon \|y(t)\|_H > R\}.$$

By virtue of condition (1.4.23) on B, for any $0 \leq s \leq \tau_R$ there exists a number $K_R > 0$ such that

$$\|B(y(s))\|^2_{\mathscr{L}_2(K_Q,H)} \leq \mu\|y(s)\|^2_H \leq K_R. \tag{3.4.10}$$

Hence, we can further derive from (3.4.8) and (3.4.9) that

$$\begin{aligned}
&\mathbb{E}\Big(\sup_{0\leq s\leq t}\|y(s\wedge\tau_R)-y_0\|^2_H\Big)\\
&\quad\leq \lambda\mathbb{E}\Big(\sup_{0\leq s\leq t}\int_0^{s\wedge\tau_R}\|y(u)\|^2_H du\Big)+\sigma\|y_0\|^p_V t\\
&\qquad+\mathbb{E}\Big(\sup_{0\leq s\leq t}\int_0^{s\wedge\tau_R}\|B(y(u))\|^2_{\mathscr{L}_2(K_Q,H)}du\Big)\\
&\qquad+2\mathbb{E}\Big(\sup_{0\leq s\leq t}\int_0^{s\wedge\tau_R}\langle y(u)-y_0, B(y(u))dW(u)\rangle_H\Big).
\end{aligned} \tag{3.4.11}$$

From (3.4.10), it follows that

$$\mathbb{E}\Big(\sup_{0\leq s\leq t}\int_0^{s\wedge\tau_R}\|B(y(u))\|^2_{\mathscr{L}_2(K_Q,H)}du\Big) \leq K_R t. \tag{3.4.12}$$

On the other hand, we compute by using the Doob inequality to obtain that

$$\begin{aligned}
&2\mathbb{E}\Big(\sup_{0\leq s\leq t}\Big|\int_0^{s\wedge\tau_R}\langle y(u)-y_0, B(y(u))dW(u)\rangle_H\Big|\Big)\\
&\quad\leq 6\mathbb{E}\Big(\int_0^{t\wedge\tau_R}\|y(u)-y_0\|^2_H\|B(y(u))\|^2_{\mathscr{L}_2(K_Q,H)}du\Big)^{1/2}\\
&\quad\leq 6\mathbb{E}\sup_{0\leq s\leq t}\|y(s\wedge\tau_R)-y_0\|_H\Big(\int_0^{t\wedge\tau_R}\|B(y(u))\|^2_{\mathscr{L}_2(K_Q,H)}du\Big)^{1/2}\\
&\quad\leq \mathbb{E}\Big\{\frac{1}{4}\sup_{0\leq s\leq t}\|y(s\wedge\tau_R)-y_0\|^2_H+36\int_0^{t\wedge\tau_R}\|B(y(u))\|^2_{\mathscr{L}_2(K_Q,H)}du\Big\}\\
&\quad\leq \frac{1}{4}\mathbb{E}\sup_{0\leq s\leq t}\|y(s\wedge\tau_R)-y_0\|^2_H+36K_R t.
\end{aligned} \tag{3.4.13}$$

Consequently, (3.4.11), (3.4.12), and (3.4.13) together imply

$$\mathbb{E}\Big(\sup_{0\leq s\leq t}\big|\|y(s\wedge\tau_R)\|_H-\|y_0\|_H\big|^2\Big) \leq \mathbb{E}\Big(\sup_{0\leq s\leq t}\|y(s\wedge\tau_R)-y_0\|^2_H\Big) \leq C_1 t,$$

for some number $C_1 = C_1(R, y_0) > 0$. Hence, one can use this relation and Chebyshev's inequality to show that for any $\delta > 0$,

$$\mathbb{P}\left\{\sup_{0\le s\le t}\Big|\|y(s\wedge\tau_R)\|_H - \|y_0\|_H\Big| > \delta\right\} \le C_1 t/\delta^2. \tag{3.4.14}$$

Let $S_R = \{x \in H: \|x\|_H \le R\}$. Since α_3 is uniformly continuous on any bounded closed set in $\mathbb{R}$, then for any $\varepsilon_2 > 0$, there exists a $\delta = \delta(\varepsilon_2) > 0$ such that

$$\big|\alpha_3(\|x\|_H) - \alpha_3(\|y\|_H)\big| \le \varepsilon_2,$$

where $x, y \in S_R$ with $|\|x\|_H - \|y\|_H| \le \delta$. On the other hand, by Corollary 3.4.2, we have for sufficiently large $R > 0$ that

$$\mathbb{P}\left\{\sup_{0\le s\le t}\|y(s)\|_H \ge R\right\} < \frac{1}{2}, \qquad \forall t \ge 0. \tag{3.4.15}$$

Thus for this number $R > 0$ and $\varepsilon_2 > 0$, we have by using (3.4.14) and (3.4.15) that

$$\begin{aligned}
&\mathbb{P}\left\{\sup_{0\le s\le t}\big|\alpha_3(\|y(s)\|_H) - \alpha_3(\|y_0\|_H)\big| > \varepsilon_2\right\}\\
&= \mathbb{P}\left\{\sup_{0\le s\le t}\big|\alpha_3(\|y(s)\|_H) - \alpha_3(\|y_0\|_H)\big| > \varepsilon_2,\ \sup_{0\le s\le t}\|y(s)\|_H < R\right\}\\
&\quad + \mathbb{P}\left\{\sup_{0\le s\le t}\big|\alpha_3(\|y(s)\|_H) - \alpha_3(\|y_0\|_H)\big| > \varepsilon_2,\ \sup_{0\le s\le t}\|y(s)\|_H \ge R\right\}\\
&\le \mathbb{P}\left\{\sup_{0\le s\le t}\Big|\|y(s)\|_H - \|y_0\|_H\Big| > \delta,\ \sup_{0\le s\le t}\|y(s)\|_H < R\right\}\\
&\quad + \mathbb{P}\left\{\sup_{0\le s\le t}\|y(s)\|_H \ge R\right\}\\
&\le C_1 t/\delta^2 + \frac{1}{2}.
\end{aligned} \tag{3.4.16}$$

In particular, by virtue of (3.4.16), for $\varepsilon_2 > 0$ there exists $t_0 = t_0(R, \varepsilon_2) \ge 0$ such that

$$\mathbb{P}\left\{\sup_{0\le s\le t_0}\big|\alpha_3(\|y(s)\|_H) - \alpha_3(\|y(0)\|_H)\big| \le \varepsilon_2\right\} \ge \frac{1}{4}. \tag{3.4.17}$$

Now let us define a sequence of stopping times,

$$\bar{\tau}_1 = \inf\{t \geq 0 \colon \alpha_3(\|y(t)\|_H) < \varepsilon_1\},$$
$$\cdots\cdots$$
$$\bar{\tau}_{2k} = \inf\{t \geq \bar{\tau}_{2k-1} \colon \alpha_3(\|y(t)\|_H) > 2\varepsilon_1\},$$
$$\bar{\tau}_{2k+1} = \inf\{t \geq \bar{\tau}_{2k} \colon \alpha_3(\|y(t)\|_H) < \varepsilon_1\}, \quad k = 1, 2, \ldots$$

Then one can find from (3.4.5) that

$$\begin{aligned} &\varepsilon_1 \sum_{k=1}^{\infty} \mathbb{E}\Big\{\mathbf{1}_{\{\bar{\tau}_{2k}<\tau_R\}}\mathbb{E}\big[(\bar{\tau}_{2k+1} - \bar{\tau}_{2k}) \mid \mathscr{F}^W_{\bar{\tau}_{2k}}\big]\Big\} \\ &\quad = \varepsilon_1 \sum_{k=1}^{\infty} \mathbb{E}\big[\mathbf{1}_{\{\bar{\tau}_{2k}<\tau_R\}}(\bar{\tau}_{2k+1} - \bar{\tau}_{2k})\big] \\ &\quad \leq \sum_{k=1}^{\infty} \mathbb{E}\left[\mathbf{1}_{\{\bar{\tau}_{2k}<\tau_R\}} \int_{\bar{\tau}_{2k}}^{\bar{\tau}_{2k+1}} \alpha_3(\|y(s)\|_H)ds\right] \\ &\quad \leq \mathbb{E}\int_0^{\infty} \alpha_3(\|y(s)\|_H)ds < \infty \end{aligned} \tag{3.4.18}$$

Since the strong solution y has strong Markov property, one can find by setting $\varepsilon_1 = 2\varepsilon_2$ that on $\{\bar{\tau}_{2k} < \tau_R\}$,

$$\begin{aligned} &\mathbb{E}\Big((\bar{\tau}_{2k+1} - \bar{\tau}_{2k}) \,\Big|\, \mathscr{F}^W_{\bar{\tau}_{2k}}\Big) \\ &\quad \geq \mathbb{E}\Big((\bar{\tau}_{2k+1} - \bar{\tau}_{2k})\mathbf{1}_{\{\sup_{0\leq t\leq t_0} |\alpha_3(\|y(t+\bar{\tau}_{2k})\|_H) - \alpha_3(\|y(\bar{\tau}_{2k})\|_H)| \leq \varepsilon_1/2\}} \,\Big|\, \mathscr{F}^W_{\bar{\tau}_{2k}}\Big) \\ &\quad \geq t_0\mathbb{P}\left\{\sup_{0\leq t\leq t_0} \Big|\alpha_3(\|y(t + \bar{\tau}_{2k})\|_H) - \alpha_3(\|y(\bar{\tau}_{2k})\|_H)\Big| \leq \frac{\varepsilon_1}{2} \,\Big|\, \mathscr{F}^W_{\bar{\tau}_{2k}}\right\} \\ &\quad = t_0\mathbb{P}\left\{\sup_{0\leq t\leq t_0} \Big|\alpha_3(\|y(t)\|_H) - \alpha_3(\|y(0)\|_H)\Big| \leq \frac{\varepsilon_1}{2}\right\} \\ &\quad \geq \frac{t_0}{4}, \qquad t_0 = t_0(R, \varepsilon_1/2). \end{aligned} \tag{3.4.19}$$

Hence, it follows from (3.4.18) and (3.4.19) that

$$\frac{t_0\varepsilon_1}{4} \sum_{k=1}^{\infty} \mathbb{P}\{\bar{\tau}_{2k} < \tau_R\} < \infty,$$

which, together with Borel–Cantelli's lemma, immediately implies

$$\mathbb{P}\{\bar{\tau}_{2k} < \tau_R \text{ for infinitely many } k \in \mathbb{N}_+\} = 0. \tag{3.4.20}$$

Note that

$$
\begin{aligned}
&\left\{\bar{\tau}_{2k} < \tau_R \text{ for infinitely many } k\right\}\\
&\quad = \left\{\bar{\tau}_{2k} < \tau_R \text{ for infinitely many } k \text{ and } \tau_R = \infty\right\}\\
&\quad \bigcup \left\{\bar{\tau}_{2k} < \tau_R \text{ for infinitely many } k \text{ and } \tau_R < \infty\right\}.
\end{aligned}
$$

Then we have from (3.4.20) that

$$\mathbb{P}\{\bar{\tau}_{2k} < \infty \text{ for infinitely many } k \text{ and } \tau_R = \infty\} = 0.$$

Hence, if we can show $\mathbb{P}\{\tau_R = \infty\} = 1$ whenever $R \to \infty$, it then follows that

$$\mathbb{P}\{\bar{\tau}_{2k} < \infty \text{ for infinitely many } k \in \mathbb{N}_+\} = 0,$$

which contradicts (3.4.6). Indeed, by analogy with (3.3.38), we can apply Itô's formula to Λ and use (ii) to obtain for any $0 \le s \le t$ that

$$\Lambda(y(t)) \le \Lambda(y(s)) + \int_s^t \langle \Lambda_y'(y(u)), B(y(u))dW(u)\rangle_H,$$

which implies that $\Lambda(y(t))$ is a supermartingale with respect to the filtration $\{\mathscr{F}_t^W\}_{t\ge 0}$. By a well-known supermartingale inequality (see Lemma (54.5), p. 154 in [198]), for any function $\alpha_1 \in \Sigma_\infty$ and $R > 0$,

$$\mathbb{P}\left\{\sup_{0\le s\le t} \Lambda(y(s)) \ge \alpha_1(R)\right\} \le \frac{12\Lambda(y_0) + 9\mathbb{E}\Lambda(y(t))}{\alpha_1(R)} \le \frac{21\Lambda(y_0)}{\alpha_1(R)}, \quad \forall t \ge 0.$$

Hence, by virtue of (i) we have for $R > 0$ that

$$
\begin{aligned}
\mathbb{P}\{\tau_R = \infty\} &\ge \mathbb{P}\left\{\sup_{t\ge 0} \|y(t)\|_H < R\right\}\\
&\ge \mathbb{P}\left\{\sup_{t\ge 0} \Lambda(y(t)) < \alpha_1(R)\right\} \ge 1 - \frac{21\Lambda(y_0)}{\alpha_1(R)}.
\end{aligned}
$$

Letting $R \to \infty$, we immediately have

$$\mathbb{P}\{\tau_R = \infty\} = 1,$$

which implies the desired limit $\lim_{t\to\infty} \alpha_3(\|y(t)\|_H) = 0$ almost surely.

Last, we show that $\lim_{t\to\infty} \|y(t)\|_H = 0$ almost surely. Indeed, if there exists a sequence $t_k \to \infty$ as $k \to \infty$ and constant $\delta > 0$ such that $\lim_{k\to\infty} \|y(t_k)\|_H \ge \delta$ with positive probability, then it follows that with positive probability,

$$\lim_{k\to\infty} \alpha_3(\|y(t_k)\|_H) \ge \inf_{\|y(t)\|_H \ge \delta} \alpha_3(\|y(t)\|_H) \ge C > 0$$

for some number $C = C(\delta) > 0$, which is a contradiction. The claim is thus true for all $y_0 \in V$. Since V is dense in H and the solution $y(t, y_0)$ is continuously dependent on initial datum, the desired result is true for all $y_0 \in H$ and thus the proof is complete. □

Example 3.4.4 Let $H = L^2(\mathcal{O})$ where $\mathcal{O}$ is an open and bounded domain in $\mathbb{R}^n$ with smooth boundary $\partial\mathcal{O}$. Let $\{e_n\}_{n\geq 1}$ be an orthonormal basis of H with $L_1 := \sup_{n\geq 1, x\in\mathcal{O}} |e_n(x)| < \infty$. Assume that $\{\lambda_n\}_{n\geq 1}$ is a sequence of nonnegative numbers satisfying $\lambda := \sum_{n\geq 1} \lambda_n < \infty$ and

$$W(t,x) := \sum_{n=1}^{\infty} \sqrt{\lambda_n} w_n(t) e_n(x)$$

is an H-valued Q-Wiener process where $Qe_n = \lambda_n e_n$, $n \geq 1$, and $w_n(t)$, $n \geq 1$, are a group of standard real-valued mutually independent Brownian motions. Let $\alpha(x)$ be a continuously differentiable function satisfying $\alpha(x) \geq \alpha_0 > 0$ on $\mathcal{O}$ and $A = \frac{\partial}{\partial x}(\alpha(x)\frac{\partial}{\partial x})$ with domain $\mathscr{D}(A) = H_0^1(\mathcal{O}) \cap H^2(\mathcal{O})$. Assume further that g satisfies a Lipschitz type of condition with $g(0) = 0$,

$$|g(x) - g(y)| \leq L_2|x - y|, \qquad x,\ y \in \mathbb{R}$$

for some $L_2 > 0$. Consider an infinite-dimensional stochastic differential equation

$$\begin{cases} du(t) = [Au(t) + F(u(t))]dt + B(u(t))dW(t), & t \geq 0, \\ u(0) = u_0 \in H, \end{cases} \tag{3.4.21}$$

where for $x \in \mathcal{O}$ and $u,\ v \in H$,

$$u(t)(x) = u(t,x), \qquad F(u)(x) = -\nu\|u\|_H u(x), \qquad \nu > 0,$$
$$B(u)(v)(x) = g(u)v(x).$$

By a simple computation, we have for $u,\ v \in H$ that

$$\|B(u) - B(v)\|^2_{\mathscr{L}_2(H_Q,H)} \leq L_1^2 L_2^2 \lambda \|u - v\|_H^2.$$

Let $\Lambda(u) = \|u\|_H^2$, $u \in H$, then it is easy to see by the Poincaré inequality that for any $u \in V = H_0^1(\mathcal{O})$,

$$\langle\langle u, Au\rangle\rangle_{V,V^*} \leq -\alpha_0\|u\|_H^2.$$

Hence, by a direct computation, we have for any $u \in V$,

$$\begin{aligned} \mathcal{L}\Lambda(u) &\leq -2\alpha_0\|u\|_H^2 - 2\nu\langle u, \|u\|_H u\rangle_H + L_1^2 L_2^2 \lambda \|u\|_H^2 \\ &= -\big(2\alpha_0 - L_1^2 L_2^2 \lambda\big)\|u\|_H^2 - 2\nu\|u\|_H^3. \end{aligned}$$

If $L := 2\alpha_0 - L_1^2 L_2^2 \lambda > 0$, then $\mathcal{L}\Lambda(u) < -L\|u\|_H^2$ and by Theorem 3.3.3, the solution is exponentially stable almost surely. If $L = 0$, then $\mathcal{L}\Lambda(u) < -2\nu\|u\|_H^3$ and by Theorem 3.4.3, the solution is globally almost sure asymptotically stable.

3.5 Lyapunov Function Characterization

Consider a linear Cauchy problem on the Hilbert space H,

$$\begin{cases} dy(t) = Ay(t)dt, & t \geq 0, \\ y(0) = y_0 \in H, \end{cases} \tag{3.5.1}$$

where A generates a C_0-semigroup e^{tA}, $t \geq 0$, on H. It is shown in Theorem 2.1.12 and Remark 2.1.13 that system (3.5.1) is exponentially stable if and only if there exists a nonnegative self-adjoint operator $P \in \mathscr{L}(H)$ such that

$$\langle Px, Ax\rangle_H + \langle Ax, Px\rangle_H = -\|x\|_H^2 \quad \text{for all} \quad x \in \mathscr{D}(A).$$

Moreover, in this case the function

$$\Lambda(x) := \langle Px, x\rangle_H = \int_0^\infty \|e^{tA}x\|_H^2 dt, \qquad x \in H,$$

could be a possible Lyapunov function satisfying the relations

$$c_2\|x\|_H^2 \leq \Lambda(x) \leq c_1\|x\|_H^2, \qquad c_2 \geq 0, \quad c_1 > 0, \tag{3.5.2}$$

and

$$\langle \Lambda'(x), Ax\rangle_H \leq -c_3\Lambda(x), \qquad c_3 > 0, \quad x \in \mathscr{D}(A). \tag{3.5.3}$$

Unfortunately, as pointed out in Remark 3.1.1, the number c_2 here is not necessarily positive, although this is always the case when H is finite dimensional. As a consequence, it may not be suitable for the function $\Lambda(\cdot)$ to play the role of a genuine Lyapunov function in handling nonlinear systems (cf. [54]). However, we can show that for an important subclass of generators A, it is possible to construct a valid Lyapunov function (see Theorem 3.5.1) with $c_2 > 0$ in (3.5.2).

We say that a linear operator $A\colon \mathscr{D}(A) \subset H \to H$ is *variational* if

(i) there exist a Hilbert space V densely embedded in H and a continuous bilinear form $a\colon V \times V \to \mathbb{R}$ such that

$$2a(v, v) \leq -\alpha\|v\|_V^2 + \lambda\|v\|_H^2, \qquad \forall v \in V, \tag{3.5.4}$$

for some $\alpha > 0$ and $\lambda \in \mathbb{R}$;

(ii) $\mathscr{D}(A) = \{u \in V : a(u,\cdot)$ is continuous in the topology of $H\}$;
(iii) $a(u,v) = \langle Au, v\rangle_H$ for any $u \in \mathscr{D}(A)$ and $v \in V$.

It is well known (see, e.g., [53]) that a variational operator A generates an analytic C_0-semigroup e^{tA}, $t \geq 0$, both in H and V^*, such that $e^{tA}: V^* \to V$ for each $t > 0$ and $\|e^{tA}\| \leq e^{\lambda t}$ for all $t \geq 0$. If a is symmetric, then A is self-adjoint. Moreover, the following equality holds:

$$\int_0^t a(e^{sA}x, e^{sA}x)ds = \frac{1}{2}\Big[\|e^{tA}x\|_H^2 - \|x\|_H^2\Big], \quad \forall t \geq 0, \quad x \in H. \quad (3.5.5)$$

Theorem 3.5.1 *Assume that A in (3.5.1) is a variational operator. If there exists a Fréchet differentiable function $\Lambda(x)$ on H such that $\Lambda'(x) \in V$ for any $x \in V$ and*

$$c_2\|x\|_H^2 \leq \Lambda(x) \leq c_1\|x\|_H^2, \quad c_1 > 0, \quad c_2 > 0, \quad \forall x \in H, \quad (3.5.6)$$

and

$$\langle\langle \Lambda'(x), Ax\rangle\rangle_{V,V^*} \leq -c_3\Lambda(x), \quad c_3 > 0, \ \forall x \in V, \quad (3.5.7)$$

then the null solution of system (3.5.1) is exponentially stable.

Conversely, assume that the null solution of (3.5.1) is exponentially stable, then there exists a continuous function Λ on H such that both (3.5.6) and (3.5.7) are satisfied. In this case, $\Lambda(x)$ can be actually taken as the form

$$\Lambda(x) = \int_0^\infty \|e^{tA}x\|_V^2 dt, \quad \forall x \in H.$$

Proof First suppose that (3.5.6) and (3.5.7) are true; then we have for any $t \geq 0$ and $x \in H$ that

$$\frac{d}{dt}\Lambda(e^{tA}x) = \langle\langle \Lambda'(e^{tA}x), Ae^{tA}x\rangle\rangle_{V,V^*} \leq -c_3\Lambda(e^{tA}x),$$

which yields the relation $(\Lambda(e^{tA}x)e^{c_3t})'_t \leq 0$ for all $t \geq 0$ and further

$$\Lambda(e^{tA}x)e^{c_3t} \leq \Lambda(x), \quad t \geq 0.$$

This inequality, in addition to (3.5.6), thus implies that

$$c_2\|e^{tA}x\|_H^2 \leq \Lambda(e^{tA}x) \leq \Lambda(x)e^{-c_3t} \leq c_1\|x\|_H^2 e^{-c_3t}, \quad \forall t \geq 0, \quad x \in H.$$

That is, the null solution of (3.5.1) is exponentially stable.

Conversely, assume that the null solution of (3.5.1) is exponentially stable, or equivalently, $\|e^{tA}\| \leq Me^{-\gamma t}$, $M \geq 1$, $\gamma > 0$ for all $t \geq 0$. From (3.5.4) and (3.5.5), we have that for all $t \geq 0$, $x \in H$,

$$\begin{aligned}\|e^{tA}x\|_H^2 &= \|x\|_H^2 + 2\int_0^t a(e^{sA}x, e^{sA}x)ds\\ &\le \|x\|_H^2 + |\lambda|\int_0^t \|e^{sA}x\|_H^2 ds - \alpha\int_0^t \|e^{sA}x\|_V^2 ds.\end{aligned}$$

This clearly implies that for all $t \ge 0$, $x \in H$,

$$\alpha\int_0^t \|e^{sA}x\|_V^2 ds \le \|e^{tA}x\|_H^2 + \alpha\int_0^t \|e^{sA}x\|_V^2 ds \le \|x\|_H^2 + |\lambda|\int_0^t \|e^{sA}x\|_H^2 ds. \tag{3.5.8}$$

By using exponential stability and letting $t \to \infty$ in (3.5.8), one thus obtains

$$\int_0^\infty \|e^{sA}x\|_V^2 ds \le \frac{2\gamma + |\lambda|M^2}{2\alpha\gamma}\|x\|_H^2, \qquad x \in H. \tag{3.5.9}$$

Now we put

$$\Lambda(x) := \int_0^\infty \|e^{tA}x\|_V^2 dt \quad \text{for all} \quad x \in H \quad \text{and} \quad c_1 = \frac{2\gamma + |\lambda|M^2}{2\alpha\gamma} > 0.$$

Then it is immediate to see that

$$\Lambda(x) = \int_0^\infty \|e^{tA}x\|_V^2 dt \le c_1\|x\|_H^2, \qquad \forall x \in H.$$

On the other hand, since a is a continuous bilinear form on $V \times V$, one can obtain

$$|a(e^{tA}x, e^{tA}x)| \le C\|e^{tA}x\|_V^2, \qquad \forall t \ge 0,$$

for some constant $C > 0$. Hence, we have, in addition to (3.5.5), that

$$\|e^{tA}x\|_H^2 - \|x\|_H^2 \ge -2C\int_0^t \|e^{sA}x\|_V^2 ds. \tag{3.5.10}$$

Letting $t \to \infty$ in (3.5.10) and using the exponential stability property of e^{tA}, $t \ge 0$, we immediately obtain

$$c_2\|x\|_H^2 \le \int_0^\infty \|e^{tA}x\|_V^2 dt = \Lambda(x) \quad \text{for all} \quad x \in H,$$

where $c_2 = (2C)^{-1} > 0$.

To conclude the proofs, it remains to show the last relation (3.5.7). To this end, we first note from (3.5.9) that for any $x \in H$, the function $t \to e^{tA}x$ is in $L^2([0,\infty); V)$. Therefore, the derivative Λ' exists, and it is continuous on H. On the other hand, recall that $\|\cdot\|_H^2 \le \beta\|\cdot\|_V^2$ for some number $\beta > 0$. Then for any $t \ge 0$, $x \in H$,

$$
\begin{aligned}
\Lambda(e^{tA}x) &= \int_0^\infty \|e^{sA}e^{tA}x\|_V^2 ds \\
&= \int_t^\infty \|e^{sA}x\|_V^2 ds \\
&= \int_0^\infty \|e^{sA}x\|_V^2 ds - \frac{1}{\beta}\int_0^t (\beta\|e^{sA}x\|_V^2 - \|e^{sA}x\|_H^2)ds \\
&\quad - \frac{1}{\beta}\int_0^t \|e^{sA}x\|_H^2 ds.
\end{aligned}
\tag{3.5.11}
$$

Since $\|e^{sA}x\|_H^2 \le \beta\|e^{sA}x\|_V^2$ for any $s \ge 0$ and $x \in H$, the function

$$
t \to \int_0^t \left(\beta\|e^{sA}x\|_V^2 - \|e^{sA}x\|_H^2\right)ds, \qquad x \in H,
$$

is nondecreasing on $[0, \infty)$. By virtue of (3.5.11) and the fact that $t \to \|e^{tA}x\|_H$ is continuous, we thus get that for all $t \ge 0$ and $x \in H$,

$$
\langle\langle \Lambda'(e^{tA}x), Ae^{tA}x\rangle\rangle_{V,V^*} = \frac{d}{dt}\Lambda(e^{tA}x) \le -\frac{1}{\beta}\|e^{tA}x\|_H^2, \tag{3.5.12}
$$

which, by putting $t = 0$ in (3.5.12), immediately yields

$$
\langle\langle \Lambda'(x), Ax\rangle\rangle_{V,V^*} \le -\frac{1}{\beta}\|x\|_H^2 \le -\frac{1}{\beta c_1}\Lambda(x), \qquad \forall x \in V.
$$

The proof is thus complete. □

In this section, we shall develop a Lyapunov function program of stochastic stability for nonlinear systems that contain as a special case the system (3.5.1) and associated Theorem 3.5.1. Precisely, we wish to consider the strong solution of the following nonautonomous stochastic differential equation in V^*

$$
\begin{cases}
dy(t) = A(t, y(t))dt + B(t, y(t))dW(t), & t \in [t_0, \infty), \\
y(t_0) = y_0 \in H,
\end{cases}
\tag{3.5.13}
$$

for some $t_0 \ge 0$. In addition to the conditions (a) through (d) in Section 1.4.2, we further impose the following ones:

(a') there exist constants $\alpha > 0$, $\gamma \ge 0$, $\mu \ge 0$, and $\lambda \in \mathbb{R}$ such that

$$
\begin{aligned}
&2\langle\langle x, A(t,x)\rangle\rangle_{V,V^*} + \|B(t,x)\|_{\mathscr{L}_2(K_Q,H)}^2 \\
&\qquad \le -\alpha\|x\|_V^2 + \lambda\|x\|_H^2 + \gamma e^{-\mu t}, \quad \forall x \in V, \ t \ge 0;
\end{aligned}
\tag{3.5.14}
$$

(b') there exists a number $c > 0$ such that for all $y \in V$, $t \ge 0$,

$$
\|A(t,y)\|_{V^*} \le c\|y\|_V; \tag{3.5.15}
$$

(c') there exists a number $L > 0$ such that for any $y, z \in V$ and $t \geq 0$,

$$\|B(t,y) - B(t,z)\|^2_{\mathscr{L}_2(K_Q,H)} \leq L\|y - z\|_V^2. \tag{3.5.16}$$

Recall that the generator $\mathcal{L}$ of the solution process $y(t)$ to (3.5.13), $t \geq t_0$, is defined by

$$(\mathcal{L}\Lambda)(t,x) = \Lambda_t'(t,x) + \langle\langle \Lambda_x'(t,x), A(t,x)\rangle\rangle_{V,V^*} + \frac{1}{2}Tr(\Lambda_{xx}''(t,x)B(t,x)Q^{1/2}(B(t,x)Q^{1/2})^*) \tag{3.5.17}$$

for any $x \in V$, $t \geq t_0$, whenever the right-hand side of (3.5.17) is meaningful.

Theorem 3.5.2 *Suppose that $y(t, y_0)$, $t \geq t_0$, is the strong solution of (3.5.13) and the coercive condition (3.5.14) holds. If there exists an Itô type of functional $\Lambda(\cdot,\cdot)\colon \mathbb{R}_+ \times H \to \mathbb{R}$ such that*

(i) $c_2\|x\|_H^2 + k_2 e^{-\mu_2 t} \leq \Lambda(t,x) \leq c_1\|x\|_H^2 + k_1 e^{-\mu_1 t}, \quad \forall x \in H, \quad t \geq t_0;$
(ii) $(\mathcal{L}\Lambda)(t,x) \leq -c_3\Lambda(t,x) + k_3 e^{-\mu_3 t}, \quad \forall x \in V, \quad t \geq t_0,$

where $c_i > 0$, $k_i \in \mathbb{R}$, $\mu_i \geq 0$, $i = 1, 2, 3$, are constants, then $y(t, y_0)$, $t \geq t_0$, satisfies

$$\mathbb{E}\|y(t,y_0)\|_H^2 \leq \alpha_1\|y_0\|_H^2 \cdot e^{-\beta_1(t-t_0)} + \alpha_2 \cdot e^{-\beta_2 t}, \quad y_0 \in H, \quad t \geq t_0, \tag{3.5.18}$$

for some constants $\alpha_1 > 0$, $\beta_1 > 0$, $\alpha_2 \geq 0$, and $\beta_2 \geq 0$.

Conversely, suppose that (3.5.18) holds and define

$$\Lambda(t_0, y_0) = \int_{t_0}^{T+t_0}\left(\int_{t_0}^{t}\mathbb{E}\|y(s,y_0)\|_V^2 ds\right)dt \quad \text{for all} \quad y_0 \in H, \quad t_0 \geq 0, \tag{3.5.19}$$

where T is a number satisfying

$$T > \max\left\{\frac{\beta}{\alpha}\left(1 + \frac{\alpha_1|\lambda|}{\beta_1}\right), \frac{\alpha_1}{\beta_1}\right\}$$

and $\beta > 0$ is the constant in $\|\cdot\|_H^2 \leq \beta\|\cdot\|_V^2$. If $\Lambda(\cdot,\cdot)$ is an Itô type of functional, then there exist constants $c_i > 0$, $k_i \in \mathbb{R}$, $\mu_i \geq 0$, $i = 1, 2, 3$, such that (i) and (ii) hold.

Proof First, suppose that the conditions (i) and (ii) are true. A direct application of Itô's formula to the strong solution $y(t, y_0)$, $t \geq t_0$, and Itô functional $\Lambda(\cdot,\cdot)$ yields that for all $t \geq s \geq t_0$,

$$\begin{aligned}\mathbb{E}\Lambda(t, y(t, y_0)) - \mathbb{E}\Lambda(s, y(s, y_0)) &= \mathbb{E}\int_s^t (\mathcal{L}\Lambda)(u, y(u, y_0))du \\ &\le \int_s^t \Big[- c_3\mathbb{E}\Lambda(u, y(u, y_0)) + k_3 e^{-\mu_3 u}\Big]du.\end{aligned}$$

Let $\phi(t) = \mathbb{E}\Lambda(t, y(t, y_0))$ and note that $\phi(t)$ is differentiable in t. We find

$$\frac{d\phi(t)}{dt} \le -c_3\phi(t) + k_3 e^{-\mu_3 t}, \qquad \forall t \ge t_0,$$

i.e.,

$$\frac{d(e^{c_3 t}\phi(t))}{dt} \le e^{c_3 t - \mu_3 t}, \qquad \forall t \ge t_0,$$

which immediately implies

$$\phi(t) \le \phi(t_0)e^{c_3 t_0} \cdot e^{-c_3 t} + k_3 e^{-c_3 t}\int_{t_0}^t e^{(c_3-\mu_3)u}du, \qquad t \ge t_0. \tag{3.5.20}$$

If $c_3 = \mu_3$, it is not difficult to deduce from (3.5.20) and (i) that

$$\phi(t) \le c_1\|y_0\|_H^2 \cdot e^{-c_3(t-t_0)} + \Big[|k_1| + \frac{2|k_3|}{c_3}\Big]e^{-\min\{\frac{c_3}{2},\, \mu_1\}t}, \qquad t \ge t_0. \tag{3.5.21}$$

Indeed, if $\mu_1 < c_3/2$, we have

$$\begin{aligned}\phi(t) &\le c_1\|y_0\|_H^2 \cdot e^{-c_3(t-t_0)} + |k_1|e^{-(\mu_1-c_3)t_0 - c_3 t} + |k_3|\cdot(t-t_0)e^{-c_3 t} \\ &\le c_1\|y_0\|_H^2 \cdot e^{-c_3(t-t_0)} \\ &\quad + e^{-\mu_1 t}\Big\{|k_1|e^{-(\mu_1-c_3)t_0 - c_3 t + \mu_1 t} + |k_3|(t-t_0)e^{-(c_3-\mu_1)t}\Big\} \\ &\le c_1\|y_0\|_H^2 \cdot e^{-c_3(t-t_0)} \\ &\quad + e^{-\mu_1 t}\Big\{|k_1|e^{(\mu_1-c_3)(t-t_0)} + |k_3|t(c_3-\mu_1)e^{-(c_3-\mu_1)t}\cdot\frac{1}{c_3-\mu_1}\Big\} \\ &\le c_1\|y_0\|_H^2 \cdot e^{-c_3(t-t_0)} + e^{-\mu_1 t}\Big\{|k_1|e^{(\mu_1-c_3)(t-t_0)} + \frac{|k_3|}{c_3-\mu_1}\Big\} \\ &\le c_1\|y_0\|_H^2 \cdot e^{-c_3(t-t_0)} + e^{-\mu_1 t}\Big\{|k_1| + \frac{2}{c_3}\cdot|k_3|\Big\}, \qquad t \ge t_0.\end{aligned} \tag{3.5.22}$$

In a similar way, we can also show (3.5.21) in the case that $\mu_1 \ge c_3/2$.

If $c_3 \ne \mu_3$, we can similarly do a direct computation, together with (3.5.20) and (i), to get

$$\phi(t) \le c_1\|y_0\|_H^2 \cdot e^{-c_3(t-t_0)} + \Big(|k_1| + \frac{2|k_3|}{|c_3-\mu_3|}\Big)e^{-\min\{\mu_3,\, c_3\}t}, \qquad t \ge t_0. \tag{3.5.23}$$

Combining the results (3.5.21) and (3.5.23), we can obtain

$$\phi(t) \leq c_1\|y_0\|_H^2 e^{-c_3(t-t_0)} + \tilde{\alpha}_2 \cdot e^{-\tilde{\beta}_2 t}, \qquad t \geq t_0, \tag{3.5.24}$$

where $\tilde{\beta}_2 = \min\{c_3/2, \mu_3, \mu_1\} \geq 0$ and

$$\tilde{\alpha}_2 = \begin{cases} |k_1| + 2|k_3|/c_3 & \text{if} \quad c_3 = \mu_3, \\ |k_1| + \dfrac{2|k_3|}{|c_3 - \mu_3|} & \text{if} \quad c_3 \neq \mu_3. \end{cases}$$

The relation (3.5.24), together with (i), immediately implies that for all $t \geq t_0$,

$$\mathbb{E}\|y(t, y_0)\|_H^2 \leq \frac{1}{c_2}\phi(t) + \frac{|k_2|}{c_2}e^{-\mu_2 t} \leq \alpha_1\|y_0\|_H^2 e^{-\beta_1(t-t_0)} + \alpha_2 \cdot e^{-\beta_2 t},$$

where $\alpha_1 = c_1/c_2 > 0$, $\alpha_2 = (\tilde{\alpha}_2 + |k_2|)/c_2 \geq 0$, $\beta_1 = c_3 > 0$ and $\beta_2 = \min\{\tilde{\beta}_2, \mu_2\} \geq 0$ are constants. The first part is proved.

Now suppose that relation (3.5.18) holds. Define

$$\Lambda(t_0, y_0) = \int_{t_0}^{T+t_0}\left(\int_{t_0}^{t} \mathbb{E}\|y(s, y_0)\|_V^2 ds\right)dt, \quad y_0 \in H, \quad t \geq t_0, \tag{3.5.25}$$

where T is a proper positive constant to be determined later on. For the sake of simplicity, suppose at present $\mu > 0$, $\beta_2 > 0$, while the corresponding results in other cases may be proved similarly.

First, applying Itô's formula to $\|y(t, y_0)\|_H^2$, $t \geq t_0$, and using the coercive condition (3.5.14), we then obtain for $t \geq t_0$ that

$$\begin{aligned} \mathbb{E}\|y(t, y_0)\|_H^2 - \|y_0\|_H^2 &= \int_{t_0}^{t} \mathbb{E}\mathcal{L}\|y(s, y_0)\|_H^2 ds \\ &\leq \lambda \int_{t_0}^{t} \mathbb{E}\|y(s, y_0)\|_H^2 ds \\ &\quad - \alpha \int_{t_0}^{t} \mathbb{E}\|y(s, y_0)\|_V^2 ds + \gamma \int_{t_0}^{t} e^{-\mu s} ds, \end{aligned} \tag{3.5.26}$$

which, in addition to (3.5.18), immediately implies

$$\begin{aligned} \int_{t_0}^{t} \mathbb{E}\|y(s, y_0)\|_V^2 ds &\leq \frac{1}{\alpha}\left(\|y_0\|_H^2 + \lambda\int_{t_0}^{t} \mathbb{E}\|y(s, y_0)\|_H^2 ds + \gamma\int_{t_0}^{t} e^{-\mu s} ds\right) \\ &\leq \frac{1}{\alpha}\left(\|y_0\|_H^2 + \alpha_1|\lambda|\|y_0\|_H^2 \int_{t_0}^{t} e^{-\beta_1(s-t_0)} ds\right. \\ &\quad \left. + |\lambda|\alpha_2 \int_{t_0}^{t} e^{-\beta_2 s} ds + \gamma\int_{t_0}^{t} e^{-\mu s} ds\right) \\ &\leq \frac{1}{\alpha}\left(\|y_0\|_H^2 + \frac{\alpha_1|\lambda|\|y_0\|_H^2}{\beta_1} + \frac{|\lambda|\alpha_2}{\beta_2}e^{-\beta_2 t_0} + \frac{\gamma}{\mu}e^{-\mu t_0}\right). \end{aligned} \tag{3.5.27}$$

Therefore, by virtue of (3.5.25) it follows that

$$\Lambda(t_0, y_0) \le \frac{1}{\alpha}\Big[\Big(T + \frac{\alpha_1 T|\lambda|}{\beta_1}\Big)\|y_0\|_H^2 + \Big(\frac{|\lambda|\alpha_2 T}{\beta_2} + \frac{\gamma T}{\mu}\Big)\cdot e^{-\min\{\beta_2,\mu\}t_0}\Big]. \tag{3.5.28}$$

On the other hand, for arbitrary $v \in V$, $t \ge t_0$,

$$(\mathcal{L}\|v\|_H^2)(t, v) = 2\langle\langle v, A(t, v)\rangle\rangle_{V,V^*} + \|B(t, v)\|^2_{\mathscr{L}_2(K_Q,H)},$$

which, together with (3.5.15) and (3.5.16), immediately implies that

$$\begin{aligned}
|(\mathcal{L}\|v\|_H^2)(t, v)| &\le 2c^2\|v\|_V^2 + \|B(t, v)\|^2_{\mathscr{L}_2(K_Q,H)}\\
&\le 2c^2\|v\|_V^2 + 2\|B(t, v) - B(t, 0)\|^2_{\mathscr{L}_2(K_Q,H)}\\
&\quad + 2\|B(t, 0)\|^2_{\mathscr{L}_2(K_Q,H)}\\
&\le 2c^2\|v\|_V^2 + 2L\|v\|_V^2 + 2\gamma e^{-\mu t}\\
&= \theta\|v\|_V^2 + 2\gamma e^{-\mu t}, \qquad t \ge t_0,
\end{aligned} \tag{3.5.29}$$

where $\theta = 2c^2 + 2L > 0$. Hence, it follows that

$$(\mathcal{L}\|v\|_H^2)(t, v) \ge -\theta\|v\|_V^2 - 2\gamma e^{-\mu t}, \qquad t \ge t_0, \qquad v \in V.$$

Now applying Itô's formula to $\|y(t, y_0)\|_V^2$, $t \ge t_0$, and taking expectation, we obtain for all $t \ge t_0$ that

$$\begin{aligned}
\mathbb{E}\|y(u, y_0)\|_H^2 - \|y_0\|_H^2 &= \int_{t_0}^t \mathbb{E}(\mathcal{L}\|y(s, y_0)\|_H^2)ds\\
&\ge -\theta\int_{t_0}^t \mathbb{E}\|y(s, y_0)\|_V^2 ds - 2\gamma\int_{t_0}^t e^{-\mu s}ds,
\end{aligned} \tag{3.5.30}$$

which, together with (3.5.18), immediately implies that for arbitrary $t \ge t_0$,

$$\begin{aligned}
\theta\int_{t_0}^t \mathbb{E}\|y(s, y_0)\|_V^2 ds &\ge \|y_0\|_H^2 - \mathbb{E}\|y(t, y_0)\|_H^2 - 2\gamma\int_{t_0}^t e^{-\mu s}ds\\
&\ge \Big(1 - \alpha_1\cdot e^{-\beta_1 t + \beta_1 t_0}\Big)\|y_0\|_H^2 - \alpha_2\cdot e^{-\beta_2 t} - \frac{2\gamma}{\mu}e^{-\mu t_0}.
\end{aligned} \tag{3.5.31}$$

Therefore, we obtain

$$\begin{aligned}
\Lambda(t_0, y_0) &= \int_{t_0}^{T+t_0}\Big(\int_{t_0}^t \mathbb{E}\|y(s, y_0)\|_V^2 ds\Big)dt\\
&\ge \frac{1}{\theta}\int_{t_0}^{T+t_0}\Big[\Big(1 - \alpha_1\cdot e^{-\beta_1 t + \beta_1 t_0}\Big)\|y_0\|_H^2 - \alpha_2\cdot e^{-\beta_2 t} - \frac{2\gamma}{\mu}e^{-\mu t_0}\Big]dt\\
&\ge \frac{1}{\theta}\Big(T - \frac{\alpha_1}{\beta_1}\Big)\|y_0\|_H^2 - \frac{1}{\theta}\Big(\frac{\alpha_2}{\beta_2} + \frac{2\gamma T}{\mu}\Big)e^{-\min\{\beta_2,\mu\}t_0}.
\end{aligned} \tag{3.5.32}$$

Now assume that $y_0 \in V$; then we have

$$
\begin{aligned}
(\mathcal{L}\Lambda)(t_0, y_0) &= \lim_{r\to 0} \frac{1}{r}\Big[\mathbb{E}\Lambda(t_0 + r, y(t_0 + r, y_0)) - \mathbb{E}\Lambda(t_0, y_0)\Big] \\
&= \lim_{r\to 0} \frac{1}{r}\bigg\{\int_{t_0+r}^{T+t_0+r}\int_{t_0+r}^{t} \mathbb{E}\|y(s, y(t_0 + r, y_0))\|_V^2 ds dt \\
&\qquad - \int_{t_0}^{T+t_0}\int_{t_0}^{t} \mathbb{E}\|y(s, y_0)\|_V^2 ds dt\bigg\} \\
&= \lim_{r\to 0} \frac{1}{r}\bigg\{\int_{t_0+r}^{T+t_0+r}\int_{t_0}^{t-r} \mathbb{E}\|y(s + r, y(t_0 + r, y_0))\|_V^2 ds dt \\
&\qquad - \int_{t_0}^{T+t_0}\int_{t_0}^{t} \mathbb{E}\|y(s, y_0)\|_V^2 ds dt\bigg\} \\
&= \lim_{r\to 0} \frac{1}{r}\bigg\{\int_{t_0}^{T+t_0}\int_{t_0}^{t} \mathbb{E}\|y(s + r, y(t_0 + r, y_0))\|_V^2 ds dt \\
&\qquad - \int_{t_0}^{T+t_0}\int_{t_0}^{t} \mathbb{E}\|y(s, y_0)\|_V^2 ds dt\bigg\} \\
&= \lim_{r\to 0} \frac{1}{r}\bigg\{\int_{t_0}^{T+t_0}\bigg[\int_{0}^{t-t_0} \mathbb{E}\|y(s + r + t_0, y(t_0 + r, y_0))\|_V^2 ds \\
&\qquad - \int_{t_0}^{t} \mathbb{E}\|y(s, y_0)\|_V^2 ds\bigg]dt\bigg\},
\end{aligned}
$$

which, by means of the Markov property of strong solutions to (3.5.13), further yields that

$$
\begin{aligned}
(\mathcal{L}\Lambda)(t_0, y_0) &= \lim_{r\to 0} \frac{1}{r}\bigg\{\int_{t_0}^{T+t_0}\int_{0}^{t-t_0} \mathbb{E}\|y(s + r + t_0, y_0)\|_V^2 ds dt \\
&\qquad - \int_{t_0}^{T+t_0}\int_{t_0}^{t} \mathbb{E}\|y(s, y_0)\|_V^2 ds dt\bigg\} \\
&= \lim_{r\to 0} \frac{1}{r}\bigg\{\int_{t_0}^{T+t_0}\int_{r}^{t-t_0+r} \mathbb{E}\|y(s + t_0, y_0)\|_V^2 ds dt \\
&\qquad - \int_{t_0}^{T+t_0}\int_{0}^{t-t_0} \mathbb{E}\|y(s + t_0, y_0)\|_V^2 ds dt\bigg\} \qquad (3.5.33) \\
&= \lim_{r\to 0} \frac{1}{r}\bigg\{\int_{t_0}^{T+t_0}\int_{t-t_0}^{t-t_0+r} \mathbb{E}\|y(s + t_0, y_0)\|_V^2 ds dt \\
&\qquad - \int_{t_0}^{T+t_0}\int_{0}^{r} \mathbb{E}\|y(s + t_0, y_0)\|_V^2 ds dt\bigg\} \\
&= \int_{t_0}^{T+t_0} \mathbb{E}\|y(t, y_0)\|_V^2 dt - T\|y_0\|_V^2.
\end{aligned}
$$

Now, substituting (3.5.27) into (3.5.33) yields

$$(\mathcal{L}\Lambda)(t_0, y_0) \le \frac{1}{\alpha}\Big(1+\frac{\alpha_1|\lambda|}{\beta_1}\Big)\|y_0\|_H^2+\Big(\frac{|\lambda|\alpha_2}{\beta_2}+\frac{\gamma}{\mu}\Big)e^{-\min\{\beta_2,\,\mu\}t_0}-\frac{T}{\beta}\|y_0\|_H^2. \tag{3.5.34}$$

Hence, by using (3.5.32) and choosing $T > 0$ so large that

$$T > \max\Big\{\frac{\beta}{\alpha}\Big(1+\frac{\alpha_1|\lambda|}{\beta_1}\Big), \frac{\alpha_1}{\beta_1}\Big\}, \tag{3.5.35}$$

then we can obtain the desired results (i) and (ii). A similar argument to the preceding applies to the case that $\mu = 0$ or $\beta_2 = 0$. The proof is now complete. □

Theorem 3.5.2 is useful in studying exponential stability and invariant measures of nonlinear stochastic systems by means of a Lyapunov function scheme and the first-order approximation method. To see this, let us first establish some useful corollaries of Theorem 3.5.2.

Consider the following autonomous stochastic differential equation in V^*:

$$\begin{cases} dy(t) = A(y(t))dt + B(y(t))dW(t), & t \in [0,\infty), \\ y(0) = y_0 \in H, \end{cases} \tag{3.5.36}$$

where $A(\cdot)$ and $B(\cdot)$ satisfy (b'), (c'), and the following coercive condition: there exist $\alpha > 0$ and $\lambda \in \mathbb{R}$ such that

$$2\langle\langle u, A(u)\rangle\rangle_{V,V^*} + \|B(u)\|^2_{\mathscr{L}_2(K_Q,H)} \le -\alpha\|u\|_V^2 + \lambda\|u\|_H^2, \quad \forall u \in V. \tag{3.5.37}$$

Corollary 3.5.3 *Assume that (b'), (c') and the coercive condition (3.5.37) hold. If there exists an Itô type of functional $\Lambda\colon H \to \mathbb{R}$ such that*

(i) $c_2\|x\|_H^2 \le \Lambda(x) \le c_1\|x\|_H^2, \quad \forall x \in H;$
(ii) $(\mathcal{L}\Lambda)(x) \le -c_3\Lambda(x), \quad \forall x \in V,$

where $c_i > 0$, $i = 1, 2, 3$, are constants, then the strong solution $y(t, y_0)$, $t \ge 0$, of (3.5.36) satisfies

$$\mathbb{E}\|y(t,y_0)\|_H^2 \le \alpha_1\|y_0\|_H^2 \cdot e^{-\beta_1 t}, \quad t \ge 0, \tag{3.5.38}$$

for some $\alpha_1 > 0$ and $\beta_1 > 0$.

Conversely, suppose that (3.5.38) holds and define

$$\Lambda(y_0) = \int_0^T\Big(\int_0^t \mathbb{E}\|y(s,y_0)\|_V^2 ds\Big)dt \quad \textit{for all} \quad y_0 \in H, \tag{3.5.39}$$

where T is some number satisfying (3.5.35). Suppose that $\Lambda(\cdot)$ is an Itô type of functional; then there exist constants $c_i > 0$, $i = 1, 2, 3$, such that (i) and (ii) hold.

The following corollary is useful in the study of invariant measures by means of a Lyapunov function method. To this end, we assume that $A(\cdot)$ and $B(\cdot)$ satisfy the following coercive condition: there exist $\alpha > 0$, $\lambda \in \mathbb{R}$, and $\gamma \geq 0$ such that

$$2\langle\langle u, A(u)\rangle\rangle_{V,V^*} + \|B(u)\|^2_{\mathscr{L}_2(K_Q,H)} \leq -\alpha\|u\|^2_V + \lambda\|u\|^2_H + \gamma, \quad \forall u \in V. \tag{3.5.40}$$

Corollary 3.5.4 *Assume that (b′), (c′), and the coercive condition (3.5.40) hold. If there exists an Itô type of functional $\Lambda\colon H \to \mathbb{R}$ such that*

(i) $c_2\|x\|^2_H + k_2 \leq \Lambda(x) \leq c_1\|x\|^2_H + k_1, \quad \forall x \in H;$
(ii) $(\mathcal{L}\Lambda)(x) \leq -c_3\Lambda(x) + k_3, \quad \forall x \in V,$

where $c_i > 0$, $k_i \in \mathbb{R}$, $i = 1, 2, 3$, are constants, then the strong solution $y(t, y_0)$, $t \geq 0$, of (3.5.36) satisfies

$$\mathbb{E}\|y(t, y_0)\|^2_H \leq \alpha_1\|y_0\|^2_H \cdot e^{-\beta_1 t} + \alpha_2, \quad t \geq 0, \tag{3.5.41}$$

for some $\alpha_1 > 0$, $\beta_1 > 0$ and $\alpha_2 \geq 0$.

Conversely, suppose that (3.5.41) holds and define

$$\Lambda(y_0) = \int_0^T \left(\int_0^t \mathbb{E}\|y(s, y_0)\|^2_V ds\right)dt \quad \textit{for all} \quad y_0 \in H, \tag{3.5.42}$$

where T satisfies (3.5.35). If $\Lambda(\cdot)$ is an Itô type of functional, then there exist constants $c_i > 0$, $k_i \in \mathbb{R}$, $i = 1, 2, 3$, such that (i) and (ii) hold.

Generally speaking, it is not obvious whether Λ defined in (3.5.19), (3.5.39), or (3.5.42) is an Itô type of functional or not. However, one can show that this is the case if equation (3.5.36) is linear. Consider a stochastic linear system given by

$$\begin{cases} dy(t) = Ay(t)dt + By(t)dW(t) & t \geq 0, \\ y(0) = y_0 \in H, \end{cases} \tag{3.5.43}$$

where $A \in \mathscr{L}(V, V^*)$ and $B \in \mathscr{L}(V, \mathscr{L}_2(K_Q, H))$ satisfy the following coercive condition: there exist $\alpha > 0$ and $\lambda \in \mathbb{R}$ such that

$$2\langle\langle u, Au\rangle\rangle_{V,V^*} + \|Bu\|^2_{\mathscr{L}_2(K_Q,H)} \leq -\alpha\|u\|^2_V + \lambda\|u\|^2_H \quad \text{for all} \quad u \in V. \tag{3.5.44}$$

Theorem 3.5.5 *Assume that the coercive condition (3.5.44) holds. If there exists an Itô type of functional* $\Lambda\colon H \to \mathbb{R}$ *such that*

(i) $c_2\|x\|_H^2 \le \Lambda(x) \le c_1\|x\|_H^2, \quad \forall x \in H;$
(ii) $(\mathcal{L}\Lambda)(x) \le -c_3\Lambda(x), \quad \forall x \in V,$

for constants $c_i > 0$, $i = 1, 2, 3$, *then the strong solution* $y(t, y_0)$, $t \ge 0$, *of (3.5.43) satisfies*

$$\mathbb{E}\|y(t, y_0)\|_H^2 \le \alpha_1\|y_0\|_H^2 \cdot e^{-\beta_1 t}, \qquad t \ge 0, \qquad y_0 \in H, \tag{3.5.45}$$

for some $\alpha_1 > 0$ *and* $\beta_1 > 0$.

Conversely, suppose that (3.5.45) holds and define

$$\Lambda(y_0) = \int_0^\infty \mathbb{E}\|y(s, y_0)\|_V^2 ds \quad \textit{for all} \quad y_0 \in H. \tag{3.5.46}$$

Then $\Lambda(\cdot)$ *is an Itô type of functional and there exist constants* $c_i > 0$, $i = 1$, 2, 3, *such that (i) and (ii) hold.*

Proof All we need prove is the latter part. By a similar argument to Theorem 3.5.2, it is easy to see that $\Lambda(\cdot)$ defined by (3.5.46) satisfies (i) and (ii) except for the fact that it is an Itô type of functional. To justify this, note that for any $x \in H$, the strong solution $y(t,x)$ of (3.5.43) with initial x is linear in x and hence

$$T(x,z) := \int_0^\infty \mathbb{E}\langle y(t,x), y(t,z)\rangle_V dt, \qquad x,\ z \in H,$$

is a bilinear form on $H \times H$. Since $y(\cdot) \in L^2(\Omega \times [0,\infty), V)$, we can use the Cauchy–Schwarz inequality to get

$$|T(x,z)| \le M\|x\|_H\|z\|_H, \qquad x,\ z \in H, \tag{3.5.47}$$

for some constant $M > 0$. Hence, there exists a continuous linear operator $P\colon H \to H$ such that $T(x,z) = \langle Px, z\rangle_H$, x, $z \in H$. In view of (3.5.46), it is immediate that

$$\Lambda'(x) = 2Px, \qquad \Lambda''(x) = 2P, \qquad x \in H. \tag{3.5.48}$$

This implies that the mappings Λ, Λ', Λ'' are locally bounded on H, and Λ, Λ' are continuous on H as

$$|\Lambda(x)| \le \|P\|\|x\|_H^2, \qquad x \in H,$$

and

$$\|\Lambda'(x)\|_H = 2\|Px\|_H \le 2\|P\|\|x\|_H, \qquad x \in H.$$

For any trace class operator D, $Tr(D\Lambda''(x)) = 2Tr(DP)$, which is clearly continuous. Lastly, to verify the condition (1.4.25), we observe that $T(x,z)$ is bilinear on $V \times V$. On the other hand, the relation (3.5.47) and continuity of the injection $V \to H$ imply that $T(\cdot,\cdot)$ is a continuous bilinear form on $V \times V$. It thus follows that there exists a continuous linear operator R on V such that

$$T(x,z) = \langle Rx, z\rangle_V \quad \forall x,\ z \in V.$$

Hence we have $\Lambda'(x) = 2Rx \in V$ for $x \in V$ and $x \to Rx$ is continuous on $V \to V$. This clearly implies (1.4.25) since

$$\|\Lambda'(x)\|_V = 2\|Rx\|_V \le 2\|R\|_{\mathscr{L}(V)}(1 + \|x\|_V) \quad \text{for all} \quad x \in V.$$

Now the proof is complete. □

Corollary 3.5.6 *Consider the stochastic linear system (3.5.43) with A and B satisfying (3.5.44). If the null solution of (3.5.43) is exponentially stable in mean square, then it is strongly stable in probability and globally asymptotically stable in sample path.*

Proof If the null solution of (3.5.43) is exponentially stable in mean square, then there exists a Lyapunov function $\Lambda(\cdot)$ that satisfies (i) and (ii) in Theorem 3.5.5. Obviously, such a function $\Lambda(\cdot)$ satisfies all the conditions in Theorems 3.4.1 and 3.4.3. The proof is thus complete. □

3.6 Two Applications

In this section, we shall employ those results derived in the last section to investigate stability and invariant measures for strong solutions of nonlinear stochastic differential equations by a method of combining a Lyapunov function scheme and the so-called first-order approximation of linear systems. In finite dimensions, this method has shown to be quite effective in dealing with large time behavior of nonlinear stochastic systems.

3.6.1 Stability in Probability

Consider the following nonlinear stochastic differential equation in V^*:

$$\begin{cases} dy(t) = \tilde{A}(y(t))dt + \tilde{B}(y(t))dW(t), & t \in [0,\infty), \\ y(0) = y_0 \in H, \end{cases} \tag{3.6.1}$$

where $\tilde{A}(\cdot)\colon V \to V^*$ and $\tilde{B}(\cdot)\colon V \to \mathscr{L}_2(K_Q, H)$ are two nonlinear measurable mappings satisfying $\|\tilde{A}(u)\|_{V^*} \le C\|u\|_V$ and $\|\tilde{B}(u)\|_{\mathscr{L}_2(K_Q,H)} \le C\|u\|_V$, $C > 0$, for all $u \in V$, respectively.

Theorem 3.6.1 *Assume that the null solution of (3.5.43) is exponentially stable in mean square. Further, suppose that for each $v \in V$, $\|v\|_H < r$, $r > 0$,*

$$\tilde{A}(v) - Av \in H$$

and

$$2\|v\|_H\|\tilde{A}(v) - Av\|_H + \|\tilde{B}(v)Q\tilde{B}(v)^* - (Bv)Q(Bv)^*\|_1 \le M(r)\|v\|_H^2 \tag{3.6.2}$$

where $\|L\|_1 := Tr((L^*L)^{1/2})$ for $L \in \mathscr{L}(H)$ and $M(r)$ is such a positive number that $M(r) \to 0$ as $r \to 0$. Then the null solution of (3.6.1) is strongly stable in probability.

Proof Recall that for any Itô type of functional Φ, we define for $v \in V$ that

$$(\mathcal{L}\Phi)(v) = \langle\langle \Phi'(v), Av\rangle\rangle_{V,V^*} + 1/2 \cdot Tr(\Phi''(v)(Bv)Q(Bv)^*),$$

$$(\tilde{\mathcal{L}}\Phi)(v) = \langle\langle \Phi'(v), \tilde{A}(v)\rangle\rangle_{V,V^*} + 1/2 \cdot Tr(\Phi''(v)\tilde{B}(v)Q\tilde{B}(v)^*).$$

Applying $\mathcal{L}$ and $\tilde{\mathcal{L}}$ to the Lyapunov function $\Lambda(\cdot)$ in Theorem 3.5.5, we find for any $v \in V$ that

$$\begin{aligned}(\tilde{\mathcal{L}}\Lambda)(v) - (\mathcal{L}\Lambda)(v) &= \langle\langle \Lambda'(v), \tilde{A}(v) - Av\rangle\rangle_{V,V^*}\\ &\quad + \frac{1}{2}\|\Lambda''(v)(\tilde{B}(v)Q\tilde{B}(v)^* - (Bv)Q(Bv)^*)\|_1\\ &= 2\langle Pv, \tilde{A}(v) - Av\rangle_H\\ &\quad + \|P(\tilde{B}(v)Q\tilde{B}(v)^* - (Bv)Q(Bv)^*)\|_1,\end{aligned}$$

where P is the operator defined in the proofs of Theorem 3.5.5. Therefore, by virtue of Theorem 3.5.5 and (3.6.2), it follows immediately that for each $v \in V$ with $\|v\|_H < r$,

$$(\tilde{\mathcal{L}}\Lambda)(v) \le -C\|v\|_H^2 + M(r)\|P\|_{\mathscr{L}(H)}\|v\|_H^2 \quad \text{for some constant} \quad C > 0.$$

Let $r \to 0$, then it is possible to choose $M(r) > 0$ such that $M(r)\|P\|_{\mathscr{L}(H)} < C$. Hence, by virtue of Theorem 3.4.1, we get the desired result. The proof is now complete. □

Example 3.6.2 Let $\mathcal{O} \subset \mathbb{R}^n$ be a bounded open domain with smooth boundary $\partial\mathcal{O}$. Assume that $H = L^2(\mathcal{O})$ and V is the standard Sobolev space $H_0^1(\mathcal{O})$. Also suppose that $\{W(t,x);\ t \ge 0,\ x \in \mathcal{O}\}$ is an H-valued Wiener process with its associated covariance operator Q, $Tr(Q) < \infty$, given by a positively definite kernel

$$q(x,y) \in L^2(\mathcal{O} \times \mathcal{O}), \qquad q(x,x) \in L^2(\mathcal{O}).$$

Let $a(\cdot,\cdot)$ be the form in $H_0^1(\mathcal{O}) \times H_0^1(\mathcal{O})$ defined by

$$a(u,v) = \int_{\mathcal{O}} \left\{ \sum_{i,\,j=1}^{n} a_{ij}(x) \frac{\partial u}{\partial x_i} \frac{\partial v}{\partial x_j} + \sum_{i=1}^{n} b_i(x) \frac{\partial u}{\partial x_i} v + c(x)uv \right\} dx, \; x \in \mathcal{O}. \tag{3.6.3}$$

Here we assume that the real-valued coefficients a_{ij}, b_i, c satisfy

$$a_{ij} = a_{ji} \in C^1(\overline{\mathcal{O}}), \qquad b_i \in C^1(\overline{\mathcal{O}}), \qquad c \in L^\infty(\overline{\mathcal{O}}), \qquad 1 \le i,\, j \le n,$$

and the uniform ellipticity

$$\sum_{i,\,j=1}^{n} a_{ij}(x) y_i y_j \ge \delta \|y\|_{\mathbb{R}^n}^2, \qquad \forall\, y = (y_1, \ldots, y_n) \in \mathbb{R}^n, \qquad x \in \mathcal{O}, \tag{3.6.4}$$

for some constant $\delta > 0$. Let A be the variational operator associated with $a(\cdot,\cdot)$. It can be shown that the coercive condition (3.5.44) is fulfilled (cf. Tanabe [203]). Then the problem (3.5.43) becomes

$$dy(t,x) = Ay(t,x)dt, \qquad t \ge 0. \tag{3.6.5}$$

In this case, let $\Lambda(u) = \|u\|_H^2$, then we get for any $u \in H_0^1(\mathcal{O})$,

$$\mathcal{L}(\|u\|_H^2) = 2\langle\langle u, Au \rangle\rangle_{V,V^*}. \tag{3.6.6}$$

On the other hand, let

$$\lambda_0 = \sup_{\substack{u \in H_0^1(\mathcal{O}) \\ \|u\|_H^2 \neq 0}} \frac{\mathcal{L}(\|u\|_H^2)}{\|u\|_H^2} = \sup_{\substack{u \in H_0^1(\mathcal{O}) \\ \|u\|_H^2 \neq 0}} \frac{2\langle\langle u, Au \rangle\rangle_{V,V^*}}{\|u\|_H^2}.$$

Then we have with the aid of Theorem 3.5.5 that the null solution of (3.6.5) is exponentially stable if $\lambda_0 < 0$.

Now let us consider a nonlinear stochastic system in $\mathcal{O}$, given by

$$dy(t,x) = \tilde{A}(y(t,x))dt + \tilde{B}(y(t,x))dW(t,x), \qquad t \ge 0, \tag{3.6.7}$$

with the boundary condition

$$y(t,x)|_{\partial\mathcal{O}} = 0, \qquad t \ge 0,$$

where

$$\tilde{A}(y(t,x)) := Ay(t,x) + \alpha_1(x, y(t,x)),$$
$$\tilde{B}(y(t,x)) := \alpha_2(x, y(t,x)),$$

and $\alpha_i : \mathcal{O} \times \mathbb{R} \to \mathbb{R}$, $i = 1, 2$, are two measurable functions satisfying the following conditions: for $i = 1, 2$, there exists a constant $c > 0$ such that

$$\sup_{x\in\mathcal{O}} |\alpha_i(x,u_2) - \alpha_i(x,u_1)| \leq c|u_2 - u_1|, \qquad \forall u_1,\ u_2 \in \mathbb{R},$$

$$\alpha_i(x,0) = 0, \qquad x \in \mathcal{O},$$

$$\sup_{x\in\mathcal{O}} |\alpha_i(x,u)| = o(|u|) \quad \text{as} \quad |u| \to 0. \tag{3.6.8}$$

Operator A is strictly elliptic in $\mathcal{O}$ so that it guarantees the fulfillment of the assumptions of coercivity and monotonicity for (3.6.7). Thus, there exists a unique strong solution of the problem. Furthermore, the last condition (3.6.8) is sufficient for (3.6.2) to be satisfied. We thus conclude, with the aid of Theorem 3.6.1, that the null solution is strongly stable in probability if $\lambda_0 < 0$.

3.6.2 Ultimate Boundedness

We shall develop a Lyapunov function program for ultimate boundedness of strong solutions in this section and study the associated problem of invariant measures in the next section. As in the last section, we shall deal with nonlinear systems by employing the first-order linear approximation method.

Definition 3.6.3 The strong solution $\{y(t,y_0),\ t \geq 0\}$ of (3.6.1) is called *exponential ultimately bounded in mean square* if the following relation holds

$$\mathbb{E}\|y(t,y_0)\|_H^2 \leq ce^{-\nu t}\|y_0\|_H^2 + M, \qquad t \geq 0,$$

for some constants $c > 0$, $M > 0$, $\nu > 0$, and any $y_0 \in H$.

Theorem 3.6.4 *Suppose that (b′), (c′), and the coercive condition (3.5.40) hold. Assume that the system (3.5.43) is exponential ultimately bounded in mean square such that*

$$\mathbb{E}\|y(t,y_0)\|_H^2 \leq ce^{-\nu t}\|y_0\|_H^2 + M \qquad \textit{for any} \qquad y_0 \in H,$$

where $c > 0$, $M > 0$, and $\nu > 0$ are constants. Further, suppose that for each $v \in V$, $\tilde{A}(v) - Av \in H$ and

$$\|\tilde{A}(v) - Av\|_H^2 + \|\tilde{B}(v)Q\tilde{B}(v)^* - (Bv)Q(Bv)^*\|_1 \leq L(1 + \|v\|_H^2)$$

for some constant $L > 0$, and for arbitrary $\varepsilon > 0$, there exists a number $K > 0$ such that

$$2\|v\|_H\|\tilde{A}(v) - Av\|_H + \|\tilde{B}(v)Q\tilde{B}(v)^* - (Bv)Q(Bv)^*\|_1 \leq \varepsilon\|v\|_H^2 + K, \tag{3.6.9}$$

whenever $\|v\|_H \geq R$ for some $R > 0$, then the strong solution of (3.6.1) is exponential ultimately bounded in mean square.

Proof We use the same notations and notions as in the proofs of Theorem 3.5.2 and Corollary 3.5.4. Define

$$\Lambda(y_0) = \int_0^T \int_0^t \mathbb{E}\|y(s, y_0)\|_V^2 ds dt \qquad \text{for all} \qquad y_0 \in H,$$

where $T > 0$ is some number satisfying (3.5.35) and y is the strong solution of (3.5.43). By an analogous argument to Theorem 3.5.5, it may be shown that $\Lambda(\cdot)$ is an Itô type of functional and satisfies

$$c_2\|y\|_H^2 + k_2 \leq \Lambda(y) \leq c_1\|y\|_H^2 + k_1, \qquad \forall y \in H,$$

for some constants $c_1 > 0$, $c_2 > 0$, and $k_1, k_2 \in \mathbb{R}$.

To conclude the proof, it remains to show that

$$\tilde{\mathcal{L}}\Lambda(v) \leq -c_3\Lambda(v) + k_3, \qquad \forall v \in V,$$

for some constants $c_3 > 0$ and $k_3 \in \mathbb{R}$. Since $\tilde{A}(v) - Av \in H$ for any $v \in V$, we have

$$\begin{aligned}\tilde{\mathcal{L}}\Lambda(v) - \mathcal{L}\Lambda(v) &= \langle \Lambda'(v), \tilde{A}(v) - Av\rangle_H \\ &\quad + \frac{1}{2}\|\Lambda''(v)(\tilde{B}(v)Q\tilde{B}^*(v) - (Bv)Q(Bv)^*)\|_1,\end{aligned}$$

with $\Lambda'(v) = 2Pv$ and $\Lambda''(v) = 2P$ for $v \in V$, where P is defined as in (3.5.48), which is a bounded positive operator on H. Hence,

$$\begin{aligned}\tilde{\mathcal{L}}\Lambda(v) - \mathcal{L}\Lambda(v) &= 2\langle Pv, \tilde{A}(v) - Av\rangle_H \\ &\quad + \|P(\tilde{B}(v)Q\tilde{B}^*(v) - (Bv)Q(Bv)^*)\|_1, \qquad v \in V.\end{aligned}$$

This further implies that for $v \in V$,

$$\begin{aligned}\tilde{\mathcal{L}}\Lambda(v) &\leq \mathcal{L}\Lambda(v) + 2\|P\|\|v\|_H\|\tilde{A}(v) - Av\|_H \\ &\quad + \|P(\tilde{B}(v)Q\tilde{B}^*(v) - (Bv)Q(Bv)^*)\|_1 \\ &\leq \mathcal{L}\Lambda(v) + \|P\|\Big\{2\|v\|_H\|\tilde{A}(v) - Av\|_H \\ &\quad + \|\tilde{B}(v)Q\tilde{B}^*(v) - (Bv)Q(Bv)^*\|_1\Big\}.\end{aligned} \tag{3.6.10}$$

Note that when $\beta_2 = 0$, $\mu = 0$ in the proofs of Theorem 3.5.2, we obtain a relation analogous to (3.5.27),

$$\int_{t_0}^{t} \mathbb{E}\|y(s,v)\|_V^2 ds \le \frac{1}{\alpha}\left(\|v\|_H^2 + \frac{\alpha_1|\lambda|\|v\|_H^2}{\beta_1} + |\lambda|\alpha_2 t + \gamma t\right), \qquad v \in V,$$

from which one further obtains the following relation analogous to (3.5.34) in the form

$$\mathcal{L}\Lambda(v) \le -\frac{\alpha_1}{\beta_1\beta}\|v\|_H^2 + (\gamma + |\lambda|\alpha_2)T, \tag{3.6.11}$$

where

$$T = \frac{\beta}{\alpha}\left(1 + \frac{\alpha_1|\lambda|}{\beta_1}\right) + \frac{\alpha_1}{\beta_1}.$$

Therefore, we have from (3.6.10) and (3.6.11) that

$$\begin{aligned}\tilde{\mathcal{L}}\Lambda(v) \le -\frac{\alpha_1}{\beta_1\beta}\|v\|_H^2 + (\gamma + |\lambda|\alpha_2)T + \|P\|\big\{2\|v\|_H\|\tilde{A}(v) - Av\|_H \\ + \|\tilde{B}(v)Q\tilde{B}^*(v) - (Bv)Q(Bv)^*\|_1\big\}.\end{aligned} \tag{3.6.12}$$

By virtue of (3.6.9), we know that for any fixed value $\kappa \in (0, \alpha_1/\|P\|\beta_1\beta)$, there exist numbers $R > 0$ and $K > 0$ such that

$$2\|v\|_H\|\tilde{A}(v) - Av\|_H + \|\tilde{B}(v)Q\tilde{B}^*(v) - (Bv)Q(Bv)\|_1 \le \kappa\|v\|_H^2 + K \tag{3.6.13}$$

for any $v \in V$ with $\|v\|_H > R$. In the meantime, for any $v \in V$ with $\|v\|_H \le R$, by assumption, we have

$$\begin{aligned}2\|v\|_H\|\tilde{A}(v) - Av\|_H &+ \|\tilde{B}(v)Q\tilde{B}^*(v) - (Bv)Q(Bv)^*\|_1 \\ &\le \|v\|_H^2 + \|\tilde{A}(v) - Av\|_H^2 \\ &\quad + \|\tilde{B}(v)Q\tilde{B}^*(v) - (Bv)Q(Bv)^*\|_1 \\ &\le \|v\|_H^2 + L(1 + \|v\|_H^2) \\ &\le L + (L+1)R^2.\end{aligned} \tag{3.6.14}$$

Therefore, (3.6.13) and (3.6.14) imply that for any $v \in V$,

$$\begin{aligned}2\|v\|_H\|\tilde{A}(v) - Av\|_H &+ \|\tilde{B}(v)Q\tilde{B}^*(v) - (Bv)Q(Bv)^*\|_1 \\ &\le \kappa\|v\|_H^2 + (L+1)R^2 + L + K.\end{aligned} \tag{3.6.15}$$

Now, it follows from (3.6.12) and (3.6.15) that

$$\begin{aligned}\tilde{\mathcal{L}}\Lambda(v) &\le \left(-\frac{\alpha_1}{\beta_1\beta}+\kappa\|P\|\right)\|v\|_H^2+(|\lambda|\alpha_2+\gamma)\left[\frac{\beta}{\alpha}\left(1+\frac{\alpha_1|\lambda|}{\beta_1}\right)+\frac{\alpha_1}{\beta_1}\right]\\&\quad+\|P\|\big[(L+1)R^2+L+K\big]\\&\le -c_3\|v\|_H^2+k_3, \qquad \forall\, v\in V,\end{aligned}$$

for some constants $c_3>0$ and $k_3\in\mathbb{R}$. The proof is now complete. □

Example 3.6.5 Consider the following stochastic differential equation:

$$dy(t)=Ay(t)dt+F(y(t))dt+B(y(t))dW(t), \qquad t\ge 0, \tag{3.6.16}$$

with initial condition $y(0)=y_0\in H$ and $Tr(Q)<\infty$. Suppose A, F, and B satisfy the following conditions:

(i) $A\colon V\to V^*$ is coercive, precisely, there exist constants $\alpha>0$ and $\lambda\in\mathbb{R}$ such that for arbitrary $v\in V$,

$$2\langle\langle v,Av\rangle\rangle_{V,V^*}\le -\alpha\|v\|_V^2+\lambda\|v\|_H^2;$$

(ii) $F\colon H\to H$ and $B\colon V\to\mathscr{L}(K,H)$ satisfy that for arbitrary $v\in V$,

$$\|F(v)\|_H+\|B(v)\|\le L(1+\|v\|_H)$$

for some constant $L>0$;

(iii) for arbitrary $u,v\in V$, there exists a constant $C>0$ such that

$$\|F(u)-F(v)\|_H^2+Tr\big[(B(u)-B(v))Q(B(u)^*-B(v)^*)\big]\le C\|u-v\|_H^2.$$

If the strong solution $\{y(t,y_0),\ t\ge 0\}$ of $dy(t)=Ay(t)dt,\ t\ge 0,$ is exponentially ultimately bounded, and for any $v\in V$, as $\|v\|_H\to\infty$,

$$\|F(v)\|_H=o(\|v\|_H), \qquad \|B(v)\|=o(\|v\|_H),$$

then the strong solution $\{y(t,y_0),\ t\ge 0\}$ of (3.6.16) is exponentially ultimately bounded in mean square.

Indeed, let $\tilde{A}(v)=Av+F(v)$ for $v\in V$. Since $F(v)\in H$,

$$\begin{aligned}&2\langle\langle v,\tilde{A}(v)\rangle\rangle_{V,V^*}+Tr[B(v)QB(v)^*]\\&\quad=2\langle\langle v,Av\rangle\rangle_{V,V^*}+2\langle F(v),v\rangle_H+Tr[B(v)QB(v)^*]\\&\quad\le -\alpha\|v\|_V^2+\lambda\|v\|_H^2+2\|v\|_H\|F(v)\|_H+Tr(Q)\|B(v)\|^2\\&\quad\le -\alpha\|v\|_V^2+M_1\|v\|_H^2+M_2\end{aligned}\tag{3.6.17}$$

for some constants $M_1\in\mathbb{R}$ and $M_2\ge 0$. Hence, under the additional assumptions (ii) and (iii), a strong solution $\{y(t,y_0),\ t\ge 0\}$ of (3.6.16) exists. Moreover, by assumption (ii) we have that for $v\in V$,

$$\|F(v)\|_H^2 + Tr[B(v)QB(v)^*] \le \|F(v)\|_H^2 + \|B(v)\|^2 Tr(Q)$$
$$\le (1 + Tr(Q))L(1 + \|v\|_H^2)$$

and since

$$\|F(v)\|_H = o(\|v\|_H), \quad Tr[B(v)QB(v)^*] \le \|B(v)\|^2 Tr(Q) = o(\|v\|_H^2)$$

as $\|v\|_H \to \infty$, the assertion follows from Theorem 3.6.4.

3.7 Invariant Measures and Ultimate Boundedness

Suppose that $y(t)$, $t \ge 0$, is a time-homogeneous Markov process defined on some probability space $(\Omega, \mathscr{F}, \{\mathscr{F}_t\}_{t\ge0}, \mathbb{P})$ with state space H. Denote the transition probability function of $y(t)$ by $\mathbb{P}(\cdot, \cdot, \cdot)$, which is defined by the conditional probability

$$\mathbb{P}(x, t, \Gamma) = \mathbb{P}\{y(t) \in \Gamma \mid y(0) = x\}, \qquad t \ge 0,$$

for $x \in H$ and $\Gamma \in \mathscr{B}(H)$, the Borel σ-field of H. A probability measure μ on $(H, \mathscr{B}(H))$ is called *invariant* for the process $y(t)$, $t \ge 0$, or more precisely, for the transition probability function $\mathbb{P}(\cdot, \cdot, \cdot)$, if it satisfies the equation

$$\mu(\Gamma) = \int_H \mathbb{P}(x, t, \Gamma)\mu(dx) \qquad \text{for any} \qquad t \ge 0, \qquad \Gamma \in \mathscr{B}(H),$$

or equivalently, the following relation holds:

$$\int_H \Psi(x)\mu(dx) = \int_H (\mathbb{P}_t\Psi)(x)\mu(dx) \qquad \text{for any} \qquad \Psi \in C_b(H), \qquad t \ge 0,$$

where the transition semigroup $\mathbb{P}_t$, $t \ge 0$, is defined by

$$(\mathbb{P}_t\Psi)(x) = \int_H \Psi(y)\mathbb{P}(x, t, dy) \qquad \text{for any} \qquad \Psi \in C_b(H), \qquad t \ge 0.$$

It is clear that the exponential decay of a system implies that this system degenerates to one with invariant measure zero. A remarkable result of the theory in the last section is that one can show a close relationship between exponentially ultimate boundedness and nontrivial invariant measures of (3.7.1).

3.7.1 Invariant Measures: Existence

Consider the following nonlinear stochastic differential equation in V^*:

$$\begin{cases} dy(t) = A(y(t))dt + B(y(t))dW(t), & t \in [0, \infty), \\ y(0) = y_0 \in H, \end{cases} \tag{3.7.1}$$

where $A(\cdot)\colon V \to V^*$ and $B(\cdot)\colon V \to \mathscr{L}_2(K_Q, H)$ are two nonlinear measurable mappings satisfying all the conditions in Section 1.4.2 so that there exists a unique strong solution to (3.7.1). In particular, it can be shown that this strong solution $y(t)$, $t \geq 0$, is a time-homogeneous Markov process and the associated transition probability function $\mathbb{P}(y_0, t, \Gamma)$ with its transition semigroup $\mathbb{P}_t$, $t \geq 0$, has the Feller property. Recall that a sequence of probability measures μ_n on $(H, \mathscr{B}(H))$ is called *weakly convergent* to a probability measure μ, denote it by $\mu_n \stackrel{w}{\Longrightarrow} \mu$, if for any $\Psi \in C_b(H)$,

$$\lim_{n\to\infty} \int_H \Psi(x)\mu_n(dx) = \int_H \Psi(x)\mu(dx).$$

To proceed further, we first state and show a general theorem about the existence of invariant measures in infinite dimensional spaces.

Theorem 3.7.1 *Let $y(t, y_0)$, $t \geq 0$, be a strong solution of (3.7.1) and $\mathbb{P}(\cdot, \cdot, \cdot)$ be its transition probability function. For some sequence $T_n \uparrow \infty$, define*

$$\mu_n(\Gamma) = \frac{1}{T_n}\int_0^{T_n} \mathbb{P}(y_0, t, \Gamma)dt, \quad \Gamma \in \mathscr{B}(H). \tag{3.7.2}$$

If μ is the weak limit of some subsequence of $\{\mu_n\}_{n\geq 1}$, then μ is an invariant measure of $\mathbb{P}(\cdot, \cdot, \cdot)$. That is, the existence of invariant measures is a consequence of the relative compactness of $\{\mu_n\}_{n\geq 1}$.

Proof Without loss of generality, we may assume $\mu_n \stackrel{w}{\Longrightarrow} \mu$, i.e., μ_n converges weakly to μ on H. By the Feller property, for each $t \geq 0$ and $\Psi \in C_b(H)$, the function $(\mathbb{P}_t\Psi)(\cdot)$ is in $C_b(H)$. It thus follows from the weak convergence of $\{\mu_n\}$ that

$$\begin{aligned}\int_H (\mathbb{P}_t\Psi)(x)\mu(dx) &= \lim_{n\to\infty} \int_H (\mathbb{P}_t\Psi)(x)\mu_n(dx)\\ &= \lim_{n\to\infty} \frac{1}{T_n}\int_0^{T_n}\int_H (\mathbb{P}_t\Psi)(x)\mathbb{P}(y_0, s, dx)ds\\ &= \lim_{n\to\infty} \frac{1}{T_n}\int_0^{T_n} (\mathbb{P}_{t+s}\Psi)(y_0)ds,\end{aligned} \tag{3.7.3}$$

where we have made use of Fubini's theorem and the Markov property of the solution process $y(t, y_0)$, $t \geq 0$. Now, for any fixed $t \geq 0$ and $y_0 \in H$, we can write

$$\begin{aligned}&\lim_{n\to\infty} \frac{1}{T_n}\int_0^{T_n} (\mathbb{P}_{t+s}\Psi)(y_0)ds\\ &= \lim_{n\to\infty} \frac{1}{T_n}\left\{\int_0^{T_n} (\mathbb{P}_s\Psi)(y_0)ds + \int_{T_n}^{T_n+t} (\mathbb{P}_s\Psi)(y_0)ds - \int_0^t (\mathbb{P}_s\Psi)(y_0)ds\right\}.\end{aligned} \tag{3.7.4}$$

Since $\|\mathbb{P}_s\Psi\| \le \|\Psi\| < \infty$ for any $s \ge 0$, by virtue of (3.7.2), (3.7.3) and (3.7.4), we obtain

$$\int_H (\mathbb{P}_t\Psi)(x)\mu(dx) = \lim_{n\to\infty} \frac{1}{T_n}\int_0^{T_n} (\mathbb{P}_{t+s}\Psi)(y_0)ds$$
$$= \lim_{n\to\infty}\int_H \Psi(x)\mu_n(dx) = \int_H \Psi(x)\mu(dx) \qquad (3.7.5)$$

according to Fubini's theorem and the weak convergence property of $\{\mu_n\}$. Hence, μ is an invariant measure. The proof is complete. □

If H is compact, it is easy to get (see, e.g., theorem 6.4, p. 45, Parthasarathy [183]) the relative compactness of $\{\mu_n(\cdot)\}$ in Theorem 3.7.1 and the existence of invariant measures thus follows. For noncompact H, we need additional conditions that lead to the following corollary to obtain the existence of invariant measures.

Corollary 3.7.2 *Assume that the embedding map $V \hookrightarrow H$ is compact. Suppose that $y(t, y_0)$, $t \ge 0$, is the strong solution of equation (3.7.1) and for some $y_0 \in H$, there exists a positive number sequence $T_n \uparrow \infty$ such that as $R \to \infty$,*

$$\frac{1}{T_n}\int_0^{T_n} \mathbb{P}\{\|y(t, y_0)\|_V > R\}dt \to 0 \text{ uniformly with respect to } n \in \mathbb{N}. \qquad (3.7.6)$$

Then equation (3.7.1) has an invariant measure on $(H, \mathscr{B}(H))$.

Proof It suffices to show that $\{\mu_n\}$ defined in (3.7.2) is relatively compact. For any $R > 0$, let $V_R = \{y \in V : \|y\|_V \le R\}$ and $V_R^{CH} = \{y \in H : y \notin V_R\}$. It is clear that V_R is compact in H and therefore by (3.7.6), for any $\varepsilon > 0$, there exists a compact set $V_R \subset H$ such that as R is sufficiently large,

$$\mu_n\{H \setminus V_R\} = \mu_n(V_R^{CH}) < \varepsilon \quad \text{for all} \quad n \ge 1.$$

Therefore, the family $\{\mu_n\}$ is tight, which, by virtue of Prokhorov's theorem (see theorem 2.3, p. 31, Da Prato and Zabczyk [53]), further implies that the family $\{\mu_n\}$ is relatively compact. □

There exists a close relationship between ultimate boundedness of a solution process and its invariant measures. For simplicity, we assume that $A(\cdot)$ and $B(\cdot)$ satisfy the following coercive condition: there exist $\alpha > 0$, $\lambda \in \mathbb{R}$, and $\gamma \ge 0$ such that

$$2\langle\langle u, A(u)\rangle\rangle_{V,V^*} + \|B(u)\|^2_{\mathscr{L}_2(K_Q,H)} \le -\alpha\|u\|_V^2 + \lambda\|u\|_H^2 + \gamma, \quad \forall u \in V. \qquad (3.7.7)$$

Theorem 3.7.3 *Assume the embedding map $V \hookrightarrow H$ is compact and coercive condition (3.7.7) is satisfied. Suppose that the strong solution $\{y(t, y_0);\ t \geq 0\}$ of (3.7.1) is exponential ultimately bounded in mean square, then there exists an invariant measure μ of $\{y(t, y_0);\ t \geq 0\}$.*

Proof Applying Itô's formula to $\Lambda(x) = \|x\|_H^2$, $x \in H$, taking expectation and using the coercive condition (3.7.7), we obtain

$$\begin{aligned}\mathbb{E}\|y(t, y_0)\|_H^2 - \|y_0\|_H^2 &= \int_0^t \mathbb{E}\mathcal{L}\|y(s, y_0)\|_H^2 ds \\ &\leq \lambda \int_0^t \mathbb{E}\|y(s, y_0)\|_H^2 ds - \alpha \int_0^t \mathbb{E}\|y(s, y_0)\|_V^2 ds + \gamma t.\end{aligned} \tag{3.7.8}$$

Hence,

$$\int_0^t \mathbb{E}\|y(s, y_0)\|_V^2 ds \leq \frac{1}{\alpha}\Big(\lambda \int_0^t \mathbb{E}\|y(s, y_0)\|_H^2 ds + \|y_0\|_H^2 + \gamma t\Big),$$

which, by virtue of Markov's inequality, immediately implies that for any $T > 0$ and $R > 0$,

$$\begin{aligned}\frac{1}{T}\int_0^T \mathbb{P}\{\|y(t, y_0)\|_V > R\} dt &\leq \frac{1}{T}\int_0^T \frac{\mathbb{E}\|y(t, y_0)\|_V^2}{R^2} dt \\ &\leq \frac{1}{\alpha R^2}\Big(\frac{|\lambda|}{T}\int_0^T \mathbb{E}\|y(t, y_0)\|_H^2 dt + \frac{\|y_0\|_H^2}{T} + \gamma\Big).\end{aligned} \tag{3.7.9}$$

Since $\{y(t, y_0), t \geq 0\}$ is exponential ultimately bounded in mean square, there exist two constants $T_0 > 0$ and $M > 0$ such that

$$\mathbb{E}\|y(t, y_0)\|_H^2 \leq M \qquad \text{for all} \qquad t \geq T_0,$$

which, in addition to (3.7.9), immediately yields

$$\begin{aligned}&\lim_{R\to\infty} \lim_{T\to\infty} \frac{1}{T}\int_0^T \mathbb{P}\{\|y(t, y_0)\|_V > R\} dt \\ &\leq \lim_{R\to\infty} \lim_{T\to\infty} \frac{|\lambda|}{\alpha R^2}\frac{1}{T}\Big(\int_0^{T_0} \mathbb{E}\|y(t, y_0)\|_H^2 dt + \int_{T_0}^T \mathbb{E}\|y(t, y_0)\|_H^2 dt\Big) \\ &\leq \lim_{R\to\infty} \lim_{T\to\infty} \frac{|\lambda|}{\alpha R^2}\frac{1}{T}\Big(\int_0^{T_0} \mathbb{E}\|y(t, y_0)\|_H^2 dt + M(T - T_0)\Big) = 0.\end{aligned}$$

Therefore, the assertion of the theorem follows. □

As an immediate consequence of Theorem 3.7.3, one may obtain information on the existence of invariant measures to the stochastic differential equation (3.7.1). For instance, the strong solution y in Example 3.6.5 is exponential ultimately bounded in mean square. If the embedding $V \hookrightarrow H$ is compact, there exists an invariant measure.

3.7.2 Invariant Measures: Uniqueness

Throughout this section, we always suppose that the embedding $V \hookrightarrow H$ is compact.

Lemma 3.7.4 *Suppose that for arbitrary $\varepsilon > 0$, $\delta > 0$ and $R > 0$, there exists $T_0 = T_0(\varepsilon, \delta, R) > 0$ such that the strong solution y of (3.7.1) satisfies*

$$\frac{1}{T}\int_0^T \mathbb{P}\{\|y(t,\xi) - y(t,\eta)\|_V \geq \delta\}dt < \varepsilon \tag{3.7.10}$$

for any ξ, $\eta \in V_R := \{y \in V : \|y\|_V \leq R\}$ and $T \geq T_0$. If there exists an invariant measure μ with support in V to system (3.7.1), then this invariant measure is unique.

Proof Suppose that μ and ν are two invariant measures with supports in V. It suffices to show that for any bounded, uniformly continuous function Ψ on H (see, e.g., theorem 5.9, p. 39, Parthasarathy [183]),

$$\int_H \Psi(x)\mu(dx) = \int_H \Psi(x)\nu(dx).$$

To this end, define for arbitrary $\Gamma \in \mathscr{B}(H)$ that

$$\mu_T^{y_0}(\Gamma) := \frac{1}{T}\int_0^T \mathbb{P}(y_0, t, \Gamma)dt, \quad y_0 \in H, \quad T > 0,$$

where $\mathbb{P}(\cdot, \cdot, \cdot)$ is the transition probability of system (3.7.1). By definition, we have for any ξ, $\eta \in H$ that

$$\begin{aligned}
&\left|\int_H \Psi(x)\mu(dx) - \int_H \Psi(x)\nu(dx)\right| \\
&= \left|\int_H\int_H \Psi(x)\Big[\mu_T^{\xi}(dx)\mu(d\xi) - \mu_T^{\eta}(dx)\nu(d\eta)\Big]\right| \\
&\leq \int_{H\times H}\left|\int_H \Psi(x)\mu_T^{\xi}(dx) - \int_H \Psi(x)\mu_T^{\eta}(dx)\right|\mu(d\xi)\nu(d\eta) \\
&= \int_{V\times V} |\Sigma(\xi,\eta)|\mu(d\xi)\nu(d\eta),
\end{aligned} \tag{3.7.11}$$

where

$$\Sigma(\xi,\eta) = \int_H \Psi(x)\mu_T^{\xi}(dx) - \int_H \Psi(x)\mu_T^{\eta}(dx). \tag{3.7.12}$$

For any $\varepsilon > 0$, let $V_R^{cV} = V \backslash V_R$ and choose a proper $R > 0$ such that

$$\mu(V_R^{cV}) + \nu(V_R^{cV}) < \varepsilon. \tag{3.7.13}$$

Then, by virtue of (3.7.12) and (3.7.13), inequality (3.7.11) yields

$$\left| \int_H \Psi(x)\mu(dx) - \int_H \Psi(x)\nu(dx) \right| \leq \int_{V_R \times V_R} |\Sigma(\xi,\eta)|\mu(d\xi)\nu(d\eta) + 4b\varepsilon + 2b\varepsilon^2, \tag{3.7.14}$$

where $b = \sup_{x \in H} |\Psi(x)| < \infty$. On the other hand, in view of (3.7.10) and the uniform continuity of Ψ, there exist positive numbers $\delta > 0$ and $T_0 = T_0(\varepsilon, \delta, R) > 0$ such that

$$\begin{aligned}
&\int_{V_R \times V_R} |\Sigma(\xi,\eta)|\mu(d\xi)\nu(d\eta) \\
&\leq \int_{V_R \times V_R} \left\{ \frac{1}{T_0} \int_0^{T_0} \mathbb{E}|\Psi(y(t,\xi)) - \Psi(y(t,\eta))|dt \right\} \mu(d\xi)\nu(d\eta) \\
&\leq 2b \sup_{u,\, v \in V_R} \frac{1}{T_0} \int_0^{T_0} \mathbb{P}\{\|y(t,u) - y(t,v)\|_V \geq \delta\}dt + \sup_{\substack{u,\, v \in V_R \\ \|u-v\|_V < \delta}} |\Psi(u) - \Psi(v)| \\
&< 2b\varepsilon + \varepsilon.
\end{aligned} \tag{3.7.15}$$

Since $\varepsilon > 0$ is arbitrary, (3.7.11) and (3.7.15) imply the desired uniqueness of invariant measures. The proof is thus complete. □

In the preceding proof, the property of the measure $\mu(\cdot)$ to have support in V is essential. Suppose that for each n, $\mu_n(\cdot)$ defined in (3.7.2) has support in the space V or $\mu_n(V) = 1$, $n = 1, 2, \ldots$ Then one can expect that the same is true for their weak limit μ. This is indeed the case as shown by the following lemma.

Lemma 3.7.5 *Suppose that a family $\{\mu_n\}$ of probability measures on H is supported in V, and for any $\varepsilon > 0$, there exists $R_0 > 0$ such that*

$$\sup_n \mu_n(V_R^{cH}) < \varepsilon, \qquad \forall R > R_0. \tag{3.7.16}$$

Then any invariant measure, as the weak limit of a subsequence of $\{\mu_n\}$, has support in V.

Proof Let $\{\mu_{n_k}\}_{k\geq 1}$ be a subsequence of $\{\mu_n\}_{n\geq 1}$ converging weakly to μ so that $\mu_{n_k}(V) = 1$ for any $k \in \mathbb{N}_+$. Since the embedding mapping $\imath$ from V into H is compact, it is immediate that the set $\imath(V_R) = \{x \in V\colon \|x\|_V \leq R\}$ in H is closed. Let $\rho(x, \imath(V_R))$, $x \in H$, be the distance from x to set $\imath(V_R)$ given by

$$\rho(x, \imath(V_R)) = \inf\Big\{\|y - x\|_H\colon y \in \imath(V_R)\Big\},$$

and define a function $\phi\colon \mathbb{R}_+ \to \mathbb{R}_+$ by $\phi(t) = t$ for $0 \leq t \leq 1$ and $\phi(t) = 1$ for $t > 1$. Now we introduce

$$\Psi_\delta^R(x) = \phi\left(\frac{1}{\delta}\rho(x, \imath(V_R))\right), \qquad \delta \in (0, 1).$$

Then $\Psi_\delta^R \in C_b(H)$ and

$$\int_H \Psi_\delta^R(x)\mu_{n_k}(dx) = \int_V \Psi_\delta^R(x)\mu_{n_k}(dx) \leq \int_V \mathbf{1}_{V_R^{cH}}(x)\mu_{n_k}(dx) = \mu_{n_k}(V_R^{cH}). \tag{3.7.17}$$

Hence, it follows from (3.7.16), (3.7.17), and the weak convergence of $\{\mu_{n_k}\}_{k\geq 1}$ that

$$\int_H \Psi_\delta^R(x)\mu(dx) = \lim_{k\to\infty} \int_H \Psi_\delta^R(x)\mu_{n_k}(dx) \leq \lim_{k\to\infty} \mu_{n_k}(V_R^{cH}) < \varepsilon, \;\; \forall\, R > R_0.$$

Thus, by the well-known Dominated Convergence Theorem, we obtain

$$\lim_{\delta\downarrow 0} \int_H \Psi_\delta^R(x)\mu(dx) = \int_H \mathbf{1}_{V_R^{cH}}(x)\mu(dx) \leq \varepsilon,$$

or

$$\mu(H\backslash V_R) = \mu(V_R^{cH}) \leq \varepsilon, \qquad \forall R > R_0. \tag{3.7.18}$$

Since $V = \bigcup_{n=1}^\infty V_n$, $V_n = \{y \in V\colon \|y\|_V \leq n\}$, it is easy to show from (3.7.18) that

$$\mu(H\backslash V) = \mu\left(H \Big\backslash \bigcup_{n=1}^\infty V_n\right) = 0,$$

as required. □

Now we are in a position to state the main result of this section on the existence and uniqueness of invariant measures to equation (3.7.1).

Theorem 3.7.6 *Assume that the strong solution $y(t, y_0)$, $t \geq 0$, of (3.7.1) satisfies (3.7.6). Then system (3.7.1) has an invariant measure μ with its support in V. Moreover, if the condition (3.7.10) holds, this invariant measure μ is unique.*

Proof Clearly, all we need to do is to show that the support $\text{supp}\{\mu\}$ of μ satisfies $\text{supp}\{\mu\} \subset V$. Define

$$\mu_T(\Gamma) = \frac{1}{T}\int_0^T \mathbb{P}\{y(t, y_0) \in \Gamma\}dt \tag{3.7.19}$$

for arbitrary $T > 0$, $\Gamma \in \mathscr{B}(H)$, which is a probability measure on $(H, \mathscr{B}(H))$. In view of Lemma 3.7.5, it suffices to show that

$$\mu_T\{H \setminus V\} = 0, \quad \forall T > 0.$$

To this end, we first note that for any $T > 0$,

$$\mathbb{E}\int_0^T \|y(t, y_0)\|_V^2 dt < \infty, \qquad y_0 \in H,$$

which further implies

$$y(t, y_0) = y(t, \omega, y_0) \in V, \quad (t, \omega) \in \Omega_T := [0, T] \times \Omega \quad \text{almost surely}.$$

Let

$$\Sigma = \{(t, \omega) \in \Omega_T \colon y(t, y_0) \in H \setminus V\}.$$

Then it is easy to see that

$$(\mathbb{L} \times \mathbb{P})\{\Sigma\} = \int_0^T \int_\Omega \mathbf{1}_\Sigma(t, \omega)\mathbb{P}(d\omega)dt = 0,$$

where $\mathbb{L}$ is the standard Lebesgue measure on $[0, T]$. Thus by the Fubini theorem, we get for any $t \in [0, T]$, $T > 0$, that

$$\mathbb{P}\{\Sigma_t\} = \int_\Omega \mathbf{1}_\Sigma(t, \omega)\mathbb{P}(d\omega) = 0, \quad t \in J \subset [0, T], \tag{3.7.20}$$

where $J = [0, T] \setminus \tilde{J}$ with $\mathbb{L}(\tilde{J}) = 0$ and

$$\Sigma_t = \{\omega \in \Omega \colon y(t, y_0) \in H \setminus V\}, \qquad t \in J.$$

Thus, by virtue of (3.7.19) and (3.7.20), we have for any $T > 0$ that

$$\mu_T\{H \setminus V\} = \frac{1}{T}\int_0^T \mathbb{P}\{\Sigma_t\}dt = \frac{1}{T}\int_0^T \mathbf{1}_J(t)\mathbb{P}\{\Sigma_t\}dt = 0$$

as desired. The proof is now complete. □

As an immediate consequence, we obtain a useful criterion that ensures a unique invariant measure of strong solutions by imposing some monotonicity conditions on the equation (3.7.1).

Corollary 3.7.7 *Assume that the condition (3.7.7) is satisfied. If there exist positive numbers $T_0 > 0$ and $M > 0$ such that*

$$\sup_{T>T_0} \left\{ \frac{1}{T} \int_0^T \mathbb{E}\|y(t, y_0)\|_H^2 dt \right\} \le M, \tag{3.7.21}$$

then system (3.7.1) has an invariant measure supported in V. This invariant measure is unique provided that the following condition holds: there exist $c > 0$ and $\delta > 0$ such that

$$\begin{aligned} &2\langle\langle u - v, A(u) - A(v)\rangle\rangle_{V,V^*} + \|B(u) - B(v)\|^2_{\mathscr{L}_2(K_Q,H)} \\ &\le -c\|u - v\|_V^\delta, \quad \forall u,\ v \in V. \end{aligned} \tag{3.7.22}$$

Proof In view of Itô's formula, we have the following energy inequality

$$\begin{aligned} \mathbb{E}\|y(t, y_0)\|_H^2 \le \|y_0\|_H^2 &+ 2\mathbb{E}\int_0^t \langle\langle y(s, y_0), A(y(s, y_0))\rangle\rangle_{V,V^*} ds \\ &+ \mathbb{E}\int_0^t \|B(y(s, y_0))\|^2_{\mathscr{L}_2(K_Q,H)} ds \end{aligned} \tag{3.7.23}$$

for any $t \ge 0$, which, in addition to the coercive condition (3.7.7), yields

$$\mathbb{E}\|y(t, y_0)\|_H^2 + \alpha\mathbb{E}\int_0^t \|y(s, y_0)\|_V^2 ds \le (\|y_0\|_H^2 + \gamma t) + \lambda\mathbb{E}\int_0^t \|y(s, y_0)\|_H^2 ds.$$

It follows that for any $T_0 > 0$,

$$\begin{aligned} \sup_{T>T_0} \frac{1}{T}\int_0^T \mathbb{E}\|y(s, y_0)\|_V^2 ds \le &\frac{|\gamma| + \|y_0\|_H^2/T_0}{\alpha} \\ &+ \frac{|\lambda|}{\alpha} \sup_{T>T_0} \frac{1}{T}\int_0^T \mathbb{E}\|y(s, y_0)\|_H^2 ds. \end{aligned} \tag{3.7.24}$$

On the other hand, $\mu_T(\cdot)$ defined by (3.7.19) is supported in V. Hence, by the well-known Markov's inequality, we have for any $T > 0$, $R > 0$ that

$$\frac{1}{T}\int_0^T \mathbb{P}\Big\{\|y(s, y_0)\|_V > R\Big\} ds \le \frac{1}{R^2} \cdot \left\{ \frac{1}{T}\int_0^T \mathbb{E}\|y(s, y_0)\|_V^2 ds \right\},$$

which, together with (3.7.24), shows that condition (3.7.21) is sufficient for the existence of an invariant measure μ on V according to Theorem 3.7.6.

To show the uniqueness of invariant measures, let $y(t, \xi)$, $y(t, \eta)$ denote two solutions with their initial data $\xi, \eta \in H$, respectively. We set $\Delta y(t) = y(t, \xi) - y(t, \eta)$. Then, by Itô's formula, we have

$$\mathbb{E}\|\Delta y(t)\|_H^2 \le \|\xi-\eta\|_H^2 + 2\mathbb{E}\int_0^t \langle\langle \Delta y(s), A(y(s,\xi)) - A(y(s,\eta))\rangle\rangle_{V,V^*} ds$$
$$+\mathbb{E}\int_0^t \|B(y(s,\xi)) - B(y(s,\eta))\|^2_{\mathscr{L}_2(K_Q,H)} ds. \tag{3.7.25}$$

In view of condition (3.7.22), we obtain

$$\mathbb{E}\|\Delta y(t)\|_H^2 \le \|\xi-\eta\|_H^2 - c\int_0^t \mathbb{E}\|\Delta y(s)\|_V^\delta ds,$$

or

$$\int_0^t \mathbb{E}\|\Delta y(s)\|_V^\delta ds \le \frac{\|\xi-\eta\|_H^2}{c}, \qquad t \ge 0.$$

This implies that condition (3.7.10) is satisfied by applying Markov's inequality

$$\mathbb{P}\Big\{\|y(t,\xi)-y(t,\eta)\|_V \ge \delta_1\Big\} \le \mathbb{E}\|y(t,\xi)-y(t,\eta)\|_V^\delta/\delta_1^\delta$$

for arbitrary $\delta_1 > 0$. The proof is now compete. □

3.8 Decay Rate

A stochastic differential equation of Itô type could be regarded as the perturbed stochastic system of some deterministic equation. Precisely, the following semilinear stochastic differential equation

$$\begin{cases} dy(t) = Ay(t)dt + F(t,y(t))dt + B(t,y(t))dW(t), & t\ge 0,\\ y(0) = y_0 \in H, \end{cases} \tag{3.8.1}$$

could be viewed as the perturbed stochastic system of the deterministic equation

$$\begin{cases} dy(t) = Ay(t)dt + F(t,y(t))dt, & t\ge 0,\\ y(0) = y_0 \in H. \end{cases} \tag{3.8.2}$$

It is known in Remark 3.3.4 that when the solution of (3.8.2) has exponential decay, it is not generally true that its stochastic perturbed system (3.8.1) has the same decay, although the solution could be, sometimes, decayable with a slower rate, e.g., polynomial decay.

3.8.1 Decay in Mean Square

Assume that $\Lambda \in C^{1,2}(\mathbb{R}_+ \times H; \mathbb{R}_+)$, the family of all functions that have continuous first-order partial derivatives in $\mathbb{R}_+$ and twice Fréchet derivatives in H. For the system (3.8.1), recall that we can introduce the operator $\mathcal{L}$ by

$$(\mathcal{L}\Lambda)(t,x) := \Lambda_t'(t,x) + \langle \Lambda_x'(t,x),\, Ax + F(t,x)\rangle_H + \frac{1}{2}\cdot Tr\Big[\Lambda_{xx}''(t,x)B(t,x)Q^{1/2}(B(t,x)Q^{1/2})^*\Big], \quad t \geq 0, \quad x \in \mathscr{D}(A).$$

Definition 3.8.1 Assume that $\lambda\colon \mathbb{R}_+ \to \mathbb{R}_+$ is a continuous function with $\lim_{t\to\infty} \lambda(t) = \infty$. The mild solution $y(t)$, $t \geq 0$, of equation (3.8.1) is said to have $\lambda(\cdot)$ *decay in mean square* if for any $y_0 \in H$, there exists a positive constant $\gamma > 0$ such that

$$\overline{\lim_{t\to\infty}} \frac{\log \mathbb{E}\|y(t,y_0)\|_H^2}{\log \lambda(t)} \leq -\gamma. \tag{3.8.3}$$

Theorem 3.8.2 *Let $\Lambda \in C^{1,2}(\mathbb{R}_+ \times H; \mathbb{R}_+)$ be an Itô functional and $\psi_1(\cdot)$, $\psi_2(\cdot)\colon \mathbb{R}_+ \to \mathbb{R}_+$ be two nonnegative, continuous functions. Assume that there exist a positive constant $m > 0$ and real numbers ν, $\theta \in \mathbb{R}$ such that*

(1) $\|x\|_H^2\lambda(t)^m \leq \Lambda(t,x)$ *for all* $(t,x) \in \mathbb{R}_+ \times H$;
(2) $(\mathcal{L}\Lambda)(t,x) \leq \psi_1(t) + \psi_2(t)\Lambda(t,x)$, $(t,x) \in \mathbb{R}_+ \times \mathscr{D}(A)$;
(3)

$$\overline{\lim_{t\to\infty}} \frac{\log \int_0^t \psi_1(s)ds}{\log \lambda(t)} \leq \nu, \qquad \overline{\lim_{t\to\infty}} \frac{\int_0^t \psi_2(s)ds}{\log \lambda(t)} \leq \theta. \tag{3.8.4}$$

If $\gamma = m - \theta - \nu > 0$, then the mild solution $y(t,y_0)$, $t \geq 0$, of equation (3.8.1) satisfies

$$\overline{\lim_{t\to\infty}} \frac{\log \mathbb{E}\|y(t,y_0)\|_H^2}{\log \lambda(t)} \leq -\gamma. \tag{3.8.5}$$

That is, the mild solution y to (3.8.1) has $\lambda(\cdot)$ decay in mean square.

Proof In order to show (3.8.5), we carry out a Yosida approximating solutions program as we did in Proposition 3.1.5 and apply Itô's formula to Λ and the solutions $y_n(t) \in \mathscr{D}(A)$, $n \in \rho(A)$, of (3.8.1) in the strong sense. Then it follows for $t \geq 0$ that

$$\Lambda(t, y_n(t)) = \Lambda(0, y_n(0)) + \int_0^t (\mathcal{L}_n\Lambda)(s, y_n(s))ds + \int_0^t \langle \Lambda_y'(s, y_n(s)), B(s, y_n(s))dW(s)\rangle_H, \tag{3.8.6}$$

where

$$(\mathcal{L}_n\Lambda)(t,x) = \Lambda_t'(t,x) + \langle \Lambda_x'(t,x),\, Ax + R(n)F(t,x)\rangle_H$$
$$+ \frac{1}{2}Tr\big[\Lambda_{xx}''(t,x)R(n)B(t,x)Q^{1/2}\big(R(n)B(t,x)Q^{1/2}\big)^*\big], \quad x \in \mathscr{D}(A), \tag{3.8.7}$$

$n \in \rho(A)$, the resolvent set of A, and $R(n) = nR(n,A)$, $R(n,A)$ is the resolvent of A. Now, for any fixed $n \in \rho(A)$, there exists a sequence of stopping times τ_m^n such that $\tau_m^n \uparrow \infty$ as $m \to \infty$ almost surely, and for $m \geq 1$,

$$\mathbb{E}\bigg(\int_0^{t\wedge\tau_m^n} \langle \Lambda_y'(s, y_n(s)),\, B(s, y_n(s))dW(s)\rangle_H\bigg) = 0, \quad t \in \mathbb{R}_+.$$

Therefore, by taking expectation on both sides of (3.8.6) and using condition (2), we get that for $t \geq 0$,

$$\begin{aligned}
&\mathbb{E}\Lambda(t\wedge\tau_m^n, y_n(t\wedge\tau_m^n))\\
&\leq \mathbb{E}\Lambda(0, y_n(0)) + \mathbb{E}\int_0^{t\wedge\tau_m^n} \big[\psi_1(s) + \psi_2(s)\Lambda(s, y_n(s))\big]ds\\
&+\mathbb{E}\int_0^{t\wedge\tau_m^n} \Big\{\langle \Lambda_y'(s, y_n(s)), (R(n) - I)F(s, y_n(s))\rangle_H\\
&+\frac{1}{2}Tr\Big[\Lambda_{yy}''(s, y_n(s))R(n)B(s, y_n(s))Q^{1/2}(R(n)B(s, y_n(s))Q^{1/2})^*\\
&- \Lambda_{yy}''(s, y_n(s))B(s, y_n(s))Q^{1/2}(B(s, y_n(s))Q^{1/2})^*\Big]\Big\}ds.
\end{aligned} \tag{3.8.8}$$

First letting $m \to \infty$, then $n \to \infty$ and applying Fatou's lemma and Corollary 1.4.6 to (3.8.8), one can obtain for $t \geq 0$,

$$\mathbb{E}\Lambda(t, y(t)) \leq \mathbb{E}\Lambda(0, y_0) + \int_0^t \big(\psi_1(s) + \psi_2(s)\mathbb{E}\Lambda(s, y(s))\big)ds.$$

Hence, by virtue of Gronwall's lemma we obtain for any $t \geq 0$

$$\mathbb{E}\Lambda(t, y(t)) \leq \bigg[\mathbb{E}\Lambda(0, y_0) + \int_0^t \psi_1(s)ds\bigg]\exp\bigg(\int_0^t \psi_2(s)ds\bigg).$$

Now, for arbitrary $\varepsilon > 0$, in view of condition (3) we have for sufficiently large $t \in \mathbb{R}_+$,

$$\log \mathbb{E}\Lambda(t, y(t)) \leq \log\big[\mathbb{E}\Lambda(0, y_0) + \lambda(t)^{\nu+\varepsilon}\big] + \log\lambda(t)^{\theta+\varepsilon},$$

which, by letting $\varepsilon \to 0$ and using condition (1), immediately implies

$$\overline{\lim_{t\to\infty}}\, \frac{\log \mathbb{E}\|y(t, y_0)\|_H^2}{\log\lambda(t)} \leq \overline{\lim_{t\to\infty}}\, \frac{\log\big[\lambda(t)^{-m}\mathbb{E}\Lambda(t, y(t, y_0))\big]}{\log\lambda(t)} \leq -\big[m - (\nu+\theta)\big] \tag{3.8.9}$$

as desired. The proof is now complete. □

Example 3.8.3 Let $\mu > 0$ be a positive number. Consider the following stochastic differential equation on the Hilbert space H,

$$\begin{cases} dy(t) = Ay(t)dt + (1+t)^{-\mu}\alpha(y(t))dW(t), & t > 0, \\ y(0) = y_0 \in H, \end{cases} \tag{3.8.10}$$

where A generates a C_0-semigroup on H and there exists a constant $\beta > 0$ such that

$$\langle u, Au\rangle_H \le -\beta\|u\|_H^2 \quad \text{for any} \quad u \in \mathscr{D}(A). \tag{3.8.11}$$

Here W is an H-valued Q-Wiener process with $Tr(Q) < \infty$ and $\alpha(\cdot): H \to \mathbb{R}$ is some bounded, Lipschitz continuous function with upper bound $M > 0$.

To proceed further, first note that the mild solution of (3.8.10) has not, in general, exponential decay, as shown in Remark 3.3.4. However, we shall show that when $\mu > 1/2$, the mild solution has polynomial decay.

Indeed, if $\mu > 1/2$, for any fixed $\varepsilon \in (0, \mu - 1/2)$, one can introduce a Lyapunov function on H of the following form:

$$\Lambda(t, u) = (1+t)^{2(\beta+\varepsilon)}\|u\|_H^2, \quad \forall u \in H, \quad t \ge 0.$$

Then a simple computation, in addition to (3.8.11), yields that for any $u \in \mathscr{D}(A)$, $t \ge 0$,

$$\begin{aligned} (\mathcal{L}\Lambda)(t,u) &= 2(\beta+\varepsilon)(1+t)^{2(\beta+\varepsilon)-1}\|u\|_H^2 + 2(1+t)^{2(\beta+\varepsilon)}\langle u, Au\rangle_H \\ &\quad + (1+t)^{2(\beta+\varepsilon)}\alpha(u)^2 Tr(Q)(1+t)^{-2\mu} \\ &\le 2(\beta+\varepsilon)(1+t)^{-1}\Lambda(t,u) - 2\beta(1+t)^{-1}\Lambda(t,u) \\ &\quad + (1+t)^{2(\beta+\varepsilon)}\alpha(u)^2 Tr(Q)(1+t)^{-2\mu} \\ &\le 2\varepsilon(1+t)^{-1}\Lambda(t,u) + (1+t)^{2(\beta+\varepsilon)-2\mu}M^2 Tr(Q). \end{aligned} \tag{3.8.12}$$

Therefore, letting $\lambda(t) = 1 + t$ and employing the same notations as in Theorem 3.8.2, we have

$$\psi_1(t) = (1+t)^{2(\beta+\varepsilon)-2\mu}M^2 Tr(Q) \quad \text{and} \quad \psi_2(t) = 2\varepsilon(1+t)^{-1}, \quad t \ge 0,$$

which yield $\nu = 2(\beta+\varepsilon) - 2\mu + 1$ and $\theta = 2\varepsilon$. Since $m = 2(\beta+\varepsilon)$, we have

$$m - \nu - \theta = 2(\beta+\varepsilon) - 2(\beta+\varepsilon) + 2\mu - 1 - 2\varepsilon = 2\mu - 2\varepsilon - 1.$$

As $\varepsilon \in (0, \mu - 1/2)$, one can always choose a value $\varepsilon > 0$ so small that $m - \nu - \theta > 0$. This fact means that when $\mu > 1/2$, the mild solution of (3.8.10) has polynomial decay in mean square. Moreover,

$$\overline{\lim_{t\to\infty}} \frac{\log \mathbb{E}\|y(t, y_0)\|_H^2}{\log(1+t)} \le -(2\mu - 1).$$

3.8.2 Pathwise Decay

For pathwise decay, the situation is much more delicate. A natural idea is to consider the mean square decay of stochastic systems in the first instance, then it is hoped that this result, probably plus some other conditions, further implies the desired pathwise decay property (cf. Theorems 3.3.1 and 3.3.3). However, it is worth pointing out that this method does not always work. For example, when one deals with exponential decay in Theorem 3.3.3, the convergence of a certain series consisting of exponential decay terms in (3.3.17) should be established. For nonexponential decay, a corresponding series like this could be divergent. In this section, we shall present a different approach to go around this difficulty.

Definition 3.8.4 Assume that $\lambda\colon \mathbb{R}_+ \to \mathbb{R}_+$ is a continuous function with $\lim_{t\to\infty}\lambda(t) = \infty$. The mild solution of equation (3.8.1) is said to have $\lambda(\cdot)$ *pathwise decay or decay in the almost sure sense* if for any $y_0 \in H$, there exists a positive constant $\gamma > 0$ such that

$$\overline{\lim_{t\to\infty}} \frac{\log \|y(t, y_0)\|_H}{\log \lambda(t)} \le -\gamma \qquad a.s.$$

In order to obtain the main results, we first state the following exponential martingale inequality.

Lemma 3.8.5 *Let $T \ge 0$. Assume that $J \in L^2([0,T]\times\Omega, H)$ and $\Phi \in \mathcal{U}^2([0,T]\times\Omega; \mathscr{L}_2(K_Q, H))$. Then for any positive numbers α and β,*

$$\begin{aligned}\mathbb{P}\Bigg\{ \sup_{0\le t\le T}\Bigg[\int_0^t \langle J(s), \Phi(s)dW(s)\rangle_H \\ - \frac{\alpha}{2}\int_0^t Tr\Big(J(s)\otimes J(s)\Phi(s)Q^{1/2}(\Phi(s)Q^{1/2})^*\Big)ds\Bigg] > \beta\Bigg\} \\ \le \exp(-\alpha\beta).\end{aligned}$$

Proof For each integer $n \ge 1$, we define a stopping time τ_n by

$$\tau_n = \begin{cases} \inf\Big\{t \ge 0\colon \Big|\int_0^t \langle J(s), \Phi(s)dW(s)\rangle_H\Big| \\ \qquad + \int_0^t Tr\Big(J(s)\otimes J(s)\Phi(s)Q^{1/2}(\Phi(s)Q^{1/2})^*\Big)ds > n\Big\}, \\ \infty \qquad \text{if the set is empty,}\end{cases}$$

and the associated process

$$x_n(t) = \alpha \int_0^{t\wedge\tau_n} \langle J(s), \Phi(s)dW(s)\rangle_H$$
$$- \frac{\alpha^2}{2}\int_0^{t\wedge\tau_n} Tr\Big(J(s)\otimes J(s)\Phi(s)Q^{1/2}(\Phi(s)Q^{1/2})^*\Big)ds.$$

Clearly, $x_n(t)$ is bounded for each $n \geq 1$ and $\tau_n \uparrow \infty$ almost surely as $n \to \infty$. Applying Itô's formula to $\exp[x_n(t)]$, we obtain

$$\begin{aligned}\exp[x_n(t)] &= 1 + \int_0^t \exp[x_n(s)]dx_n(s)\\ &\quad + \frac{\alpha^2}{2}\int_0^{t\wedge\tau_n} \exp[x_n(s)]Tr\big(J(s)\otimes J(s)\Phi(s)Q^{1/2}(\Phi(s)Q^{1/2})^*\big)ds\\ &= 1 + \alpha\int_0^{t\wedge\tau_n} \exp[x_n(s)]\langle J(s), \Phi(s)dW(s)\rangle_H.\end{aligned} \tag{3.8.13}$$

Therefore, one can easily see that for each $n \geq 1$, $\exp[x_n(t)]$ is a continuous nonnegative martingale over $\mathbb{R}_+$ with $\mathbb{E}\big(\exp[x_n(t)]\big) = 1$, $t \geq 0$. Moreover, by Theorem 1.2.9, it follows that

$$\mathbb{P}\Big\{\sup_{0\leq t\leq T} \exp[x_n(t)] \geq e^{\alpha\beta}\Big\} \leq e^{-\alpha\beta}\mathbb{E}\big(\exp[x_n(T)]\big) = e^{-\alpha\beta}.$$

That is,

$$\mathbb{P}\Big\{\sup_{0\leq t\leq T}\Big[\int_0^{t\wedge\tau_n} \langle J(s), \Phi(s)dW(s)\rangle_H - \frac{\alpha}{2}\int_0^{t\wedge\tau_n} Tr\Big(J(s)\otimes J(s)\Phi(s)Q^{1/2}(\Phi(s)Q^{1/2})^*\Big)ds\Big] > \beta\Big\} \leq \exp(-\alpha\beta). \tag{3.8.14}$$

Now the required result follows by letting $n \to \infty$ and using Fatou's lemma. □

In connection with the equation (3.8.1), we define for each $\Lambda \in C^{1,2}(\mathbb{R}_+ \times H; \mathbb{R}_+)$ the following mapping:

$$(\mathcal{Q}\Lambda)(t,x) = Tr\big[(\Lambda_x'\otimes\Lambda_x')(t,x)B(t,x)Q^{1/2}(B(t,x)Q^{1/2})^*\big], \quad t \geq 0, \quad x \in H. \tag{3.8.15}$$

Moreover, we impose the following condition on the decay function $\lambda(t)$:

(H1) $\log\lambda(t)$ is uniformly continuous on $[t_0, \infty)$ for sufficiently large $t_0 \geq 0$;

(H2) there exists a nonnegative constant $\kappa \geq 0$ such that

$$\overline{\lim_{t\to\infty}} \frac{\log\log t}{\log\lambda(t)} \leq \kappa. \tag{3.8.16}$$

It is easy to see that functions e^t, $1+t$, and $\log(1+t)$, $t \geq 0$, satisfy (H1) and (H2), and correspond respectively to exponential, polynomial, and logarithmic decay.

Theorem 3.8.6 *Let λ be a continuous function satisfying (H1) and (H2). Let $\Lambda \in C^{2,1}(\mathbb{R}_+ \times H; \mathbb{R}_+)$ be an Itô functional and $\psi_1(\cdot)$, $\psi_2(\cdot)$ be two nonnegative continuous functions. Suppose that there exist a positive constant $m > 0$ and numbers ν, $\theta \in \mathbb{R}$ such that*

(1) $\|x\|_H^2\lambda(t)^m \leq \Lambda(t,x)$, $(t,x) \in \mathbb{R}_+ \times H$;
(2) $(\mathcal{L}\Lambda)(t,x) + (\mathcal{Q}\Lambda)(t,x) \leq \psi_1(t) + \psi_2(t)\Lambda(t,x)$, $x \in \mathscr{D}(A)$, $t \in \mathbb{R}_+$;
(3)

$$\overline{\lim_{t\to\infty}} \frac{\log\left(\int_0^t \psi_1(s)ds\right)}{\log\lambda(t)} \leq \nu, \qquad \overline{\lim_{t\to\infty}} \frac{\int_0^t \psi_2(s)ds}{\log\lambda(t)} \leq \theta.$$

If $\gamma = m - \theta - \kappa \vee \nu > 0$ where $\kappa \vee \nu = \max\{\kappa, \nu\}$, then

$$\overline{\lim_{t\to\infty}} \frac{\log\|y(t,y_0)\|_H}{\log\lambda(t)} \leq -\frac{\gamma}{2} \qquad a.s.$$

That is, the mild solution of equation (3.8.1) has $\lambda(\cdot)$ decay in the almost sure sense.

Proof In a similar manner to the proofs of Theorem 3.8.2, we apply Itô's formula to $\Lambda(\cdot,\cdot)$ and the approximating solution $y_n(t)$ of (3.8.1) in the strong sense to obtain that for any $t \geq 0$,

$$\begin{aligned}
\Lambda(t, y_n(t)) &= \Lambda(0, y_n(0)) + \int_0^t (\mathcal{L}_n\Lambda)(s, y_n(s))ds \\
&\quad + \int_0^t \langle \Lambda'_y(s, y_n(s)), R(n)B(s, y_n(s))dW(s)\rangle_H \\
&=: \Lambda(0, y_n(0)) + \int_0^t (\mathcal{L}\Lambda)(s, y_n(s))ds \\
&\quad + \int_0^t \langle \Lambda'_y(s, y(s)), B(s, y(s))dW(s)\rangle_H + I_1(t,n) + I_2(t,n),
\end{aligned} \tag{3.8.17}$$

where

$$I_1(t,n) = \int_0^t (\mathcal{L}_n\Lambda)(s, y_n(s))ds - \int_0^t (\mathcal{L}\Lambda)(s, y_n(s))ds,$$
$$I_2(t,n) = \int_0^t \langle \Lambda'_y(s, y_n(s)), [R(n) - I]B(s, y_n(s))dW(s)\rangle_H, \tag{3.8.18}$$

and

$$(\mathcal{L}_n\Lambda)(t,y) := \Lambda'_t(t,y) + \langle \Lambda'_y(t,y), Ay + R(n)F(t,y)\rangle_H$$
$$+ \frac{1}{2}Tr\Big[\Lambda''_{yy}(t,y)R(n)B(t,y)Q^{1/2}\big(R(n)B(t,y)Q^{1/2}\big)^*\Big], \quad y \in \mathscr{D}(A), \quad t \geq 0. \tag{3.8.19}$$

By virtue of (H1), for any $\varepsilon > 0$ there exist positive integers $N = N(\varepsilon)$ and $k_1 = k_1(\varepsilon)$ such that if $\frac{k-1}{2^N} \leq t \leq \frac{k}{2^N}$, $k \geq k_1(\varepsilon)$, then

$$\left|\log\lambda\left(\frac{k}{2^N}\right) - \log\lambda(t)\right| \leq \varepsilon.$$

On the other hand, in view of Lemma 3.8.5, we have

$$\mathbb{P}\left\{\omega\colon \sup_{0\leq t\leq w}\left[\int_0^t \langle \Lambda'_y(s, y(s)), B(s, y(s))dW(s)\rangle_H - \int_0^t \frac{u}{2}(\mathcal{Q}\Lambda)(s, y(s))ds\right] > v\right\} \leq e^{-uv}$$

for any positive constants u, v, and w. In particular, by putting

$$u = 2, \quad v = \log\left(\frac{k-1}{2^N}\right), \quad w = \frac{k}{2^N}, \quad k = 2, 3, \ldots,$$

we can apply the Borel–Cantelli lemma to obtain a random integer $k_0(\varepsilon,\omega) > 0$ such that

$$\int_0^t \langle \Lambda'_y(s, y(s)), B(s, y(s))dW(s)\rangle_H \leq \log\left(\frac{k-1}{2^N}\right) + \int_0^t (\mathcal{Q}\Lambda)(s, y(s))ds \qquad a.s.$$

for all $0 \leq t \leq \frac{k}{2^N}$, $k(\omega) \geq k_0(\varepsilon,\omega)$. Substituting this into (3.8.17) and using condition (3), we see that for almost all $\omega \in \Omega$,

$$\begin{aligned}\Lambda(t, y_n(t)) \leq\ & \log\left(\frac{k-1}{2^N}\right) + \Lambda(0, y_n(0)) + \int_0^t (\mathcal{L}\Lambda)(s, y_n(s))ds \\ & + \int_0^t (\mathcal{Q}\Lambda)(s, y_n(s))ds + I_1(t,n) + I_2(t,n) + I_3(t,n) \\ \leq\ & \log\left(\frac{k-1}{2^N}\right) + \Lambda(0, y_n(0)) + \int_0^t \big(\psi_1(s) + \psi_2(s)\Lambda(s, y_n(s))\big)ds \\ & + I_1(t,n) + I_2(t,n) + I_3(t,n),\end{aligned} \tag{3.8.20}$$

where

$$I_3(t,n) := \int_0^t \Big((\mathcal{Q}\Lambda)(s, y(s)) - (\mathcal{Q}\Lambda)(s, y_n(s))\Big)ds$$

for all $t \in [0, \frac{k}{2^N}]$, $k \geq k_0(\varepsilon, \omega) \vee k_1(\varepsilon)$. Hence, by using Gronwall's lemma, we obtain

$$\begin{aligned}\Lambda(t, y_n(t)) \leq & \left[\Lambda(0, y_n(0)) + \log\left(\frac{k-1}{2^N}\right) + |I_1(t,n)| + |I_2(t,n)|\right. \\ & \left. + |I_3(t,n)| + \int_0^t \psi_1(s)ds\right] \cdot \exp\left(\int_0^t \psi_2(s)ds\right) \qquad a.s.\end{aligned} \tag{3.8.21}$$

for all $0 \leq t \leq \frac{k}{2^N}$, $k \geq k_0(\varepsilon, \omega) \vee k_1(\varepsilon)$. By virtue of Corollary 1.4.6, there exists a subsequence of $y_n(t)$, still denote it by $y_n(t)$, such that $y_n(t) \to y(t)$ almost surely, as $n \to \infty$ uniformly with respect to $t \in [0, \frac{k}{2^N}]$. Hence, we have that $|I_i(t,n)| \to 0$ almost surely, $i = 1, 2, 3$, as $n \to \infty$ for all $0 \leq t \leq \frac{k}{2^N}$, $k \geq k_0(\varepsilon, \omega) \vee k_1(\varepsilon)$. Hence, by letting $n \to \infty$ in (3.8.21), one can have

$$\begin{aligned}\Lambda(t, y(t)) \leq & \left[\Lambda(0, y_0) + \log\left(\frac{k-1}{2^N}\right)\right. \\ & \left. + \int_0^t \psi_1(s)ds\right] \exp\left(\int_0^t \psi_2(s)ds\right) \qquad a.s.\end{aligned}$$

for all $0 \leq t \leq \frac{k}{2^N}$, $k \geq k_0(\varepsilon, \omega) \vee k_1(\varepsilon)$. By using the conditions (H1) and (3), for the previous $\varepsilon > 0$ there exists a positive integer $k_2(\varepsilon, \omega) > 0$ such that

$$\begin{aligned}\log \Lambda(t, y(t)) \leq\ & \log\left[\Lambda(0, y_0) + \lambda(t)^{(\nu+\varepsilon)} + \log\left(\frac{k-1}{2^N}\right)\right] \\ & + (\theta + \varepsilon)\log \lambda(t) \qquad a.s.\end{aligned}$$

for all $\frac{k-1}{2^N} \leq t \leq \frac{k}{2^N}$, $k \geq k_0(\varepsilon, \omega) \vee k_1(\varepsilon) \vee k_2(\varepsilon, \omega)$, which, letting $\varepsilon \to 0$ and using the conditions (1) and (H2), immediately implies

$$\overline{\lim_{t\to\infty}} \frac{\log \|y(t,y_0)\|_H}{\log \lambda(t)} \le \overline{\lim_{t\to\infty}} \frac{1}{2} \frac{\log\left[\lambda(t)^{-m}\Lambda(t,y(t))\right]}{\log \lambda(t)}$$
$$\le -\frac{m-[\theta+\nu\vee\kappa]}{2} \qquad a.s.$$

as required. The proof is complete. □

Example 3.8.7 (Example 3.8.3 revisited) As an immediate consequence, we may apply Theorem 3.8.6 to the equation (3.8.10) in Example 3.8.3 to obtain that if $\mu > \frac{1}{2} + \beta$, the solution of (3.8.10) has polynomial decay in the almost sure sense.

Indeed, we can introduce a Lyapunov function of the form

$$\Lambda(t,u) = (1+t)^{2\beta}\|u\|_H^2, \quad \forall u \in H, \quad t \ge 0.$$

Then a similar computation to (3.8.12) yields that for any $u \in \mathscr{D}(A)$, $t \ge 0$,

$$(\mathcal{L}\Lambda)(t,u) + (\mathcal{Q}\Lambda)(t,u) \le (1+t)^{2\beta-2\mu}M^2Tr(Q)$$
$$+ 4M^2Tr(Q)(1+t)^{-2\mu+2\beta}\Lambda(t,u).$$

Therefore, by letting $\lambda(t) = 1+t$ and employing the same notations as in Theorem 3.8.6, we have

$$\psi_1(t) = (1+t)^{2\beta-2\mu}M^2Tr(Q) \quad \text{and} \quad \psi_2(t) = 4M^2Tr(Q)(1+t)^{-2\mu+2\beta},$$

which yield $\nu = 2\beta - 2\mu + 1 < 0$ and $\theta = 0$. It is clear that $\kappa = 0$ in (3.8.16). Since $m = 2\beta$, we have

$$m - \theta - \nu \vee \kappa = 2\beta.$$

This means that the mild solution y of (3.8.10) has polynomial decay almost surely. Moreover,

$$\overline{\lim_{t\to\infty}} \frac{\log \|y(t,y_0)\|_H}{\log(1+t)} \le -\beta \qquad a.s.$$

3.9 Stabilization of Systems by Noise

Let $r_0 \in \mathbb{R}$ and consider a deterministic system

$$\begin{cases} \dfrac{\partial y(t,x)}{\partial t} = \dfrac{\partial^2 y(t,x)}{\partial x^2} + r_0 y(t,x), \quad t \ge 0, \quad 0 \le x \le \pi, \\ y(t,0) = y(t,\pi) = 0, \quad t \ge 0, \\ y(0,x) = y_0(x), \quad 0 \le x \le \pi. \end{cases} \tag{3.9.1}$$

Recall that (see Example 2.5.1) this equation has an explicit solution

$$y(t,x) = \sum_{n=1}^{\infty} a_n e^{(r_0-n^2)t} \sin nx, \qquad t \geq 0, \qquad x \in [0,\pi],$$

where $\{a_n\}_{n=1}^{\infty}$ is a sequence satisfying $\sum_{n=1}^{\infty} a_n^2 < \infty$. Therefore, it is immediate to obtain exponential stability when $r_0 < 1$, while for $r_0 \geq 1$ the null solution is generally unstable.

When $r_0 \geq 1$, we can add, however, a noise term to (3.9.1) to obtain a stochastic perturbed partial differential equation

$$\begin{cases} dy(t,x) = \Big(\dfrac{\partial^2}{\partial x^2} + r_0\Big) y(t,x)dt + r_1 y(t,x)dw(t), \\ \qquad\qquad t \geq 0, \quad 0 \leq x \leq \pi, \\ y(t,0) = y(t,\pi) = 0, \quad t \geq 0; \quad y(0,x) = y_0(x), \quad 0 \leq x \leq \pi, \end{cases} \tag{3.9.2}$$

where $r_1 \in \mathbb{R}$ and $w(t)$ is a standard one-dimensional Brownian motion. Because a new noise term is introduced into the stochastic system (3.9.2), a natural question is: when $r_0 \geq 1$, is it possible to obtain any stability results for the stochastic system (3.9.2)? More precisely, is it possible to stabilize the deterministic system (3.9.1) by choosing a proper noise source such as that one in (3.9.2)? The answer is affirmative (see Example 3.9.4) as shown in this section.

3.9.1 Linear and Commutative Case

Consider the following deterministic equation in a Hilbert space H,

$$\begin{cases} dy(t) = Ay(t)dt, \qquad t \geq 0, \\ y(0) = y_0 \in H. \end{cases} \tag{3.9.3}$$

We shall prove that as long as A generates a C_0-semigroup, this system can be stabilized by appropriate noise sources. For arbitrary $y_0 \in H$ and the corresponding mild solution y of (3.9.3), we define the so-called Lyapunov exponent of (3.9.3) by

$$\lambda_d(y_0) = \overline{\lim_{t\to\infty}} \frac{\log \|y(t,y_0)\|_H}{t}.$$

Lemma 3.9.1 *Suppose that operator A generates a C_0-semigroup e^{tA}, $t \geq 0$, on H, then for any initial value $y_0 \in H$,*

$$\lambda_d(y_0) \leq \lambda < \infty,$$

for some $\lambda \in \mathbb{R}$ that is independent of $y_0 \in H$.

Proof This is deduced from the fact that $y(t, y_0) = e^{tA}y_0$, $t \geq 0$, is the unique mild solution of (3.9.3) and for the C_0-semigroup e^{tA}, $t \geq 0$, there exist constants $\lambda \in \mathbb{R}$ and $M \geq 1$ such that

$$\|e^{tA}\| \leq M \cdot \exp(\lambda t), \qquad \forall t \geq 0.$$

The proof is thus complete. □

Now let us consider the following stochastic differential equation with multiplicative noise

$$\begin{cases} dy(t) = Ay(t)dt + By(t)dw(t), & t \geq 0, \\ y(0) = y_0 \in H, \end{cases} \tag{3.9.4}$$

where $A\colon \mathscr{D}(A) \subset H \to H$ generates a C_0-semigroup e^{tA}, $t \geq 0$, $B \in \mathscr{L}(H)$ and $w(t)$, $t \geq 0$, is a standard real Brownian motion. Assume further that e^{tA} and B commute for any $t \geq 0$. In this case, there exists a unique mild solution to the problem (3.9.4) and the solution is explicitly given by

$$y(t, y_0) = \exp\left(w(t)B - \frac{1}{2}tB^2\right)e^{tA}y_0, \qquad t \geq 0. \tag{3.9.5}$$

For arbitrary $y_0 \in H$, the random Lyapunov exponent of the system (3.9.4) is similarly defined as

$$\lambda_s(y_0)(\omega) = \overline{\lim_{t\to\infty}} \frac{\log \|y(t, \omega, y_0)\|_H}{t} \qquad a.s.,$$

where $y(t, \omega, y_0)$ is the mild solution of (3.9.4) with initial $y_0 \in H$. In particular, from Lemma 3.9.1 and the next theorem we have that $\lambda_s(y_0)(\omega) < \infty$ almost surely.

Theorem 3.9.2 *Let $y_0 \in H$ and $y(t, y_0)$, $t \geq 0$, be the mild solution of (3.9.4). Then the following inequality holds:*

$$\lambda_s(y_0)(\omega) \leq \lambda_d(y_0) + \alpha \qquad a.s.,$$

where $\lambda_d(y_0)$ is the Lyapunov exponent of (3.9.3) and

$$\alpha = \overline{\lim_{t\to\infty}} \frac{1}{t} \log \left\| \exp[-(1/2)tB^2] \right\| < \infty.$$

Proof We have by virtue of (3.9.5) that

$$\begin{aligned}
\lambda_s(y_0)(\omega) &= \overline{\lim_{t\to\infty}} \frac{1}{t} \log \left\| \exp\left(w(t)B - \frac{1}{2}tB^2\right)e^{tA}y_0\right\|_H \\
&\le \overline{\lim_{t\to\infty}} \frac{1}{t} \log \| \exp[(-1/2)tB^2]\| \, \| \exp(w(t)B)\| \\
&\quad + \overline{\lim_{t\to\infty}} \frac{1}{t} \log \|e^{tA}y_0\|_H \\
&\le \lambda_d(y_0) + \alpha + \overline{\lim_{t\to\infty}} \frac{1}{t} \log \| \exp(w(t)B)\|.
\end{aligned} \tag{3.9.6}$$

By virtue of the well-known strong law of large numbers, we have that $\overline{\lim}_{t\to\infty} |w(t)|/t = 0$ almost surely. This implies that

$$\overline{\lim_{t\to\infty}} \frac{1}{t} \log \| \exp(w(t)B)\| \le \overline{\lim_{t\to\infty}} \frac{1}{t} \log \exp(\|B\||w(t)|) = 0 \qquad a.s. \tag{3.9.7}$$

Therefore, by substituting (3.9.7) into (3.9.6), we obtain immediately the desired result. □

Generally speaking, the commutative requirement between B and e^{tA} is rather restrictive. However, this limitation could be less serious when one deals with stochastic stabilization problems by seeking simple noise sources to stabilize unstable systems. For example, we can consider the following stochastic system in H

$$\begin{cases} dy(t) = Ay(t)dt + by(t)dw(t), & t \ge 0, \\ y(0) = y_0 \in H, \end{cases} \tag{3.9.8}$$

where $b \in \mathbb{R}$. A similar argument to Theorem 3.9.2 can be applied to this system to yield the following.

Corollary 3.9.3 *Let $y_0 \in H$ and $y(t, y_0)$, $t \ge 0$, be the mild solution of (3.9.8). Then the following inequality holds:*

$$\lambda_s(y_0)(\omega) \le \lambda_d(y_0) - \frac{b^2}{2} \qquad a.s.$$

In particular, the deterministic system (3.9.3) can be always stabilized by noise when b is sufficiently large.

Example 3.9.4 Let us apply Corollary 3.9.3 to the linear system described at the beginning of this section. Indeed, we may formulate this problem by setting $H = L^2([0, \pi])$, $Ay = \partial^2 y/\partial x^2 + r_0 y$, and $By = r_1 y$. It is obvious that this equation has a unique mild solution for any $y_0 \in H$ and it is easy to see that $\lambda_d(y_0) \le r_0 - 1$. Thus, it follows that the mild solution of (3.9.2) satisfies

$$\overline{\lim_{t\to\infty}} \frac{\log \|y(t,y_0)\|_H}{t} \le -\left[\frac{r_1^2}{2} - (r_0 - 1)\right] \qquad a.s.$$

for any $y_0 \in H$. That is, we obtain pathwise exponential stability with probability one if $r_1^2 > 2(r_0 - 1)$. In particular, for any $r_0 \in \mathbb{R}$, the preceding result shows that it is always possible to find an appropriate multiplicative noise source to stabilize the unperturbed system (3.9.1).

3.9.2 Nonlinear and Noncommutative Case

We shall employ a variational approach to consider the strong solution of the following deterministic nonlinear equation in V^*,

$$\begin{cases} dy(t) = A(t, y(t))dt, & t \ge 0, \\ y(0) = y_0 \in H, \end{cases} \tag{3.9.9}$$

where $A(t,\cdot)\colon V \to V^*$ with $A(t,0) = 0$, $t \in \mathbb{R}_+$, is a measurable mapping satisfying the following:

(i) there exist a local integrable function ν and a real number $\nu_0 \in \mathbb{R}$ such that

$$2\langle\langle u, A(t,u)\rangle\rangle_{V,V^*} \le \nu(t)\|u\|_H^2 \quad \text{for all} \quad t \ge 0, \quad u \in V, \tag{3.9.10}$$

and

$$\overline{\lim_{t\to\infty}} \frac{1}{t}\int_0^t \nu(s)\,ds \le \nu_0.$$

Now we are concerned with the following question. If the system (3.9.9) is not exponentially stable, is it possible to stabilize it by using a stochastic perturbation, for example, a noise of the form $B(t, y(t))\dot{w}(t)$? Here $B(t,\cdot)\colon H \to H$ with $B(t,0) = 0$ is a measurable mapping, not necessarily commutative with A, but satisfying the following condition:

(ii) there exist a local integrable function λ and a number $\lambda_0 \ge 0$ such that

$$\|B(t,u) - B(t,v)\|_H^2 \le \lambda(t)\|u - v\|_H^2, \quad t \in \mathbb{R}_+, \quad u,\ v \in V,$$

and

$$\overline{\lim_{t\to\infty}} \frac{1}{t}\int_0^t \lambda(s)ds \le \lambda_0.$$

The answer to this question is affirmative if we choose an appropriate $B(\cdot,\cdot)$ satisfying (ii). In other words, we intend to consider the following stochastic perturbed differential equation on H,

$$\begin{cases} dy(t) = A(t, y(t))dt + B(t, y(t))dw(t), & t \geq 0, \\ y(0) = y_0 \in H. \end{cases} \tag{3.9.11}$$

Theorem 3.9.5 *Let $y_0 \neq 0$ and $y(t, y_0)$ be a strong solution of (3.9.11). In addition to the conditions (i) and (ii), assume further that*

$$\langle B(t,u), u\rangle_H^2 \geq \rho(t)\|u\|_H^4 \quad \textit{for all} \quad t \in \mathbb{R}_+, \quad u \in H,$$

where $\rho(t)$, $t \geq 0$, is a local integrable function such that

$$\varliminf_{t\to\infty} \frac{1}{t}\int_0^t \rho(s)\,ds \geq \rho_0 \quad \textit{for some} \quad \rho_0 \in \mathbb{R}.$$

Then, the inequality

$$\varlimsup_{t\to\infty} \frac{1}{t}\log\|y(t, y_0)\|_H \leq -\left(\rho_0 - \frac{\nu_0 + \lambda_0}{2}\right)$$

holds almost surely. In particular, if $2\rho_0 > \nu_0 + \lambda_0$ the null solution of equation (3.9.11) is almost surely exponentially stable.

Proof Since $y_0 \neq 0$, it is easy to see, by the uniqueness of solutions, that $y(t, y_0) \neq 0$ almost surely. Then we apply Itô's formula to function $\log\|\cdot\|_H^2$ to obtain

$$\begin{aligned} \log\|y(t)\|_H^2 \leq \log\|y_0\|_H^2 &+ \int_0^t \frac{2}{\|y(s)\|_H^2}\langle y(s), B(s, y(s))\rangle_H dw(s) \\ &+ \int_0^t \left(\frac{2\langle\langle y(s), A(s, y(s))\rangle\rangle_{V,V^*} + \|B(s, y(s))\|_H^2}{\|y(s)\|_H^2}\right. \\ &\left. - \frac{2\langle y(s), B(s, y(s))\rangle_H^2}{\|y(s)\|_H^4}\right)ds. \end{aligned} \tag{3.9.12}$$

On the other hand, by using Lemma 3.8.5, we obtain

$$\begin{aligned} \mathbb{P}\Bigg\{\omega\colon \sup_{0\leq t\leq w}\Bigg[&\int_0^t \frac{2}{\|y(s)\|_H^2}\langle y(s), B(s, y(s))\rangle_H dw(s) \\ &- \int_0^t \frac{2u}{\|y(s)\|_H^4}\langle y(s), B(s, y(s))\rangle_H^2 ds\Bigg] > v\Bigg\} \leq e^{-uv} \end{aligned} \tag{3.9.13}$$

for any positive constants u, v, and w. For arbitrarily given $\varepsilon > 0$, let

$$u = \delta < 1, \quad v = 2\delta^{-1}\log k, \quad w = k\varepsilon, \qquad k = 1, 2, 3, \ldots,$$

in (3.9.13), then one can apply the well-known Borel–Cantelli lemma to show that there exists an integer $k_0(\varepsilon, \omega) > 0$ for almost all $\omega \in \Omega$ such that

$$\int_0^t \frac{2}{\|y(s)\|_H^2} \langle y(s), B(s, y(s))\rangle_H dw(s)$$
$$\leq 2\delta^{-1} \log k + \delta \int_0^t \frac{2}{\|y(s)\|_H^4} \langle y(s), B(s, y(s))\rangle_H^2 ds$$

for all $0 \leq t \leq k\varepsilon$, $k \geq k_0(\varepsilon, \omega)$. Substituting this into (3.9.12) and using the conditions in this theorem, we see that for the preceding $\varepsilon > 0$, there exists a positive integer $k_1(\varepsilon) > 0$ such that

$$\begin{aligned}\frac{\log \|y(t)\|_H}{t} &\leq \frac{1}{2t}\bigg(\log \|y_0\|_H^2 + 2\delta^{-1} \log k \\ &\qquad + \int_0^t (\nu(s) + \lambda(s))ds - 2(1-\delta) \int_0^t \rho(s)ds \bigg) \\ &\leq \frac{1}{2t}\Big(\log \|y_0\|_H^2 + 2\delta^{-1} \log k + \big(\nu_0 + \lambda_0 + \varepsilon\big)t \\ &\qquad - 2(1-\delta)\big(\rho_0 + \varepsilon\big)t \Big) \qquad a.s.\end{aligned}$$

for all $(k-1)\varepsilon \leq t \leq k\varepsilon$, $k \geq k_0(\varepsilon, \omega) \vee k_1(\varepsilon)$, which immediately implies

$$\overline{\lim_{t\to\infty}} \frac{\log \|y(t)\|_H}{t} \leq \frac{1}{2}\big[(\nu_0 + \lambda_0 + \varepsilon) - 2(1-\delta)(\rho_0 + \varepsilon)\big] \qquad a.s.$$

Therefore, by letting $\varepsilon \to 0$ and $\delta \to 0$, we obtain

$$\overline{\lim_{t\to\infty}} \frac{\log \|y(t, y_0)\|_H}{t} \leq -\left(\rho_0 - \frac{\nu_0 + \lambda_0}{2}\right) \qquad a.s.$$

as desired. The proof is thus complete. □

Example 3.9.6 Let $\mathcal{O}$ be an open, bounded subset in $\mathbb{R}^n$ with regular boundary and $p > 2$ be an integer. Consider a Sobolev space $V = W_0^{1,p}(\mathcal{O})$, $H = L^2(\mathcal{O})$ with its usual inner product, and the monotone operator $A\colon V \to V^*$ defined by

$$\begin{aligned}\langle\!\langle v, A(u)\rangle\!\rangle_{V,V^*} = &-\sum_{i=1}^n \int_{\mathcal{O}} \left|\frac{\partial u(x)}{\partial x_i}\right|^{p-2} \frac{\partial u(x)}{\partial x_i}\frac{\partial v(x)}{\partial x_i} dx \\ &+ a\int_{\mathcal{O}} u(x)v(x)dx, \quad \forall u,\, v \in V,\end{aligned} \tag{3.9.14}$$

where $a \in \mathbb{R}$. We also put $B(t, u) = bu$, $u \in H$, where $b \in \mathbb{R}$ and $w(t)$ is a standard real Brownian motion.

In this case, it is easy to see that the condition (A) in Section 3.3 becomes

$$2\langle\langle u, A(t,u)\rangle\rangle_{V,V^*} + \|B(t,u)\|^2 \le -2\|u\|_V^p + 2a\|u\|_H^2 + b^2\|u\|_H^2, \quad u \in V.$$

The condition (ii) of Theorem 3.3.1 requires $2a + b^2 < 0$, i.e., $a < 0$ and $b^2 < -2a$. Therefore, Theorem 3.3.1 guarantees almost surely exponential stability of the null solution only for those values of a and b for which the deterministic system $dy(t) = A(y(t))dt$ is exponentially stable and the random perturbation is small enough. However, it is possible to show that Theorem 3.9.5 is applicable to this example to secure exponential stability of the null solution for sufficiently large perturbations when the corresponding deterministic system is unstable. Indeed, in this case, it is not difficult to see that

$$2\langle\langle u, A(u)\rangle\rangle_{V,V^*} = -2\|u\|_V^p + 2a\|u\|_H^2 \le 2a\|u\|_H^2, \qquad u \in V.$$

Therefore, by virtue of Theorem 3.9.5 it immediately follows that the strong solution y satisfies

$$\overline{\lim_{t\to\infty}} \frac{1}{t} \log \|y(t, y_0)\|_H \le -(b^2/2 - a)$$

for any $y_0 \neq 0$. Consequently, we get the almost surely exponential stability if $b^2 > 2a$.

3.10 Notes and Comments

The main results in Section 3.1 are taken from Ichikawa [94]. A Lyapunov function of the form (3.1.3) appeared in the work of Pritchard and Zabczyk [191] associated with a stability analysis of linear deterministic hyperbolic equations. This method was also applied to stochastic beam and hyperbolic equations in Brzeźniak et al [21] and Maslowski et al [170]. The comparison approach for stability of finite-dimensional differential equations in Section 3.2 has been extensively studied since it appears in some feedback control systems (cf. Hahn [82]). The systems (3.2.7) and (3.2.8) are often called stochastic differential equations of Lue's type. In finite-dimensional cases, Morozan [179] showed that the system (3.2.8) is absolutely stable in the class Σ_α given in Section 3.2 and that the null solution is exponentially stable in mean square for each $g \in \Sigma_\alpha$ if and only if the linear equation (3.2.7) is stable. The main material in Section 3.2 is due to Ichikawa [92].

After the work of Pardoux [184], who established some fundamental results on the existence and uniqueness of strong solutions for stochastic nonlinear partial differential equations of monotone type, Chow [40] showed that under some circumstances a coercive type of condition can play the role of a stability criterion. The main results in Section 3.3, which are due to Caraballo and Liu [31], improved and generalized the corresponding results in Chow [40] to a nonautonomous situation. Stochastic stability of partial differential equations by using finite-dimensional approximations was considered in Yavin [222, 223].

In the history of (asymptotic) stability of stochastic systems, the Lyapunov function method is probably the most influential one (cf. Khas'minskii [103]). The basic technique is to construct, as the first step, a proper Lyapunov function and then deal with the stability property of nonlinear systems by means of a first-order approximation scheme. In infinite-dimensional cases, the initial research in this direction was done by Khas'minskii and Mandrekar [105]. The general nonautonomous version presented in Theorem 3.5.2 is taken from Liu [134]. The relation between ultimate boundedness in the mean square sense and invariant measures of stochastic differential equations has been pointed out by Miyahara [173, 174] in finite dimensions. But the basic idea goes back at least to Wonham [218] and Zakai [230, 231]. The corresponding investigation in the infinite-dimensional case was carried out in Liu and Mandrekar [153] by a variational method and Liu and Mandrekar [154] for semilinear systems. The generalizations in Theorem 3.7.3 and Theorem 3.7.6 of finite-dimensional systems are presented in Chow and Khas'minskii [42]. A closely related topic to ultimate boundedness is stability in distribution of stochastic systems, which is frequently based on a tight set analysis of some transition probabilities. In an infinite-dimensional space, this method might fail when one tries to locate a compact set, although this difficulty could be overcome somehow based on the fact that any bounded set in a Hilbert space has compact closure in the weak topology.

There is a rich theory on invariant measures and related topics of SPDEs in the existing literature. A systematic statement in this respect can be found in Da Prato and Zabczyk [52, 53]. In this chapter, our main intention is to show the power of a Lyapunov function program in dealing with this topic. In addition, it is worth mentioning the work by Hairer and Mattingly [83], whose treatment of the uniqueness of invariant measures for equations with degenerate noise is enlightening. The reader can also find an interesting method in Ethier and Kurtz [71] to deal with the existence of invariant measures by constructing a Lyapunov function and solving some relevant martingale problem.

The research about decay rates of finite-dimensional Itô's stochastic differential equations goes back at least to Mao [164], in which polynomial decay was investigated. The material in Section 3.8 is based on Liu [131]. One of the most important topics in stability theory is the so-called stabilization by noise sources of deterministic (stochastic) systems. In this respect, there is a systematic statement in Khas'minskii [103] for finite-dimensional systems. In Duncan et al. [67], stabilization by a fractional Brownian motion was considered somehow. It turns out that fractional white noise indexed by various Hurst parameters can stabilize or destabilize the stochastic systems under investigation. The presentation in Section 3.9 is mainly taken from Caraballo et al. [33]. On the other hand, some research works explored the stabilization problem by noise in a somewhat different but interesting way. For instance, although it is plausible that the Lyapunov exponent of the deterministic equation (3.9.3) is quite different from its random versions of the perturbed Itô equation (3.9.4), it may be shown, however, that this is not the case if we focus on the following Stratonovich version of (3.9.4):

$$\begin{cases} dy(t) = Ay(t)dt + By(t) \circ dw(t), & t \geq 0, \\ y(0) = y_0 \in H. \end{cases} \tag{3.10.1}$$

It was shown in Caraballo and Langa [28] that the null solution of equation (3.9.3) is exponentially stable if and only if the null solution of (3.10.1) is exponentially stable in the almost sure sense. In finite-dimensional cases, Arnold [5] proved that the deterministic system (3.9.3) can be stabilized by a suitable Stratonovich linear noise as in (3.10.1) if and only if $Tr(A) < 0$. An infinite-dimensional generalization of this result was considered in Caraballo and Robinson [37]. Last, we like to mention some works on stabilization of finite-dimensional systems, Arnold et al. [6], Mao [165] and Scheutzow [200], among others.

4
Stability of Stochastic Functional Differential Equations

The aim of this chapter is to investigate stability of stochastic retarded functional differential equations in infinite dimensions. To make this material easily accessible to various researchers, we present a theory focusing on bounded delay operators. We begin our statement by extending the standard C_0-semigroup concept to establish a time-delay version, i.e., (retarded) Green operator. The stability of linear systems driven by an additive and multiplicative noise is handled through Green operators and lift-up solution methods. For nonlinear stochastic systems, we first introduce a coercive condition approach to capture the exponential decay of strong solutions. Afterward, Lyapunov function, fixed-point principle, Razumikhin function, and comparison methods are introduced to study various stability properties of these systems.

4.1 Deterministic Systems

Let $r > 0$ and X be a Banach space with norm $\|\cdot\|_X$. Let $L_r^2 = L^2([-r,0],X)$ and define a product Banach space $\mathcal{X} = X \times L_r^2$, equipped with its Banach space norm

$$\|\phi\|_{\mathcal{X}} := \left(\|\phi_0\|_X^2 + \int_{-r}^{0} \|\phi_1(\theta)\|_X^2 d\theta\right)^{1/2} \quad \text{for any} \quad \phi = (\phi_0,\phi_1) \in \mathcal{X}.$$

If $X = H$, a Hilbert space, then $\mathcal{X}$ is a Hilbert space whose inner product is given by

$$\langle \phi,\psi\rangle_{\mathcal{X}} := \langle \phi_0,\psi_0\rangle_H + \int_{-r}^{0} \langle \phi_1(\theta),\psi_1(\theta)\rangle_H d\theta \quad \text{for any} \quad \phi = (\phi_0,\phi_1),\ \psi = (\psi_0,\psi_1) \in \mathcal{X}.$$

Consider the following Cauchy problem in X,

$$\begin{cases} dy(t) = Ay(t)dt + \int_{-r}^{0} d\eta(\theta)y(t+\theta)dt, & t \geq 0, \\ y(0) = \phi_0, \ y(\theta) = \phi_1(\theta), \ \theta \in [-r,0], & \phi = (\phi_0,\phi_1) \in \mathcal{X}, \end{cases} \tag{4.1.1}$$

or its integral form,

$$\begin{cases} y(t) = e^{tA}\phi_0 + \int_0^t e^{(t-s)A} \int_{-r}^{0} d\eta(\theta)y(s+\theta)ds, & t \geq 0, \\ y(0) = \phi_0, \ y(\theta) = \phi_1(\theta), \ \theta \in [-r,0], & \phi = (\phi_0,\phi_1) \in \mathcal{X}. \end{cases} \tag{4.1.2}$$

Here $A\colon \mathscr{D}(A) \subset X \to X$ generates a C_0-semigroup e^{tA}, $t \geq 0$, in X and the Stieltjes delay measure η is given by

$$\int_{-r}^{0} d\eta(\theta)\varphi(\theta) = A_1\varphi(-r) + \int_{-r}^{0} A_2(\theta)\varphi(\theta)d\theta, \qquad \forall \varphi \in C([-r,0];X),$$

where $A_1 \in \mathscr{L}(X)$ and $A_2(\cdot) \in L^2([-r,0];\mathscr{L}(X))$.

4.1.1 Green Operator and Semigroup

Definition 4.1.1 A family $\{G(t)\}_{t\in\mathbb{R}}$ of $\mathscr{L}(X)$ is called a *(retarded) Green operator* or *fundamental solution* of (4.1.1) if

(i) for any $x \in X$, $G(t)x$ is continuous on $\mathbb{R}_+$ and $G(0) = I$, $G(\theta) = 0$ for $\theta \in [-r,0)$;
(ii) $G(t)\mathscr{D}(A) \subset \mathscr{D}(A)$ for each $t \geq 0$, and for any $x \in \mathscr{D}(A)$, $AG(t)x$ is continuous and $G(t)x$ is continuously differentiable on $\mathbb{R}_+$;
(iii) for all $x \in \mathscr{D}(A)$ and $t \geq 0$,

$$\begin{aligned} \frac{dG(t)x}{dt} &= AG(t)x + A_1G(t-r)x + \int_{-r}^{0} A_2(\theta)G(t+\theta)xd\theta \\ &= G(t)Ax + G(t-r)A_1x + \int_{-r}^{0} G(t+\theta)A_2(\theta)xd\theta \qquad a.e. \end{aligned} \tag{4.1.3}$$

Theorem 4.1.2 *For the equation (4.1.1), it has at most one fundamental solution.*

Proof Suppose that $G_1(t)$ and $G_2(t)$, $t \in \mathbb{R}$, are two fundamental solutions. Let $x \in \mathscr{D}(A)$, then one finds by using (4.1.3) that for any $t \geq 0$,

$$\begin{aligned}
G_1(t)x - G_2(t)x &= \int_0^t \frac{d}{ds}\{G_2(t-s)G_1(s)\}xds \\
&= \int_0^t \{G_2(t-s)A_1G_1(s-r)x - G_2(t-s-r)A_1G_1(s)x\}ds \\
&\quad + \int_0^t \Big\{G_2(t-s)\int_{-r}^0 A_2(\theta)G_1(s+\theta)xd\theta \\
&\qquad - \int_{-r}^0 G_2(t-s+\theta)A_2(\theta)G_1(s)xd\theta\Big\}ds \\
&= \int_0^t \Big\{G_2(t-s)\int_{-r}^0 A_2(\theta)G_1(s+\theta)xd\theta \\
&\qquad - \int_{-r}^0 G_2(t-s+\theta)A_2(\theta)G_1(s)xd\theta\Big\}ds.
\end{aligned} \tag{4.1.4}$$

On the other hand, by definition we have for any $x \in \mathscr{D}(A)$ and $t \geq 0$ that

$$\int_0^t G_2(t-s)A_1G_1(s-r)xds = \int_0^t G_2(t-s-r)A_1G_1(s)xds, \tag{4.1.5}$$

and since $G_2(t-s)A_2(\theta)G_1(s+\theta)x$ is integrable on $[0,t] \times [-r,0]$ for each $t \geq 0$, we have by using Fubini's theorem that for each $t \geq 0$,

$$\begin{aligned}
&\int_0^t \int_{-r}^0 G_2(t-s)A_2(\theta)G_1(s+\theta)xd\theta ds \\
&= \int_0^t \int_{-r}^0 G_2(t-s+\theta)A_2(\theta)G_1(s)xd\theta ds.
\end{aligned} \tag{4.1.6}$$

Hence, (4.1.4), (4.1.5), and (4.1.6) together imply that $G_1(t)x = G_2(t)x$ for any $t \geq 0$ and $x \in \mathscr{D}(A)$. Since $\mathscr{D}(A)$ is dense in X and $G_1(t)$, $G_2(t)$ are bounded for each $t \geq 0$, we thus obtain $G_1(t) = G_2(t)$ for each $t \geq 0$. The proof is complete. □

Definition 4.1.3 Let $T \geq 0$. A function $y\colon [-r,T] \to X$ is called a *classical solution* of (4.1.1) if $y(t)$ is continuously differentiable on $(0,T)$, $y(t) \in \mathscr{D}(A)$, $Ay(t)$ is continuous on all $t \in [0,T]$, and equation (4.1.1) holds on $[-r,T]$.

Theorem 4.1.4 *Suppose that equation (4.1.1) admits a fundamental solution $G(t)$, $t \in \mathbb{R}$. Then for each $\phi = (\phi_0, \phi_1) \in \mathcal{X}$, the initial value problem (4.1.1) has at most one classical solution. In particular, if (4.1.1) has a classical solution, this solution is given by*

$$y(t) = G(t)\phi_0 + \int_{-r}^{0} U(t,\theta)\phi_1(\theta)d\theta, \qquad t \in [0,T], \tag{4.1.7}$$

where

$$U(t,\theta) = G(t-\theta-r)A_1 + \int_{-r}^{\theta} G(t-\theta+s)A_2(s)ds, \qquad t \geq 0, \qquad \theta \in [-r,0].$$

Proof Let y be a classical solution of (4.1.1), then the function $x(s) = G(t-s)y(s)$ is differentiable for $0 < s < t < T$ and

$$\begin{aligned}\frac{dx(s)}{ds} &= -G(t-s)Ay(s) - \int_{-r}^{0} G(t-s+\theta)d\eta(\theta)y(s) + G(t-s)Ay(s) \\ &\quad + \int_{-r}^{0} G(t-s)d\eta(\theta)y(s+\theta) \\ &= -\int_{-r}^{0} G(t-s+\theta)d\eta(\theta)y(s) + \int_{-r}^{0} G(t-s)d\eta(\theta)y(s+\theta).\end{aligned} \tag{4.1.8}$$

By integrating both sides of (4.1.8) from 0 to t, we obtain that for $t \in [0,T]$,

$$\begin{aligned}y(t) = G(t)\phi_0 + \int_0^t \Big[&-\int_{-r}^{0} G(t-s+\theta)d\eta(\theta)y(s) \\ &+ \int_{-r}^{0} G(t-s)d\eta(\theta)y(s+\theta)\Big]ds.\end{aligned} \tag{4.1.9}$$

On the other hand, by using Fubini's theorem and noting $G(\theta) = 0$ for $\theta < 0$, we easily get for $t \in [0,T]$ that

$$\begin{aligned}&\int_0^t \Big[-\int_{-r}^{0} G(t-s+\theta)d\eta(\theta)y(s) + \int_{-r}^{0} G(t-s)d\eta(\theta)y(s+\theta)\Big]ds \\ &= -\int_{-r}^{0}\int_0^t G(t-s+\theta)dsd\eta(\theta)y(s) \\ &\quad + \int_{-r}^{0}\int_{\theta}^{t+\theta} G(t-s+\theta)dsd\eta(\theta)y(s) \\ &= \int_{-r}^{0}\int_{\theta}^{0} G(t-s+\theta)dsd\eta(\theta)y(s).\end{aligned} \tag{4.1.10}$$

Using the initial condition in (4.1.1) and Fubini's theorem, one can immediately see

$$\int_{-r}^{0}\int_{-r}^{s} G(t-s+\theta)d\eta(\theta)\phi_1(s)ds = \int_{-r}^{0} U(t,s)\phi_1(s)ds, \tag{4.1.11}$$

which, together with (4.1.9) and (4.1.10), immediately yields the desired form (4.1.7). □

The function y given by (4.1.7) is called a *mild solution* of (4.1.1). Next, we justify the existence of Green operator G. The relation (4.1.7) suggests that if there exists a unique mild solution y of (4.1.1), then the Green operator G should be given by $G(t)x = y(t,\phi)$, where $\phi = (x,0)$ for $x \in X$.

Theorem 4.1.5 *Let $T \geq 0$ and $\phi \in \mathcal{X}$. Then there exists a unique solution $y(\cdot) = y(\cdot,\phi) \in C([0,T],X)$ of (4.1.2) such that $\|y(t)\|_X \leq C\|\phi\|_{\mathcal{X}}e^{\gamma t}$, $t \in [0,T]$, where $C > 0$ and $\gamma > 0$ are constants depending only on η, r, and A.*

Proof See Appendix A. □

By virtue of Theorems 4.1.2 and 4.1.5, it is immediate that there exists a unique fundamental solution G, given by $G(t)x = y(t,\phi)$ with $\phi = (x,0)$ for any $x \in X$ and $t \geq 0$, for (4.1.1). Moreover, there exist constants $C \geq 1$ and $\gamma > 0$ such that

$$\|G(t)\| \leq Ce^{\gamma t}, \qquad t \geq 0. \tag{4.1.12}$$

Note that the relation (4.1.2) implies that for any $t \geq 0$ and $x \in X$,

$$G(t)x = e^{tA}x + \int_0^t \int_{-r}^0 e^{(t-s)A}d\eta(\theta)G(t+\theta)xds. \tag{4.1.13}$$

For any measurable function $\varphi(t)$ from $[-r,0]$ to X, we define its right extension function $\vec{\varphi}(t)$ by

$$\vec{\varphi}\colon [-r,\infty) \to X, \quad \vec{\varphi}(t) = \begin{cases} \varphi(t), & -r \leq t \leq 0, \\ 0, & 0 < t < \infty. \end{cases} \tag{4.1.14}$$

It is useful to introduce the so-called *structure operator* S on the space $L^2([-r,0],X)$ by

$$(S\varphi)(\theta) = \int_{-r}^0 d\eta(\tau)\vec{\varphi}(\tau-\theta), \quad \theta \in [-r,0], \quad \text{a.e.} \quad \forall\, \varphi(\cdot) \in L^2([-r,0];X). \tag{4.1.15}$$

It is easy to see that S is a linear and bounded operator on $L^2([-r,0];X)$. Indeed, we have by virtue of (A.2) that there exists a constant $C > 0$ such that

$$\begin{aligned}\int_{-r}^0 \|S\varphi(\theta)\|_X^2 d\theta &= \int_0^r \left\| \int_{-r}^0 d\eta(\tau)\vec{\varphi}(\tau+\theta)\right\|_X^2 d\theta \leq C\int_{-r}^r \|\vec{\varphi}(t)\|_X^2 dt \\ &= C\int_{-r}^0 \|\varphi(\theta)\|_X^2 d\theta, \qquad \forall\, \varphi(\cdot) \in L^2([-r,0];X).\end{aligned}$$

In terms of the operator S, the variation of constants formula (4.1.7) may be rewritten as

$$y(t) = G(t)\phi_0 + \int_{-r}^{0} G(t+\theta)(S\phi_1)(\theta)d\theta, \qquad t \geq 0. \tag{4.1.16}$$

The fundamental solution $G(t)$, $t \in \mathbb{R}_+$, is generally not a semigroup on X, although it is a "quasisemigroup" in the sense that

$$G(t+s)x = G(t)G(s)x + \int_{-r}^{0} G(t+\theta)[SG(s+\cdot)x](\theta)d\theta$$
$$\text{for all} \quad s,\ t \geq 0, \quad x \in X. \tag{4.1.17}$$

Indeed, consider the integral equation (4.1.2) with $\phi = (x, 0)$, $x \in X$, then we have by (4.1.7) that for any t, $s \geq 0$ and $x \in X$,

$$\begin{aligned} G(t+s)x = y(t+s,\phi) &= y(t,(y(s,\phi), y_s(\phi))) \\ &= G(t)y(s,\phi) + \int_{-r}^{0} U(t,\theta)y(s+\theta,\phi)d\theta \\ &= G(t)G(s)x + \int_{-r}^{0} U(t,\theta)G(s+\theta)xd\theta \\ &= G(t)G(s)x + \int_{-r}^{0} G(t+\theta)[SG(s+\cdot)x](\theta)d\theta. \end{aligned}$$

Let $y(t,\phi)$ denote the unique solution of system (4.1.2) and define a family of operators $\mathcal{S}(t)\colon \mathcal{X} \to \mathcal{X}$, $t \geq 0$, by

$$\mathcal{S}(t)\phi = (y(t,\phi), y(t+\cdot,\phi)) \quad \text{for any} \quad \phi \in \mathcal{X}. \tag{4.1.18}$$

Proposition 4.1.6 *For any $t \geq s \geq 0$ and $\phi \in \mathcal{X}$, the following relation holds:*

$$\mathcal{S}(t-s)(y(s,\phi), y(s+\cdot,\phi)) = (y(t,\phi), y(t+\cdot,\phi)), \tag{4.1.19}$$

that is, $\mathcal{S}(t-s)\mathcal{S}(s)\phi = \mathcal{S}(t)\phi$. Moreover, $\mathcal{S}(t)$ is a C_0-semigroup of bounded linear operators on $\mathcal{X}$.

Proof The linearity of $\mathcal{S}(t)$ is obvious. Strong continuity of $\mathcal{S}(t)$ on $\mathcal{X}$ follows from the fact that $y(t,\phi) \to \phi_0$ in X as $t \to 0+$ by virtue of (4.1.2), and it is not hard to get that $y(t+\cdot,\phi) \to \phi_1(\cdot)$ in $L^2([-r,0], X)$ as $t \to 0+$. In order to show the semigroup property (4.1.19), let $t \geq s$ and from (4.1.2), it is easy to verify that

$$
\begin{aligned}
y(t-s, &\mathcal{S}(s)\phi) \\
&= e^{(t-s)A}y(s,\phi) + \int_0^{t-s} e^{(t-s-u)A}\int_{-r}^0 d\eta(\theta)y(u+\theta,\mathcal{S}(s)\phi)du \\
&= e^{(t-s)A}\Big[e^{sA}\phi_0 + \int_0^s e^{(s-u)A}\int_{-r}^0 d\eta(\theta)y(u+\theta,\phi)du\Big] \\
&\quad + \int_s^t e^{(t-u)A}\int_{-r}^0 d\eta(\theta)y(u-s+\theta,\mathcal{S}(s)\phi)du \\
&= e^{tA}\phi_0 + \int_0^s e^{(t-u)A}\int_{-r}^0 d\eta(\theta)y(u+\theta,\phi)du \\
&\quad + \int_s^t e^{(t-u)A}\int_{-r}^0 d\eta(\theta)y(u-s+\theta,\mathcal{S}(s)\phi)du.
\end{aligned}
$$

On the other hand, for $t \geq s$,

$$
\begin{aligned}
y(t,\phi) = e^{tA}\phi_0 &+ \int_0^s e^{(t-u)A}\int_{-r}^0 d\eta(\theta)y(u+\theta,\phi)du \\
&+ \int_s^t e^{(t-u)A}\int_{-r}^0 d\eta(\theta)y(u+\theta,\phi)du.
\end{aligned}
\tag{4.1.20}
$$

Thus, by the uniqueness of solutions of (4.1.2), it follows that

$$
y(t-s,\mathcal{S}(s)\phi) = y(t,\phi), \qquad t \geq s. \tag{4.1.21}
$$

Hence, $[\mathcal{S}(t-s)\mathcal{S}(s)\phi]_0 = [\mathcal{S}(t)\phi]_0$ and similarly $[\mathcal{S}(t-s)\mathcal{S}(s)\phi]_1 = [\mathcal{S}(t)\phi]_1$. The semigroup property (4.1.19) is thus proved and the proof is now complete. □

The following proposition fully describes the infinitesimal generator $\mathcal{A}$ of $\mathcal{S}(t)$.

Proposition 4.1.7 *The infinitesimal generator $\mathcal{A}$ of $\mathcal{S}(t)$ is given by*

$$
\mathscr{D}(\mathcal{A}) = \big\{\phi = (\phi_0,\phi_1) \in \mathcal{X}\colon \phi_1 \in W^{1,2}([-r,0],X),\ \phi_1(0) = \phi_0 \in \mathscr{D}(A)\big\}, \tag{4.1.22}
$$

$$
\mathcal{A}\phi = \left(A\phi_0 + \int_{-r}^0 d\eta(\theta)\phi_1(\theta), \frac{d\phi_1(\theta)}{d\theta}\right) \quad \text{for } \phi = (\phi_0,\phi_1) \in \mathscr{D}(\mathcal{A}). \tag{4.1.23}
$$

Proof See Appendix B. □

4.1.2 Stability and Spectrum

First, note that C_0-semigroup $e^{t\mathcal{A}}$, $t \geq 0$, allows us to transfer the stability problem of the time-delay system (4.1.2) to that of a non–time-delay system. Actually, we can rewrite (4.1.2) as a Cauchy problem (see, e.g., [11]),

$$\begin{cases} dY(t) = \mathcal{A}Y(t)dt, & t \geq 0, \\ Y(0) = \phi \in \mathcal{X}, \end{cases} \tag{4.1.24}$$

where $Y(t) = (y(t), y(t+\cdot))$ is the *lift-up* function of $y(t)$, $t \geq 0$. It turns out that the exponential stability property of (4.1.2) is equivalent to that of system (4.1.24). Indeed, if the null solution of (4.1.2) is exponentially stable, i.e., there exist constants $C > 0$, $\nu > 0$ such that $\|y(t,\phi)\|_X^2 \leq Ce^{-\nu t}\|\phi\|_{\mathcal{X}}^2$, $t \geq 0$, then it follows that

$$\begin{aligned} \|Y(t,\phi)\|_{\mathcal{X}}^2 &\leq \|y(t,\phi)\|_X^2 + r \cdot \max_{-r\leq\theta\leq 0} \|y(t+\theta,\phi)\|_X^2 \\ &< (C + rCe^{\nu r})e^{-\nu t}\|\phi\|_{\mathcal{X}}^2, \qquad t \geq 0. \end{aligned}$$

Conversely, if the null solution of (4.1.24) is exponentially stable, then there exist constants $M > 0$ and $\mu > 0$ such that

$$\begin{aligned} \|y(t,\phi)\|_X^2 &\leq \|y(t,\phi)\|_X^2 + \int_{-r}^0 \|y(t+\theta,\phi)\|_X^2 d\theta \\ &= \|Y(t,\phi)\|_{\mathcal{X}}^2 \leq Me^{-\mu t}\|\phi\|_{\mathcal{X}}^2, \quad t \geq 0. \end{aligned}$$

Owing to this equivalent relation, it is plausible to consider stability of the non–time-delay system (4.1.24) rather than the time-delay one (4.1.2). To establish the exponential stability property of C_0-semigroup $e^{t\mathcal{A}}$, $t \geq 0$, one may intend to develop a dissipative operator method by using Proposition 1.1.18. However, the following example shows some difficulties in association with this scheme.

Let $\alpha > 0$, $\beta \in \mathbb{R}$ and consider a real-valued differential equation

$$\begin{cases} du(t) = -\alpha u(t)dt + \beta u(t-1)dt, & t \geq 0, \\ u(\theta) = 1,\ \theta \in [-1,0]. \end{cases} \tag{4.1.25}$$

If $\beta = 0$, it is clear that $u(t) = e^{-\alpha t}$, $t \geq 0$, which implies the exponential stability of (4.1.25). If $\beta \neq 0$ and we take the preceding viewpoint to consider system (4.1.24) in the Hilbert space $\mathcal{X} := \mathbb{R} \times L^2([-1,0],\mathbb{R})$, it turns out in this case that for any number $a > 0$ and $\phi \in \mathscr{D}(\mathcal{A})$,

$$\begin{aligned} \langle \mathcal{A}\phi, \phi\rangle_{\mathcal{X}} + a\|\phi\|_{\mathcal{X}}^2 &= (a-\alpha)\phi_0^2 + \beta\phi_0\phi_1(-1) \\ &\quad + \frac{1}{2}\phi_1^2(0) - \frac{1}{2}\phi_1^2(-1) + a\int_{-1}^0 \phi_1^2(\theta)d\theta. \end{aligned}$$

It is obvious that one can always find a nonzero element $\phi \in \mathscr{D}(\mathcal{A})$ with $\phi_0 = \phi_1(0) = \phi_1(-1) = 0$ such that

$$\langle \mathcal{A}\phi, \phi\rangle_{\mathcal{X}} + a\|\phi\|_{\mathcal{X}}^2 > 0.$$

In other words, there does not exist a constant $a > 0$, according to Proposition 1.1.18, such that $\|e^{t\mathcal{A}}\| \le e^{-at}$, $t \ge 0$, although $\|e^{tA}\| \le e^{-\alpha t}$, $t \ge 0$. However, it is possible to overcome this difficulty by introducing an equivalent norm $|\cdot|_{\mathcal{X}}$ to the canonical one $\|\cdot\|_{\mathcal{X}}$ in $\mathcal{X}$ to establish the exponential stability property of semigroup $e^{t\mathcal{A}}$, $t \ge 0$.

Theorem 4.1.8 *Let $X = H$ be a real Hilbert space and $\mathcal{X} = H \times L^2([-r,0], H)$ with their inner products $\langle\cdot,\cdot\rangle_H$ and $\langle\cdot,\cdot\rangle_{\mathcal{X}}$, respectively. Suppose that $\langle x, Ax\rangle_H \le -\alpha\|x\|_H^2$ for all $x \in \mathscr{D}(A)$ and some $\alpha > 0$. If, further,*

$$\lambda < \alpha - \left(\big(|\eta|(0) - |\eta|(-r)\big)\int_{-r}^0 e^{-2\lambda\theta} d|\eta|(\theta)\right)^{1/2} \quad \textit{for some} \quad 0 < \lambda < \alpha, \tag{4.1.26}$$

where $|\eta|(\theta)$ is the total variation on $[-r,\theta]$, $\theta \in [-r,0]$, of the bounded variation function η, then $\|e^{t\mathcal{A}}\| \le e^{-\lambda t}$ for all $t \ge 0$.

Proof We show that there exists an equivalent inner product $(\cdot,\cdot)_{\mathcal{X}}$ to the canonical one $\langle\cdot,\cdot\rangle_{\mathcal{X}}$ in $\mathcal{X}$ such that

$$(\mathcal{A}\phi, \phi)_{\mathcal{X}} \le -\lambda\|\phi\|_{\mathcal{X}}^2 \quad \text{for all} \quad \phi \in \mathscr{D}(\mathcal{A}).$$

Here $(\cdot,\cdot)_{\mathcal{X}}$ is defined by

$$(\phi, \psi)_{\mathcal{X}} := \langle \phi_0, \psi_0\rangle_H + \int_{-r}^0 \gamma(\theta)\langle \phi_1(\theta), \psi_1(\theta)\rangle_H d\theta, \qquad \phi,\ \psi \in \mathcal{X}, \tag{4.1.27}$$

where $\gamma: [-r,0] \to \mathbb{R}_+$ is given by

$$\gamma(\theta) = e^{2\lambda\theta}\left[\alpha - \lambda - \frac{|\eta|(0) - |\eta|(-r)}{\alpha - \lambda}\int_\theta^0 e^{-2\lambda\tau} d|\eta|(\tau)\right], \qquad \theta \in [-r,0]. \tag{4.1.28}$$

First, note that under the conditions of Theorem 4.1.8, bilinear form (4.1.27) does define an inner product. Indeed, both (4.1.26) and (4.1.28) imply the lower boundedness of $\gamma(\cdot)$,

$$\gamma(\theta) \ge e^{-2\lambda r}\left[\alpha - \lambda - \frac{|\eta|(0) - |\eta(-r)|}{\alpha - \lambda}\int_{-r}^0 e^{-2\lambda\tau} d|\eta|(\tau)\right]$$
$$\text{for any} \quad \theta \in [-r,0].$$

This implies that $(\cdot,\cdot)_{\mathcal{X}}$ defines an inner product on $\mathcal{X}$. In particular, it is easy to see that for any $\phi \in \mathcal{X}$,

$$
\begin{aligned}
(\phi, \phi)_{\mathcal{X}} &= \|\phi_0\|_H^2 + \int_{-r}^{0} \gamma(\theta)\|\phi_1(\theta)\|_H^2 d\theta \\
&\le [1 + e^{2\lambda r}(\alpha - \lambda)]\Big(\|\phi_0\|_H^2 + \int_{-r}^{0} \|\phi_1(\theta)\|_H^2 d\theta \Big)
\end{aligned}
$$

which implies that the inner product $(\cdot, \cdot)_{\mathcal{X}}$ is also equivalent to canonical inner product $\langle \cdot, \cdot \rangle_{\mathcal{X}}$ in $\mathcal{X}$. On the other hand,

$$
\begin{aligned}
(\phi, (\mathcal{A} + \lambda)\phi)_{\mathcal{X}} &= \Big\langle \phi_0, A\phi_0 + \lambda\phi_0 + A_1\phi_1(-r) + \int_{-r}^{0} A_2(\theta)\phi_1(\theta) d\theta \Big\rangle_H \\
&\quad + \int_{-r}^{0} \gamma(\theta)\langle \phi_1(\theta), \dot{\phi}_1(\theta) + \lambda\phi_1(\theta)\rangle_H d\theta.
\end{aligned}
$$

Since $\langle x, Ax\rangle_H \le -\alpha\|x\|_H^2$ for any $x \in \mathscr{D}(A)$, it follows for any $\phi \in \mathscr{D}(\mathcal{A})$ that

$$
\begin{aligned}
(\phi, (\mathcal{A} + \lambda)\phi)_{\mathcal{X}} &= \langle \phi_0, A\phi_0\rangle_H + \Big\langle \phi_0, A_1\phi_1(-r) + \int_{-r}^{0} A_2(\theta)\phi_1(\theta) d\theta \Big\rangle_H \\
&\quad + \int_{-r}^{0} \gamma(\theta)\langle \dot{\phi}_1(\theta), \phi_1(\theta)\rangle_H d\theta + \lambda\|\phi_0\|_H^2 \\
&\quad + \lambda \int_{-r}^{0} \gamma(\theta)\|\phi_1(\theta)\|_H^2 d\theta \\
&\le (\lambda - \alpha)\|\phi_0\|_H^2 + \|\phi_0\|_H \int_{-r}^{0} \|\phi_1(\theta)\|_H d|\eta|(\theta) \\
&\quad + \int_{-r}^{0} \gamma(\theta)\Big(\frac{1}{2}\frac{d}{d\theta}\|\phi_1(\theta)\|_H^2 + \lambda\|\phi_1(\theta)\|_H^2 \Big) d\theta.
\end{aligned}
\tag{4.1.29}
$$

By using integration by parts, one can further obtain from (4.1.26) and (4.1.29) that for $\phi \in \mathscr{D}(\mathcal{A})$,

$$
\begin{aligned}
(\phi, (\mathcal{A} + \lambda)\phi)_{\mathcal{X}} &\le (\lambda - \alpha)\|\phi_0\|_H^2 + \|\phi_0\|_H \int_{-r}^{0} \|\phi_1(\theta)\|_H d|\eta|(\theta) \\
&\quad + \frac{1}{2}(\alpha - \lambda)\|\phi_0\|_H^2 - \frac{1}{2}\gamma(-r)\|\phi_1(-r)\|_H^2 \\
&\quad - \frac{|\eta|(0) - |\eta|(-r)}{2(\alpha - \lambda)} \int_{-r}^{0} \|\phi_1(\theta)\|_H^2 d|\eta|(\theta) \\
&\le -\frac{1}{2}(\alpha - \lambda)\|\phi_0\|_H^2 + \|\phi_0\|_H \int_{-r}^{0} \|\phi_1(\theta)\|_H d|\eta|(\theta) \\
&\quad - \frac{|\eta|(0) - |\eta|(-r)}{2(\alpha - \lambda)} \int_{-r}^{0} \|\phi_1(\theta)\|_H^2 d|\eta|(\theta).
\end{aligned}
\tag{4.1.30}
$$

If $\|\phi_0\|_H = 0$ or $|\eta|(0) = 0$, i.e., η is constant, it is immediate from (4.1.30) that

$$(\phi, (\mathcal{A}+\lambda)\phi)_{\mathcal{X}} \le 0 \quad \text{for all} \quad \phi \in \mathscr{D}(\mathcal{A}).$$

If $\|\phi_0\|_H \neq 0$ and $|\eta|(0) > 0$ for $\phi \in \mathscr{D}(\mathcal{A})$, we have from (4.1.30) that

$$\begin{aligned}(\phi, (\mathcal{A}+\lambda)\phi)_{\mathcal{X}} &\le \|\phi_0\|_H^2 \int_{-r}^0 \Bigg[-\frac{\alpha-\lambda}{2(|\eta|(0)-|\eta|(-r))} + \frac{\|\phi_1(\theta)\|_H}{\|\phi_0\|_H} \\ &\quad - \frac{|\eta|(0)-|\eta|(-r)}{2(\alpha-\lambda)} \frac{\|\phi_1(\theta)\|_H^2}{\|\phi_0\|_H^2} \Bigg] d|\eta|(\theta) \\ &= \frac{\|\phi_0\|_H^2}{2} \cdot \frac{|\eta|(0)-|\eta|(-r)}{\lambda-\alpha} \\ &\quad \times \int_{-r}^0 \left(\frac{\|\phi_1(\theta)\|_H}{\|\phi_0\|_H} - \frac{\alpha-\lambda}{|\eta|(0)-|\eta|(-r)} \right)^2 d|\eta|(\theta) \\ &\le 0.\end{aligned}$$

Hence, it follows that $(\phi, (\mathcal{A}+\lambda)\phi)_{\mathcal{X}} \le 0$ for all $\phi \in \mathscr{D}(\mathcal{A})$. Due to the equivalence between $\langle \cdot, \cdot \rangle_{\mathcal{X}}$ and $(\cdot, \cdot)_{\mathcal{X}}$, this further implies, in addition to Proposition 1.1.18, that C_0-semigroup $e^{t\mathcal{A}}$, $t \ge 0$, is exponentially stable. The proof is now complete. □

Corollary 4.1.9 *Suppose that $\|e^{tA}\| \le e^{-\alpha t}$ for some $\alpha > 0$ and all $t \ge 0$. Further, if*

$$\lambda < \alpha - \left((|\eta|(0) - |\eta|(-r)) \int_{-r}^0 e^{-2\alpha\theta} d|\eta|(\theta) \right)^{1/2} \quad \textit{for some} \quad \lambda > 0, \tag{4.1.31}$$

then

$$\|e^{t\mathcal{A}}\| \le e^{-\lambda t} \quad \textit{for all} \quad t \ge 0.$$

Proof Note that $e^{-2\alpha\theta} \ge e^{-2\lambda\theta}$ for all $\theta \in [-r, 0]$, from which the desired result is immediate.

□

Next we present an equivalent result whose proof is referred to Appendix C regarding exponential stability between the Green operator $G(t)$ and C_0-semigroup $e^{t\mathcal{A}}$, $t \ge 0$, on a Banach space X.

Proposition 4.1.10 *For the Green operator $G(t)$, $t \in \mathbb{R}$, and C_0-semigroup $e^{t\mathcal{A}}$, $t \ge 0$, the following relations are equivalent.*

(i) For any $\phi \in \mathcal{X}$, $\displaystyle\int_0^\infty \|e^{t\mathcal{A}}\phi\|_{\mathcal{X}}^2 dt < \infty.$

(ii) *There exist constants* $M > 0$ *and* $\mu > 0$ *such that* $\|e^{t\mathcal{A}}\| \le Me^{-\mu t}$, $t \ge 0$.

(iii) *For any* $x \in X$, $\int_0^\infty \|G(t)x\|_X^2 dt < \infty$.

(iv) *There exist constants* $M > 0$ *and* $\mu > 0$ *such that* $\|G(t)\| \le Me^{-\mu t}$, $t \ge 0$.

For each $\lambda \in \mathbb{C}$, we define a densely defined, closed linear operator $\Delta(\lambda, A, \eta)$ by

$$\Delta(\lambda, A, \eta) = \lambda I - A - \int_{-r}^{0} e^{\lambda\theta} d\eta(\theta).$$

The retarded resolvent set $\rho(A, \eta)$ is defined as the set of all values λ in $\mathbb{C}$ for which the operator $\Delta(\lambda, A, \eta)$ has a bounded inverse, $\Delta(\lambda, A, \eta)^{-1}$, in X. Operator $\Delta(\lambda, A, \eta)^{-1}$ is called the *retarded resolvent* of (A, η).

Definition 4.1.11 The *retarded spectrum* is defined to be $\sigma(A, \eta) = \mathbb{C} \setminus \rho(A, \eta)$. In particular, the *retarded point spectrum* is defined to be the set

$$\sigma_p(A, \eta) = \{\lambda \in \mathbb{C}\colon \Delta(\lambda, A, \eta) \text{ is not injective}\},$$

and the *retarded continuous spectrum* is

$$\sigma_c(A, \eta) = \big\{\lambda \in \mathbb{C}\colon \Delta(\lambda, A, \eta) \text{ is injective } \mathscr{R}(\Delta(\lambda, A, \eta)) \neq X \text{ and } \overline{\mathscr{R}(\Delta(\lambda, A, \eta))} = X\big\}.$$

The *retarded residual spectrum* is defined by

$$\sigma_r(A, \eta) = \big\{\lambda \in \mathbb{C}\colon \Delta(\lambda, A, \eta) \text{ is injective and } \overline{\mathscr{R}(\Delta(\lambda, A, \eta))} \neq X\big\}.$$

Note that $\lambda \in \sigma_p(A, \eta)$ if and only if there exists a nonzero $x \in \mathscr{D}(A)$ such that $\Delta(\lambda, A, \eta)x = 0$. The value $\lambda \in \sigma_p(A, \eta)$ is often called a *characteristic value* of (A, η). It is obvious that $\sigma(A, \eta) = \sigma_p(A, \eta) \cup \sigma_c(A, \eta) \cup \sigma_r(A, \eta)$.

The following proposition can be used to establish a relation between the resolvent operator of $\mathcal{A}$ and the retarded resolvent $\Delta(\lambda, A, \eta)^{-1}$.

Proposition 4.1.12 *Let* $\lambda \in \mathbb{C}$ *and* $\psi = (\psi_0, \psi_1) \in \mathcal{X}$. *If* $\phi = (\phi_1(0), \phi_1) \in \mathscr{D}(\mathcal{A})$ *satisfies*

$$\lambda\phi - \mathcal{A}\phi = \psi, \tag{4.1.32}$$

then

$$\phi_1(\theta) = e^{\lambda\theta}\phi_1(0) + \int_{\theta}^{0} e^{\lambda(\theta-\tau)}\psi_1(\tau)d\tau, \qquad -r \le \theta \le 0, \tag{4.1.33}$$

and, by letting $\phi_0 = \phi_1(0)$, *we have*

$$\begin{aligned}\Delta(\lambda, A, \eta)\phi_1(0) = &\int_{-r}^{0} e^{\lambda(-r-\tau)} A_1 \psi_1(\tau) d\tau \\ &+ \int_{-r}^{0}\int_{\theta}^{0} e^{\lambda(\theta-\tau)} A_2(\theta)\psi_1(\tau) d\tau d\theta + \psi_0.\end{aligned} \tag{4.1.34}$$

Conversely, if $\phi_1(0) \in \mathscr{D}(A)$ *satisfies the equation (4.1.34), and if* $\phi_1(0) = \phi_0$,

$$\phi_1(\theta) = e^{\lambda\theta}\phi_1(0) + \int_{\theta}^{0} e^{\lambda(\theta-\tau)}\psi_1(\tau)d\tau, \qquad -r \le \theta \le 0, \tag{4.1.35}$$

then $\phi_1 \in W^{1,2}([-r,0], X)$, $\phi = (\phi_1(0), \phi_1) \in \mathscr{D}(\mathcal{A})$ *and* ϕ *satisfies (4.1.32).*

Proof The equation (4.1.32) can be equivalently written as

$$\lambda\phi_1(0) - A\phi_1(0) - \int_{-r}^{0} d\eta(\theta)\phi_1(\theta) = \psi_0, \tag{4.1.36}$$

and

$$\lambda\phi_1(\theta) - d\phi_1(\theta)/d\theta = \psi_1(\theta) \quad \text{ for } \theta \in [-r, 0]. \tag{4.1.37}$$

It is easy to see that (4.1.37) is equivalent to (4.1.33). Hence, if (4.1.32) holds, we find that $\phi_1(0) \in \mathscr{D}(A)$ and (4.1.34) is true according to (4.1.36) and (4.1.33).

Conversely, if $\phi_1(0) \in \mathscr{D}(A)$, it is easy to see by a simple calculation that ϕ_1, defined by (4.1.35), belongs to $W^{1,2}([-r,0], X)$. In addition, let $\phi_1(0) = \phi_0$ and assume that (4.1.34) holds true; then from (4.1.35), we have

$$\begin{aligned}&\lambda\phi_1(0) - A\phi_1(0) \\ &= \int_{-r}^{0} d\eta(\theta) e^{\lambda\theta}\phi_1(0) + \psi_0 + \int_{-r}^{0} d\eta(\theta)\Big(\int_{\theta}^{0} e^{\lambda(\theta-\tau)}\psi_1(\tau)d\tau\Big) \\ &= \int_{-r}^{0} d\eta(\theta) e^{\lambda\theta}\phi_1(0) + \psi_0 \\ &\quad + \int_{-r}^{0} d\eta(\theta)\Big(\int_{\theta}^{0} e^{\lambda(\theta-\tau)}\Big[\lambda\phi_1(\tau) - \frac{d\phi_1(\tau)}{d\tau}\Big]d\tau\Big) \\ &= \psi_0 + \int_{-r}^{0} d\eta(\theta)\phi_1(\theta),\end{aligned} \tag{4.1.38}$$

which is the first coordinate relation of (4.1.32). The second coordinate equality of (4.1.32) is obvious. The proof is thus complete. □

An immediate result of Proposition 4.1.12 is the following spectrum relations between $G(t)$ and $e^{t\mathcal{A}}$, $t \ge 0$.

Corollary 4.1.13 *Let $\mathcal{A}$ be the linear operator given in (4.1.22) and (4.1.23), then*

$$\sigma_p(\mathcal{A}) = \sigma_p(A,\eta), \qquad \sigma_c(\mathcal{A}) = \sigma_c(A,\eta), \qquad \sigma_r(\mathcal{A}) = \sigma_r(A,\eta),$$

and thus $\sigma(\mathcal{A}) = \sigma(A,\eta)$.

Proof By using Proposition 4.1.12 with $\psi = 0$, we immediately have that $\mathscr{K}(\lambda - \mathcal{A}) = \{0\}$ is equivalent to $\mathscr{K}(\Delta(\lambda, A,\eta)) = \{0\}$, and hence $\mathscr{K}(\lambda - \mathcal{A}) \neq \{0\}$ if and only if $\mathscr{K}(\Delta(\lambda, A,\eta)) \neq \{0\}$. This implies, by definition, $\sigma_p(\mathcal{A}) = \sigma_p(A,\eta)$. The remaining equality $\sigma_c(A,\eta) = \sigma_c(\mathcal{A})$ and $\sigma_r(A,\eta) = \sigma_r(\mathcal{A})$ can be similarly proved. □

To proceed further, we present a result whose proof is referred to Appendix D, which is important in its own right.

Proposition 4.1.14 *Let A generate a C_0-semigroup e^{tA}, $t \geq 0$, on a Banach space X. If e^{tA}, $t \geq 0$, is compact for all $t > 0$, then $e^{t\mathcal{A}}$ is compact for all $t > r$.*

Example 4.1.15 Consider the following retarded partial differential equation

$$\begin{cases} \dfrac{\partial y(t,x)}{\partial t} = \dfrac{\partial^2 y(t,x)}{\partial x^2} + ay(t,x) + by(t-r,x), \quad t > 0, \quad x \in [0,\pi], \\ y(t,0) = y(t,\pi) = 0, \ t \geq 0; \ y(t,x) = \varphi(t,x), \ 0 \leq x \leq \pi, \quad -r \leq t \leq 0, \end{cases} \tag{4.1.39}$$

where $r > 0$, a and b are real numbers.

Let $A = \partial^2/\partial x^2$, which generates a compact C_0-semigroup on $H = L^2(0,\pi)$. By virtue of Propositions 4.1.7 and 4.1.14, operator $\mathcal{A}$ given by (4.1.23) generates a C_0-semigroup $e^{t\mathcal{A}}$, $t \geq 0$, such that $e^{t\mathcal{A}}$, $t > r$, are compact. Hence, this implies by Corollary 4.1.13 and Theorem 1.1.26 that the spectrum $\sigma(\mathcal{A})$ consists of the countable point spectrum $\sigma_p(\mathcal{A})$, determined by the equation

$$\Delta(\lambda, A,\eta)x = [\lambda - A - a - be^{-\lambda r}]x = 0, \qquad x \in \mathscr{D}(A) - \{0\}.$$

Since the eigenvalues of A are $-n^2$ with $Ae_n = -n^2 e_n$ for some orthonormal basis $\{e_n\} \subset \mathscr{D}(A)$, $n = 1, 2, \ldots$, in H. This implies that for exponential stability, it suffices to consider the relation $\sup\{Re\,\lambda\colon \lambda \in \sigma_p(A,\eta)\} < 0$, where

$$\sigma_p(A,\eta) = \big\{\lambda \in \mathbb{C}\colon \lambda + n^2 - a - be^{-\lambda r} = 0 \ \text{ for all } \ n = 1, 2, \ldots\big\},$$

which is equivalent to the condition that the roots of all the equations

$$\lambda = -n^2 + a + be^{-\lambda r}, \qquad n \in \mathbb{N}_+,$$

have negative real parts.

4.2 Linear Systems with Additive Noise

Consider the following stochastic retarded differential equation with additive noise on a Hilbert space H,

$$\begin{cases} dy(t) = Ay(t)dt + \int_{-r}^{0} d\eta(\theta)y(t+\theta)dt + BdW(t), & t \geq 0, \\ y(0) = \phi_0, \; y_0 = \phi_1, \quad \phi = (\phi_0, \phi_1) \in \mathcal{H} := H \times L^2([-r,0], H), & r > 0, \end{cases} \tag{4.2.1}$$

where A and η are given as in the last section, $B \in \mathscr{L}_2(K_Q, H)$ and W is a Q-Wiener process in some Hilbert space K. The following theorem gives some conditions under which there exists a unique stationary solution to the stochastic functional differential equation (4.2.1) of retarded type.

Theorem 4.2.1 *Suppose that the Green operator $G(\cdot)$ is exponentially stable, i.e., there exist constants $M \geq 1$ and $\mu > 0$ such that*

$$\|G(t)\| \leq Me^{-\mu t} \quad \textit{for all} \quad t \geq 0, \tag{4.2.2}$$

then there exists a unique stationary solution to (4.2.1). Moreover, this stationary solution is a zero mean Gaussian process with its covariance operator K given by

$$K(t,s) = \int_0^\infty G(t-s+u)BQ^{1/2}(G(u)BQ^{1/2})^* du, \qquad t \geq s \geq 0. \tag{4.2.3}$$

Proof The proofs are similar to those in Theorem 2.4.3, thus we only sketch them here. By assumption, we can define a process

$$U(t) = \int_{-\infty}^{t} G(t-s)Bd\overline{W}(s), \qquad t \geq -r, \tag{4.2.4}$$

where $\overline{W}$ is a two-sided Q-Wiener process in K on the whole real axis $\mathbb{R}$. It is easy to see that $\mathbb{E}U(t) = 0$ for any $t \geq -r$ and process U is stationary. Moreover, U is a Gaussian process with the covariance operator $K(t,s)$, $t \geq s$, given by

$$K(t,s) = \int_0^\infty G(t-s+u)BQ^{1/2}(G(u)BQ^{1/2})^* du.$$

Thus, the covariance operator $K(t,s)$ is a map of $t-s$ only.

Now we show that $U(t)$, $t \geq -r$, in (4.2.4) is a mild solution of (4.2.1). To this end, first note that by virtue of (4.1.3), we have for any $h \in \mathscr{D}(A^*)$ that

$$\frac{dG^*(t)}{dt}h = G^*(t)A^*h + \int_{-r}^{0} G^*(t+\theta)d\eta^*(\theta)h, \qquad t \geq 0, \tag{4.2.5}$$

where

$$\int_{-r}^{0} G^*(t+\theta)d\eta^*(\theta)h = G^*(t-r)A_1^*h + \int_{-r}^{0} G^*(t+\theta)A_2^*(\theta)hd\theta, \qquad t \geq 0.$$

Therefore, by using the well-known stochastic Fubini's theorem and (4.2.5) we have for any $t \geq 0$ and $h \in \mathscr{D}(A^*)$ that

$$\begin{aligned}
&\int_0^t \langle A^*h, U(s)\rangle_H ds + \langle h, U(0)\rangle_H \\
&\quad = \int_0^t \left\langle A^*h, \int_{-\infty}^{s} G(s-u)Bd\overline{W}(u)\right\rangle_H ds + \left\langle h, \int_{-\infty}^{0} G(-u)Bd\overline{W}(u)\right\rangle_H \\
&\quad = \int_{-\infty}^{0} \left\langle \int_0^t \frac{d}{ds}B^*G^*(s-u)hds, d\overline{W}(u)\right\rangle_H \\
&\qquad + \int_0^t \left\langle \int_u^t \frac{d}{ds}B^*G^*(s-u)hds, d\overline{W}(u)\right\rangle_H + \left\langle h, \int_{-\infty}^{0} G(-u)Bd\overline{W}(u)\right\rangle_H \\
&\qquad - \int_0^t \left\langle h, \int_{-\infty}^{s}\int_{-r}^{0} d\eta(\theta)G(s-u+\theta)Bd\overline{W}(u)\right\rangle_H ds \\
&\quad = \int_{-\infty}^{t} \left\langle B^*G^*(t-u)h, d\overline{W}(u)\right\rangle_H - \int_{-\infty}^{0} \left\langle B^*G^*(-u)h, d\overline{W}(u)\right\rangle_H \\
&\qquad - \int_0^t \langle B^*h, d\overline{W}(u)\rangle_H + \left\langle h, \int_{-\infty}^{0} G(-u)Bd\overline{W}(u)\right\rangle_H \\
&\qquad - \int_0^t \left\langle h, \int_{-\infty}^{s}\int_{-r}^{0} d\eta(\theta)G(s-u+\theta)Bd\overline{W}(u)\right\rangle_H ds \\
&\quad = \left\langle h, U(t) - \int_0^t Bd\overline{W}(u)\right\rangle_H \\
&\qquad - \int_0^t \left\langle h, \int_{-r}^{0} d\eta(\theta)\int_{-\infty}^{s+\theta} G(s-u+\theta)Bd\overline{W}(u)\right\rangle_H ds \\
&\quad = \langle h, U(t)\rangle_H - \left\langle h, \int_0^t Bd\overline{W}(s)\right\rangle_H - \left\langle h, \int_0^t\int_{-r}^{0} d\eta(\theta)U(s+\theta)ds\right\rangle_H .
\end{aligned}$$

Hence, U is a weak solution that is also a mild solution of (4.2.1).

Last, we show the uniqueness of stationary solutions. To this end, first note that for any initial $\psi = (\psi_0, \psi_1) \in \mathcal{H}$, the mild solution $y(t, \psi)$ of (4.2.1) is represented by

$$y(t) = \tilde{y}(t) + \int_0^t G(t-s)BdW(s), \qquad t \geq 0,$$

where

$$\tilde{y}(t) = G(t)\psi_0 + \int_{-r}^{0} G(t+\theta)S\psi(\theta)d\theta, \qquad t \geq 0.$$

It suffices to show that $\tilde{y}(t) \to 0$ in probability as $t \to \infty$. Indeed, since $\|G(t)\| \to 0$ as $t \to \infty$, it follows that

$$\begin{aligned}\mathbb{E}\|\tilde{y}(t)\|_H^2 &\leq 2\mathbb{E}\|\psi_0\|_H^2 \cdot \|G(t)\|^2 + 2\mathbb{E}\left\| \int_{-r}^{0} G(t+\theta)S\psi(\theta)d\theta \right\|_H^2 \\ &\leq 2\|G(t)\|^2\mathbb{E}\|\psi_0\|_H^2 + 2\|S\|^2 r\mathbb{E}\|\psi\|_{L_r^2}^2 \int_{-r}^{0} \|G(t+\theta)\|^2 d\theta \\ &\to 0 \quad \text{as} \quad t \to \infty.\end{aligned}$$

This implies that $\tilde{y}(t) \to 0$ in probability as $t \to \infty$. □

Let $\phi^* \in \mathcal{H}$ be such an initial datum that the corresponding solution $y(\cdot, \phi^*)$ of (4.2.1) is stationary. The following corollary shows that this unique stationary solution of equation (4.2.1) is exponentially stable in mean square.

Corollary 4.2.2 *Suppose that (4.2.2) holds. Then the unique stationary solution $y(\cdot, \phi^*)$ of (4.2.1) is exponentially stable in mean square such that for any mild solution $y(\cdot, \phi)$ of (4.2.1), there exist constants $M > 0$, $\mu > 0$ such that*

$$\mathbb{E}\|y(t,\phi) - y(t,\phi^*)\|_H^2 \leq Me^{-\mu t}\mathbb{E}\|\phi - \phi^*\|_{\mathcal{H}}^2, \qquad t \geq 0.$$

Proof The equation (4.2.1) can be lifted up into an equivalent equation

$$\begin{cases} dY(t) = \mathcal{A}Y(t)dt + \mathcal{B}dW(t), & t \geq 0, \\ Y(0) = \phi, \quad \phi \in \mathcal{H}, \end{cases} \tag{4.2.6}$$

where $Y(t) = (y(t), y_t)$ and $y_t(\theta) := y(t+\theta)$, $t \geq 0$, $\theta \in [-r, 0]$. Here

$$\mathcal{A}\phi = \left(A\phi_0 + \int_{-r}^{0} d\eta(\theta)\phi_1(\theta), \frac{d\phi_1(\theta)}{d\theta} \right) \quad \text{for } \phi = (\phi_0, \phi_1) \in \mathscr{D}(\mathcal{A}),$$

and $\mathcal{B} \in \mathscr{L}_2(K_Q, \mathcal{H})$ is given by

$$\mathcal{B}x = (Bx, 0), \qquad x \in K_Q. \tag{4.2.7}$$

For the equation (4.2.6), we know from (4.2.4), Proposition 4.1.10, and Theorem 2.4.3 that

$$Y(t, \phi^*) = \int_{-\infty}^{t} e^{(t-s)\mathcal{A}}\mathcal{B}d\overline{W}(s), \qquad t \geq 0,$$

is the unique stationary solution with initial $\phi^* \in \mathcal{H}$, where $\overline{W}$ is the two-sided Q-Wiener process given in (2.4.5) on $\mathbb{R}$. Let $Y(t,\phi)$ be an arbitrary mild solution of (4.2.6) with initial $\phi \in \mathcal{H}$. Then for any $t \geq 0$, it follows that

$$\begin{aligned} Y(t,\phi) - Y(t,\phi^*) &= e^{t\mathcal{A}}\phi + \int_0^t e^{(t-s)\mathcal{A}}\mathcal{B}d\overline{W}(s) - \int_{-\infty}^t e^{(t-s)\mathcal{A}}\mathcal{B}d\overline{W}(s) \\ &= e^{t\mathcal{A}}\phi - \int_{-\infty}^0 e^{(t-s)\mathcal{A}}\mathcal{B}d\overline{W}(s) = e^{t\mathcal{A}}(\phi - \phi^*). \end{aligned} \tag{4.2.8}$$

Hence, by virtue of Proposition 4.1.10 and condition (4.2.2) there exist constants $M > 0$ and $\mu > 0$ such that

$$\begin{aligned} \mathbb{E}\|y(t,\phi) - y(t,\phi^*)\|_H^2 &\leq \mathbb{E}\|Y(t,\phi) - Y(t,\phi^*)\|_{\mathcal{H}}^2 \\ &\leq Me^{-\mu t}\mathbb{E}\|\phi - \phi^*\|_{\mathcal{H}}^2, \qquad t \geq 0. \end{aligned}$$

The proof is now complete. □

Example 4.2.3 Consider a stochastic version of equation (4.1.39) in the form

$$\begin{cases} dy(t,x) = \dfrac{\partial^2 y(t,x)}{\partial x^2}dt + by(t-r,x)dt + f(x)dw(t), \quad t > 0, \quad x \in [0,\pi], \\ y(t,0) = y(t,\pi) = 0, \ t \geq 0; \ y(t,x) = \varphi(t,x), \ 0 \leq x \leq \pi, \quad -r \leq t \leq 0, \end{cases} \tag{4.2.9}$$

where $r > 0$, b are real numbers, $f(\cdot)$, $\varphi(t,\cdot) \in L^2(0,\pi)$, and $w(t)$ is the standard one-dimensional Brownian motion on $[0,\infty)$.

By virtue of Example 4.1.15, it is known that the spectrum set of $\mathcal{A}$ consists only of the point spectrum determined by

$$\sigma_p(A,\eta) = \{\lambda \in \mathbb{C}\colon \lambda + n^2 - be^{-\lambda r} = 0 \ \text{ for } \ n = 1, 2, \ldots\},$$

and the condition $\sup\{Re\,\lambda\colon \lambda \in \sigma_p(A,\eta)\} < 0$ ensures the exponential stability of $G(t)$, $t \geq 0$. It is thus possible (see Theorem A.5, p. 416 in [85]) to show that

(i) if $b \geq 1$, there is not a unique stationary solution;
(ii) if $-1 < b < 1$, there is a unique stationary solution for any $r > 0$;
(iii) if $b \leq -1$, one can always find a number $r > 0$ small enough such that the relation

$$\cos\rho - \frac{\rho\sin\rho}{r} < b \leq -1 \tag{4.2.10}$$

holds where $\rho \in (\pi/2,\pi)$ is the root of $\rho = -r\tan\rho$. This implies that for such a small value $r > 0$, there exists a unique stationary solution.

4.3 Linear Systems with Multiplicative Noise

In this section, we shall consider exponential stability of stochastic linear systems driven by multiplicative noise. Let $\mathcal{H} = H \times L^2([-r,0],H)$ and consider the following stochastic delay differential equation in H,

$$\begin{cases} dy(t) = Ay(t)dt + \int_{-r}^{0} d\eta(\theta)y(t+\theta)dt + By(t)dW(t), & t \geq 0, \\ y(0) = \phi_0, \quad y(\theta) = \phi_1(\theta), \quad \theta \in [-r,0], \quad \phi \in \mathcal{H}, \end{cases} \tag{4.3.1}$$

where η is the delay term given in Section 4.1, $B \in \mathscr{L}(H, \mathscr{L}_2(K_Q, H))$ and W is a Q-Wiener process in Hilbert space K.

Let $\mathcal{A}$ be the generator of the delay semigroup $e^{t\mathcal{A}}$, $t \geq 0$, on $\mathcal{H}$ with

$$\mathcal{A}\phi = \left(A\phi_0 + \int_{-r}^{0} d\eta(\theta)\phi_1(\theta), \frac{d\phi_1(\theta)}{d\theta} \right) \qquad \text{for } \phi \in \mathscr{D}(\mathcal{A}).$$

Further, define a linear operator $\mathcal{B} \in \mathscr{L}(\mathcal{H}, \mathscr{L}_2(K_Q, \mathcal{H}))$ by

$$\mathcal{B}\phi = (B\phi_0, 0) \qquad \text{for any} \quad \phi \in \mathcal{H}. \tag{4.3.2}$$

Then equation (4.3.1) can be lifted up into an equivalent equation

$$\begin{cases} dY(t) = \mathcal{A}Y(t)dt + \mathcal{B}Y(t)dW(t), & t \geq 0, \\ Y(0) = \phi, \quad \phi \in \mathcal{H}, \end{cases} \tag{4.3.3}$$

where $Y(t) = (y(t), y_t)$, $t \geq 0$. In most situations, it is more effective to consider stability of the linear system (4.3.3), instead of (4.3.1), by using the theory established in the previous chapters.

4.3.1 Systems with Coercive Condition

Assume that $A\colon V \to V^*$ is a bounded linear operator and there exist constants $\alpha > 0$, $\lambda \in \mathbb{R}$ such that

$$2\langle\langle x, Ax\rangle\rangle_{V,V^*} + \langle x, \Delta(I)x\rangle_H \leq -\alpha\|x\|_V^2 + \lambda\|x\|_H^2, \qquad \forall x \in V, \tag{4.3.4}$$

where $\langle x, \Delta(I)x\rangle_H = Tr\{B(x)Q^{1/2}(B(x)Q^{1/2})^*\}$. On this occasion, let us focus on the point delay term to illustrate our ideas. That is,

$$\int_{-r}^{0} d\eta(\theta)\varphi(\theta) = A_1\varphi(-r), \quad A_1 \in \mathscr{L}(H), \quad \varphi \in C([-r,0];H).$$

In this case, it is easy to see the existence of a unique strong solution $y(t)$, $t \geq -r$, of equation (4.3.1) through a step-by-step argument on $[-r,0]$, $[0,r]$, $[r,2r], \ldots$

Suppose that

$$\|A_1\|^2 + \lambda + 1 < \alpha/\beta, \tag{4.3.5}$$

where $\beta > 0$ is the constant satisfying $\|x\|_H^2 \leq \beta\|x\|_V^2$ for any $x \in V$, then there exists a number $\delta > 0$ such that

$$\mathbb{E}\|y(t,\phi)\|_H^2 \leq C\|\phi\|_{\mathcal{H}}^2 e^{-\delta t}, \qquad \forall t \geq 0, \tag{4.3.6}$$

for some constant $C > 0$. Indeed, let $\nu = \alpha/\beta - \lambda - 1$. Since $\nu > \|A_1\|^2 \geq 0$, it is possible to find a number $\delta \in (0, \nu)$ such that

$$\|A_1\|^2 e^{\delta r} \leq \nu - \delta. \tag{4.3.7}$$

Hence, by applying Itô's formula to the strong solution $y(t,\phi) \in V$, $t \geq 0$, we find that for $t \geq 0$,

$$\begin{aligned} e^{\delta t}\|y(t)\|_H^2 - \|\phi_0\|_H^2 = \delta \int_0^t e^{\delta s}\|y(s)\|_H^2 ds \\ &+ 2\int_0^t e^{\delta s}\langle\langle y(s), Ay(s) + A_1 y(s-r)\rangle\rangle_{V,V^*} ds \\ &+ 2\int_0^t e^{\delta s}\langle y(s), By(s)dW(s)\rangle_H \\ &+ \int_0^t e^{\delta s} Tr(By(s)Q^{1/2}(By(s)Q^{1/2})^*)ds. \end{aligned} \tag{4.3.8}$$

As before, there exists a sequence $\{\tau_n\}_{n\geq 0}$ of stopping times such that $\tau_n \uparrow \infty$, as $n \to \infty$, and for any $n \geq 1$,

$$\mathbb{E}\left(\int_0^{t\wedge\tau_n} e^{\delta s}\langle y(s), By(s)dW(s)\rangle_H\right) = 0, \quad t \in \mathbb{R}_+. \tag{4.3.9}$$

Hence, we have from (4.3.8) and (4.3.9) that for all $t \geq 0$ and large $n \in \mathbb{N}$,

$$\begin{aligned} \mathbb{E}e^{\delta(t\wedge\tau_n)}\|y(t\wedge\tau_n)\|_H^2 &\leq \|\phi_0\|_H^2 - (\nu-\delta)\mathbb{E}\int_0^{t\wedge\tau_n} e^{\delta s}\|y(s)\|_H^2 ds \\ &\quad + \|A_1\|^2\mathbb{E}\int_0^{t\wedge\tau_n} e^{\delta s}\|y(s-r)\|_H^2 ds \\ &= \|\phi_0\|_H^2 - (\nu-\delta)\mathbb{E}\int_0^{t\wedge\tau_n} e^{\delta s}\|y(s)\|_H^2 ds \\ &\quad + \|A_1\|^2 e^{\delta r}\mathbb{E}\int_0^{t\wedge\tau_n} e^{\delta(s-r)}\|y(s-r)\|_H^2 ds. \end{aligned} \tag{4.3.10}$$

By virtue of (4.3.7) and (4.3.10), we have

$$\begin{aligned}\mathbb{E}e^{\delta(t\wedge\tau_n)}\|y(t\wedge\tau_n)\|_H^2 &\le \|\phi_0\|_H^2 - (\nu-\delta)\mathbb{E}\int_0^{t\wedge\tau_n} e^{\delta s}\|y(s)\|_H^2 ds\\ &\quad + \|A_1\|^2 e^{\delta r}\mathbb{E}\int_0^{t\wedge\tau_n} e^{\delta s}\|y(s)\|_H^2 ds\\ &\quad + \|A_1\|^2 e^{\delta r}\int_{-r}^0 e^{\delta\theta}\|\phi_1(\theta)\|_H^2 d\theta\\ &\le (1+\|A_1\|^2 e^{\nu r})\|\phi\|_{\mathcal{H}}^2.\end{aligned} \tag{4.3.11}$$

Letting n tend to infinity and using Fatou's lemma in (4.3.11), we have

$$\mathbb{E}\|y(t,\phi)\|_H^2 \le C\|\phi\|_{\mathcal{H}}^2 e^{-\delta t}, \qquad t\ge 0,$$

where $C = 1+\|A_1\|^2 e^{\nu r} > 0$, $\delta > 0$. By carrying out a similar scheme to that in Theorem 3.3.3, we can also obtain almost sure pathwise stability of equation (4.3.1).

4.3.2 Systems with Small Time Delay

(I) **Mean Square Stability.**

Let us consider equation (4.3.1) and its lift-up system (4.3.3). Clearly, we have by virtue of Theorem 2.2.8 that the null solution of (4.3.3) is mean square exponentially stable if there exists a constant $\mu > 0$ such that

$$\|e^{t\mathcal{A}}\| \le e^{-\mu t} \qquad \text{for all} \qquad t \ge 0, \tag{4.3.12}$$

and

$$\left\|\int_0^\infty e^{t\mathcal{A}^*}\Delta(\mathcal{I})e^{t\mathcal{A}}dt\right\| < 1, \tag{4.3.13}$$

where $\mathcal{I}$ is the identity operator in $\mathcal{H}$ and $\Delta(\mathcal{I})$ is defined by the relation

$$\langle\phi,\Delta(\mathcal{I})\phi\rangle_{\mathcal{H}} = Tr\{\mathcal{B}(\phi)Q^{1/2}(\mathcal{B}(\phi)Q^{1/2})^*\}, \qquad \phi\in\mathcal{H}.$$

Note that $\|\mathcal{B}\| = \|B\|$. Indeed, for any $\phi\in\mathcal{H}$,

$$\|\mathcal{B}\phi\|_{\mathcal{H}}^2 = \|(B\phi_0,0)\|_{\mathcal{H}}^2 = \|B\phi_0\|_H^2 \le \|B\|^2\|\phi_0\|_H^2 \le \|B\|^2\|\phi\|_{\mathcal{H}}^2,$$

i.e., $\|\mathcal{B}\| \le \|B\|$. On the other hand, by definition it is immediate that

$$\|\mathcal{B}\phi\|_{\mathcal{H}}^2 \le \|\mathcal{B}\|^2\|\phi\|_{\mathcal{H}}^2 \qquad \text{for any} \qquad \phi\in\mathcal{H}. \tag{4.3.14}$$

In particular, let $\phi = (\phi_0, 0)$, $\phi_0\in H$, then we have from (4.3.14) that

$$\|B\phi_0\|_H^2 \le \|\mathcal{B}\|^2\|\phi_0\|_H^2, \qquad \phi_0\in H,$$

that is, $\|B\| \leq \|\mathcal{B}\|$. For simplicity, let us focus on the case that $Tr(Q) < \infty$, $\eta(\tau) = -\mathbf{1}_{(-\infty,-r]}(\tau)A_1$, $A_1 \in \mathscr{L}(H)$ to illustrate our ideas. In this case, if (4.3.12) holds, then

$$\left\| \int_0^\infty e^{t\mathcal{A}^*}\Delta(\mathcal{I})e^{t\mathcal{A}}dt \right\| \leq \|\Delta(\mathcal{I})\| \int_0^\infty \|e^{t\mathcal{A}}\|^2 dt \leq \frac{Tr(Q)\cdot\|B\|^2}{2\mu}. \tag{4.3.15}$$

Therefore, by virtue of (4.3.12), (4.3.13), and Theorem 4.1.8, we have whenever there exists $\alpha > 0$ such that

$$\|e^{tA}\| \leq e^{-\alpha t}, \quad \frac{Tr(Q)\cdot\|B\|^2}{2\lambda} < 1,$$
$$\|A_1\|^2 e^{2\lambda r} < (\alpha-\lambda)^2 \quad \text{for some } \lambda \in (0,\alpha), \tag{4.3.16}$$

the null solution of (4.3.3) (thus, the null solution of (4.3.1)) is mean square exponentially stable. In this case, if $A_1 \neq 0$, the delay parameter $r > 0$ satisfies

$$r < \frac{1}{\lambda}\ln\left(\frac{\alpha-\lambda}{\|A_1\|}\right). \tag{4.3.17}$$

Example 4.3.1 (Example in (2.5.2) revisited) Consider the following linear stochastic partial differential equation,

$$\begin{cases} dy(t,x) = \Delta y(t,x)dt + \alpha_0 y(t,x)dt + \alpha_1 y(t,x)dw(t), & t \geq 0, \quad x \in [0,\pi], \\ y(t,0) = y(t,\pi) = 0, & t \geq 0, \\ y(0,\cdot) = \phi_0(\cdot) \in H = L^2(0,\pi), \end{cases} \tag{4.3.18}$$

where $\Delta = \partial^2/\partial x^2$, $\alpha_0 > 0$, $\alpha_1 \in \mathbb{R}$ and w is a standard real Brownian motion.

It is known by Example 2.5.1 that if

$$\alpha_1^2 < 2(1-\alpha_0), \tag{4.3.19}$$

the null solution of (4.3.18) is exponentially stable in mean square. Further, let $r > 0$ and consider a time-delay version of (4.3.18) in the form

$$\begin{cases} dy(t,x) = \Delta y(t,x)dt + \alpha_0 y(t-r,x)dt + \alpha_1 y(t,x)dw(t), \ t \geq 0, \ x \in [0,\pi], \\ y(t,0) = y(t,\pi) = 0, \quad t \geq 0, \\ y(0,\cdot) = \phi_0(\cdot) \in L^2(0,\pi), \ y_0(\cdot,\cdot) = \phi_1(\cdot,\cdot) \in L^2([-r,0];L^2(0,\pi)). \end{cases} \tag{4.3.20}$$

A natural question is whether one can secure, under the condition (4.3.19), the same exponential stability of (4.3.20), at least for sufficiently small $r > 0$. The answer turns out to be affirmative.

Indeed, let $H = L^2(0,\pi)$ and $\mathcal{A}$ be the generator of the associated delay semigroup $e^{t\mathcal{A}}$, $t \geq 0$, in $\mathcal{H}$ with

$$\mathcal{A}\phi = (\Delta\phi_0 + \alpha_0\phi_1(-r), d\phi_1(\theta)/d\theta) \qquad \text{for } \phi = (\phi_0, \phi_1) \in \mathscr{D}(\mathcal{A}).$$

Also, let $\mathcal{B} \in \mathscr{L}(\mathcal{H})$ be given by $\mathcal{B}\phi = (\alpha_1\phi_0, 0)$ for $\phi = (\phi_0, \phi_1) \in \mathcal{H}$. By virtue of (4.3.16), if there exists a number $\lambda \in (0,1)$ such that

$$\|e^{t\Delta}\| \leq e^{-t}, \qquad \alpha_1^2 < 2\lambda, \qquad \alpha_0^2 e^{2\lambda r} < (1-\lambda)^2, \tag{4.3.21}$$

then the null solution of (4.3.20) is exponentially stable in mean square. It may be verified that $\lambda = \frac{1}{2}\alpha_1^2 + \varepsilon$ with $\varepsilon > 0$ sufficiently small can satisfy the condition (4.3.21). In this case, the third inequality in (4.3.21) is actually reduced to

$$\alpha_0^2 \exp(r(\alpha_1^2 + 2\varepsilon)) < \left(1 - \frac{\alpha_1^2}{2} + \varepsilon\right)^2.$$

If $\alpha_1 \neq 0$, then this inequality is, by letting $\varepsilon \to 0$, reduced to

$$r < \frac{2}{\alpha_1^2} \ln\left(\frac{2-\alpha_1^2}{2\alpha_0}\right). \tag{4.3.22}$$

That is, in addition to the condition $\alpha_1^2 < 2(1-\alpha_0)$, which ensures the mean square exponential stability of equation (4.3.18), if one further assumes that the relation (4.3.22) is satisfied, then the null solution of (4.3.20) is exponentially stable in mean square. In other words, the stability at the moment is not sensitive to small delays.

(II) **Almost Sure Stability**.
Once again, suppose that $B \in \mathscr{L}(H)$, $K = \mathbb{R}$, $\eta(\tau) = -\mathbf{1}_{(-\infty,-r]}(\tau)A_1$, $A_1 \in \mathscr{L}(H)$, and $W(t) = w(t)$, $t \geq 0$, is a standard real-valued Brownian motion. In this case, we notice in the formulation (4.3.3) of equation (4.3.1) that for any $\beta \in \mathbb{R}$,

$$(\mathcal{A} + \beta\mathcal{B})\phi = \big((A + \beta B)\phi_0 + A_1\phi_1(-r), d\phi_1(\theta)/d\theta\big), \qquad \phi \in \mathscr{D}(\mathcal{A}).$$

Hence, by virtue of Proposition 2.2.11 and Theorem 4.1.8, if there exists a number $\beta \in \mathbb{R}$ such that $A + \beta B$ generates an exponentially stable C_0-semigroup $e^{t(A+\beta B)}$, $t \geq 0$, and

$$\begin{aligned} &\|e^{t(A+\beta B)}\| \leq e^{-\alpha t}, \quad \frac{\beta^2}{2} + \|B\|^2 \leq 2\alpha, \\ &\|A_1\|^2 e^{2\lambda r} < (\alpha - \lambda)^2 \quad \text{for some} \quad \alpha > \lambda > 0, \end{aligned} \tag{4.3.23}$$

then the null solution of (4.3.3) (thus, the null solution of (4.3.1)) is exponentially stable in the almost sure sense. In particular, if $A_1 \neq 0$, the delay parameter $r > 0$ satisfies

$$r < \frac{1}{\lambda} \ln\left(\frac{\alpha - \lambda}{\|A_1\|}\right). \tag{4.3.24}$$

Example 4.3.2 (Example 4.3.1 revisited) Consider the following linear stochastic partial differential equation,

$$\begin{cases} dy(t,x) = \Delta y(t,x)dt + \alpha_0 y(t,x)dt + \alpha_1 y(t,x)dw(t), \quad t \geq 0, \quad x \in [0,\pi], \\ y(t,0) = y(t,\pi) = 0, \quad t \geq 0, \\ y(0,\cdot) = \phi_0(\cdot) \in H = L^2(0,\pi), \end{cases} \tag{4.3.25}$$

where $\Delta = \partial^2/\partial x^2$, $\alpha_0 > 0$, $\alpha_1 \in \mathbb{R}$, and w is a standard real Brownian motion.

It is known by Example 2.2.12 that if

$$\alpha_0 - 1 < \alpha_1^2/2, \tag{4.3.26}$$

the null solution of (4.3.25) is exponentially stable almost surely. Further, let $r > 0$ and consider a time-delay version of (4.3.25) in the form

$$\begin{cases} dy(t,x) = \Delta y(t,x)dt + \alpha_0 y(t-r,x)dt \\ \qquad\qquad + \alpha_1 y(t,x)dw(t), \quad t \geq 0, \quad x \in [0,\pi], \\ y(t,0) = y(t,\pi) = 0, \quad t \geq 0, \\ y(0,\cdot) = \phi_0(\cdot) \in L^2(0,\pi), \; y_0(\cdot,\cdot) = \phi_1(\cdot,\cdot) \in L^2([-r,0]; L^2(0,\pi)). \end{cases} \tag{4.3.27}$$

By using the same notations and notions as in Example 4.3.1, we have

$$(\mathcal{A} + \beta\mathcal{B})\phi = (\Delta\phi_0 + \alpha_0\phi_1(-r) + \beta\alpha_1\phi_0, d\phi_1(\theta)/d\theta), \qquad \phi \in \mathscr{D}(\mathcal{A}).$$

By virtue of (4.3.23), if there exists a number $\beta \in \mathbb{R}$ such that for some $\lambda > 0$,

$$\lambda < 1 - \beta\alpha_1, \qquad \frac{\beta^2}{2} + \alpha_1^2 \leq 2(1 - \beta\alpha_1), \qquad \alpha_0^2 e^{2\lambda r} < (1 - \beta\alpha_1 - \lambda)^2, \tag{4.3.28}$$

then the null solution of (4.3.27) is exponentially stable in the almost sure sense. It may be verified that $\beta = -2\alpha_1$ and $\lambda = \frac{3}{2}\alpha_1^2 + \varepsilon$ with $\varepsilon > 0$ sufficiently small can satisfy the condition (4.3.28). If $\alpha_1 \neq 0$, the third inequality in (4.3.28) is, by letting $\varepsilon \to 0$, reduced to

$$r < \frac{2}{3\alpha_1^2} \ln\left(\frac{\frac{1}{2}\alpha_1^2 + 1}{\alpha_0}\right). \tag{4.3.29}$$

That is, in addition to the condition $\alpha_0 - 1 < \alpha_1^2/2$, which ensures the pathwise exponential stability of equation (4.3.25), if we further assume that the relation (4.3.29) is satisfied, then the null solution of (4.3.27) is exponentially stable in the almost sure sense.

4.3.3 Systems with Time Delay in Diffusion Terms

In contrast with equation (4.3.1), let us consider at present the stability property of a stochastic differential equation with time delay appearing in its diffusion term:

$$\begin{cases} dy(t) = Ay(t)dt + By(t-r)dW(t), \quad t \geq 0, \\ y(0) = \phi_0, \quad y(\theta) = \phi_1(\theta), \quad \theta \in [-r, 0], \quad \phi \in \mathcal{H}, \end{cases} \tag{4.3.30}$$

where $r > 0$. If we intend to follow the same idea as in the previous sections to transfer the stability problem of (4.3.30) to the corresponding one of some system without time delay, it is natural to define an operator $\mathcal{B}$ by $\mathcal{B}\phi = (B\phi_1(-r), 0)$ for $\phi \in \mathcal{H} = H \times L^2([-r, 0], H)$. This implies that $\mathcal{B} \notin \mathscr{L}(\mathcal{H})$, i.e., $\mathcal{B}$ is unbounded, in spite of the fact that $B \in \mathscr{L}(H)$. To avoid this difficulty, we will proceed by considering the strong solutions of (4.3.30) and employing a variational approach.

Suppose that $A \in \mathscr{L}(V, V^*)$ satisfies a coercive condition

$$2\langle\langle x, Ax\rangle\rangle_{V,V^*} \leq -\alpha\|x\|_V^2 + \lambda\|x\|_H^2, \qquad \forall x \in V, \tag{4.3.31}$$

for some $\alpha > 0$ and $\lambda \in \mathbb{R}$. Recall that for arbitrary $P \in \mathscr{L}(H)$, we define a unique operator $\Delta(P) \in \mathscr{L}(H)$ by the relation

$$\langle \Delta(P)x, y\rangle_H = Tr[PB(x)Q^{1/2}(B(y)Q^{1/2})^*], \qquad x,\ y \in H. \tag{4.3.32}$$

It is not difficult to see that under condition (4.3.31), there exists a unique strong solution to (4.3.30). Moreover, A generates a C_0-semigroup e^{tA}, $t \geq 0$, on H and the strong solution is also a mild solution of (4.3.30).

Theorem 4.3.3 *Suppose that*

(i) there exist constants $M > 0$ and $\mu > 0$ such that

$$\|e^{tA}\| \leq Me^{-\mu t}, \qquad t \geq 0;$$

(ii)

$$\left\| \int_0^\infty \Delta(e^{tA^*}e^{tA})dt \right\| < 1.$$

Then the null solution of (4.3.30) is exponentially stable in mean square. That is, there exist positive constants $C > 0$ and $\nu > 0$ such that

$$\mathbb{E}\|y(t,\phi)\|_H^2 \le C\|\phi\|_{\mathcal{H}}^2 e^{-\nu t} \quad \textit{for all} \quad t \ge 0.$$

Proof Since $y(t,\phi)$ is also the mild solution of (4.3.30), it follows that

$$y(t,\phi) = e^{tA}\phi_0 + \int_0^t e^{(t-s)A}By(s-r,\phi)dW(s), \quad t \ge 0,$$

which immediately implies that

$$\|y(t,\phi)\|_H^2 = \|e^{tA}\phi_0\|_H^2 + \left\|\int_0^t e^{(t-s)A}By(s-r,\phi)dW(s)\right\|_H^2 + 2\left\langle e^{tA}\phi_0, \int_0^t e^{(t-s)A}By(s-r,\phi)dW(s)\right\rangle_H, \quad t \ge 0. \tag{4.3.33}$$

Hence, it is easy to see that for any $t \ge 0$,

$$\mathbb{E}\|y(t,\phi)\|_H^2 = \|e^{tA}\phi_0\|_H^2 + \mathbb{E}\left\|\int_0^t e^{(t-s)A}By(s-r,\phi)dW(s)\right\|_H^2, \tag{4.3.34}$$

since the stochastic integral in the inner product $\langle\cdot,\cdot\rangle_H$ of (4.3.33) is a martingale, and consequently for each $t \ge 0$,

$$\mathbb{E}\left\langle e^{tA}\phi_0, \int_0^t e^{(t-s)A}By(s-r,\phi)dW(s)\right\rangle_H = 0.$$

On the other hand, note that

$$Tr\big[PGQG^*\big] = Tr\big[G^*PGQ\big] \ge 0,$$

for any $G \in \mathscr{L}(K,H)$ and self-adjoint operator $P \in \mathscr{L}^+(H)$. We therefore have by virtue of (4.3.32) that

$$\begin{aligned}
&\mathbb{E}\left\|\int_0^t e^{(t-s)A}By(s-r,\phi)dW(s)\right\|_H^2 \\
&= \int_0^t \mathbb{E}\Big\{Tr\Big[e^{(t-s)A}By(s-r,\phi)Q^{1/2}(e^{(t-s)A}By(s-r,\phi)Q^{1/2})^*\Big]\Big\}ds \\
&= \int_0^t \mathbb{E}\langle\Delta(e^{(t-s)A^*}e^{(t-s)A})y(s-r,\phi), y(s-r,\phi)\rangle_H ds.
\end{aligned} \tag{4.3.35}$$

Let $\nu \in (0, 2\mu)$ be a number that will be determined later on. From (4.3.34) and (4.3.35), it follows that

$$\int_0^\infty e^{\nu t}\mathbb{E}\|y(t,\phi)\|_H^2 dt = \int_0^\infty e^{\nu t}\|e^{tA}\phi_0\|_H^2 dt + \int_0^\infty e^{\nu t}\int_0^t \mathbb{E}\langle\Delta(e^{(t-s)A^*}e^{(t-s)A})y(s-r,\phi), y(s-r,\phi)\rangle_H ds dt. \tag{4.3.36}$$

Evaluating the first term on the right hand side of (4.3.36), we obtain by using condition (i) that

$$\int_0^\infty e^{\nu t}\|e^{tA}\phi_0\|_H^2 dt \le \frac{M^2}{2\mu - \nu}\|\phi\|_{\mathcal{H}}^2. \tag{4.3.37}$$

On the other hand, by applying Fubini's theorem and the change of variables $s - r \to s$ to the second term on the right-hand side of (4.3.36), one can have

$$\begin{aligned}
&\int_0^\infty e^{\nu t}\int_0^t \mathbb{E}\langle\Delta(e^{(t-s)A^*}e^{(t-s)A})y(s-r,\phi), y(s-r,\phi)\rangle_H ds dt\\
&= \int_0^\infty e^{\nu s}\int_0^\infty e^{\nu t}\mathbb{E}\langle\Delta(e^{tA^*}e^{tA})y(s-r,\phi), y(s-r,\phi)\rangle_H dt ds\\
&\le \left\|\int_0^\infty e^{\nu t}\Delta(e^{tA^*}e^{tA})dt\right\|\int_0^\infty e^{\nu s}\mathbb{E}\|y(s-r,\phi)\|_H^2 ds\\
&\le f(\nu)e^{\nu r}\|\phi\|_{\mathcal{H}}^2 + f(\nu)e^{\nu r}\int_0^\infty e^{\nu s}\mathbb{E}\|y(s,\phi)\|_H^2 ds,
\end{aligned} \tag{4.3.38}$$

where

$$f(\nu) = \left\|\int_0^\infty e^{\nu t}\Delta(e^{tA^*}e^{tA})dt\right\| \quad \text{and} \quad \nu > 0.$$

Hence, by employing (4.3.36) and (4.3.38), we can further have

$$\begin{aligned}
&\int_0^\infty e^{\nu t}\int_0^t \mathbb{E}\langle\Delta(e^{(t-s)A^*}e^{(t-s)A})y(s-r,\phi), y(s-r,\phi)\rangle_H ds dt\\
&\le f(\nu)e^{\nu r}\left(1 + \frac{M^2}{2\mu - \nu}\right)\|\phi\|_{\mathcal{H}}^2 + f(\nu)e^{\nu r}\\
&\quad\cdot\int_0^\infty e^{\nu t}\int_0^t \mathbb{E}\langle\Delta(e^{(t-s)A^*}e^{(t-s)A})y(s-r,\phi), y(s-r,\phi)\rangle_H ds dt.
\end{aligned} \tag{4.3.39}$$

Note that by virtue of condition (ii) and the continuity of $f(\nu)$ at $\nu = 0$, we have the relation $\lim_{\nu\to 0^+} f(\nu)e^{\nu r} < 1$. Hence, we can choose a number ν

such that $0 < \nu < 2\mu$ and $f(\nu)e^{\nu r} < 1$. Consequently, there exists a constant $C_1 = C_1(\nu) > 0$ such that

$$\int_0^\infty e^{\nu t}\int_0^t \mathbb{E}\langle \Delta(e^{(t-s)A^*}e^{(t-s)A})y(s-r,\phi), y(s-r,\phi)\rangle_H dsdt \le C_1(\nu)\|\phi\|_{\mathcal{H}}^2. \tag{4.3.40}$$

From (4.3.36), (4.3.37), and (4.3.40), it follows that for sufficiently small $\nu > 0$, there exists a positive constant $C_2 = C_2(\nu)$ such that

$$\int_0^\infty e^{\nu t}\mathbb{E}\|y(t,\phi)\|_H^2 dt \le C_2\|\phi\|_{\mathcal{H}}^2. \tag{4.3.41}$$

Now we can apply Itô's formula to the process $e^{\nu t}\|y(t,\phi)\|_H^2$, $t \ge 0$, to obtain

$$\begin{aligned} e^{\nu t}\|y(t,\phi)\|_H^2 &= \|\phi_0\|_H^2 + \nu\int_0^t e^{\nu s}\|y(s,\phi)\|_H^2 ds \\ &\quad + 2\int_0^t e^{\nu s}\langle\langle y(s,\phi), Ay(s,\phi)\rangle\rangle_{V,V^*}ds \\ &\quad + \int_0^t e^{\nu s}(\Delta(I)y(s-r,\phi), y(s-r,\phi)\rangle_H ds \\ &\quad + 2\int_0^t e^{\nu s}\langle y(s,\phi), By(s-r,\phi)dW(s)\rangle_H. \end{aligned} \tag{4.3.42}$$

Applying the coercive condition (4.3.31) to (4.3.42), we thus have for any $t \ge 0$ that

$$\begin{aligned} e^{\nu t}\mathbb{E}\|y(t,\phi)\|_H^2 &\le \|\phi\|_{\mathcal{H}}^2 + (\nu + |\lambda|)\int_0^t e^{\nu s}\mathbb{E}\|y(s,\phi)\|_H^2 ds \\ &\quad + \|\Delta(I)\|\int_0^t e^{\nu s}\mathbb{E}\|y(s-r,\phi)\|_H^2 ds. \end{aligned}$$

By virtue of (4.3.41) and the change of variables used before, it follows that for some constant $C > 0$,

$$e^{\nu t}\mathbb{E}\|y(t,\phi)\|_H^2 \le C\|\phi\|_{\mathcal{H}}^2, \qquad \forall t \ge 0.$$

The proof is now complete. □

Theorem 4.3.4 *Under the same conditions as in Theorem 4.3.3, there exist positive constants M and μ and a random time $T(\omega) > 0$ such that*

$$\|y(t,\phi)\|_H \le M\|\phi\|_{\mathcal{H}}e^{-\mu t}, \qquad t \ge T(\omega),$$

almost surely.

Proof The proofs are similar to those in Theorem 2.2.17, and are omitted. □

Corollary 4.3.5 *Assume that $Tr(Q) < \infty$ and y is a strong solution of (4.3.30) on the Hilbert space H. If there exist constants $M > 0$ and $\mu > 0$ such that*

$$\|e^{tA}\| \le Me^{-\mu t}, \qquad t \ge 0,$$

and

$$\frac{M^2\|B\|^2 Tr(Q)}{2\mu} < 1,$$

then the null solution of (4.3.30) is exponentially stable in both the mean square and almost sure sense.

Proof For any $L \in \mathscr{L}(H)$, define a sesquilinear form φ on $H \times H$ by

$$\varphi(x, y) = Tr(L(Bx)Q^{1/2}(L(By)Q^{1/2})^*), \qquad x,\ y \in H,$$

then it is easy to see that φ is Hermitian, i.e., $\varphi(x, y) = \overline{\varphi(y, x)}$ and

$$|\varphi(x, y)| \le Tr(Q)\|B\|^2\|L\|^2\|x\|_H\|y\|_H,$$

for any $x,\ y \in H$. Hence, for any $t \ge 0$, the operator $\Delta(e^{tA^*}e^{tA}) \in \mathscr{L}(H)$ generated by φ is self-adjoint on H. Therefore, we have

$$\begin{aligned}
\left\|\int_0^\infty \Delta(e^{tA^*}e^{tA})dt\right\| &\le \int_0^\infty \sup_{\|x\|_H=1} \left|\langle\Delta(e^{tA^*}e^{tA})x, x\rangle_H\right|dt \\
&= \int_0^\infty \sup_{\|x\|_H=1} \left|Tr(e^{tA}(Bx)Q^{1/2}(e^{tA}(Bx)Q^{1/2})^*)\right|dt \\
&\le Tr(Q)\cdot\|B\|^2\int_0^\infty \|e^{tA}\|^2 dt \\
&\le \frac{Tr(Q)\|B\|^2M^2}{2\mu} < 1.
\end{aligned}$$

Hence, by virtue of Theorems (4.3.3) and (4.3.4), we obtain the desired result. □

4.4 Stability of Nonlinear Systems

In this section, we shall introduce some useful methods to deal with stochastic stability of nonlinear stochastic delay differential equations (SDDEs) in the Hilbert space H.

4.4.1 Lyapunov Function Method

One of the most important methods in dealing with stability of time-delay systems is Lyapunov's direct method. Let us illustrate the main ideas by considering the mild solutions of a class of semilinear retarded stochastic evolution equations.

Let $r > 0$ and consider the following stochastic system in H,

$$\begin{cases} dy(t) = Ay(t) + F(y(t-r))dt + B(y(t-r))dW(t), & \forall t \geq 0, \\ y(\theta) = \varphi(\theta) \in H, \quad \theta \in [-r,0], \end{cases} \tag{4.4.1}$$

where $\varphi \in C([-r,0],H)$; A is the infinitesimal generator of some C_0-semigroup e^{tA}, $t \geq 0$, over H; and F and B are two nonlinear measurable mappings with $F(0) = 0$, $B(0) = 0$ from H to H and $\mathscr{L}_2(K_Q, H)$, respectively, which satisfy the usual Lipschitz and linear growth conditions. We introduce the following Yosida approximating systems of (4.4.1),

$$\begin{cases} dy(t) = Ay(t)dt + R(n)F(y(t-r))dt \\ \qquad\qquad + R(n)B(y(t-r))dW(t), \quad \forall t \geq 0, \\ y(\theta) = R(n)\varphi(\theta) \in \mathscr{D}(A), \quad \theta \in [-r,0], \end{cases} \tag{4.4.2}$$

where $n \in \rho(A)$, the resolvent set of A, and $R(n) = nR(n,A)$. Here $R(n,A)$ is the resolvent of A at $n \in \rho(A)$.

Let Λ be a nonnegative Itô functional on H, and the infinitesimal generators $\mathcal{L}$ and $\mathcal{L}_n$ in correspondence with (4.4.1) and (4.4.2) are defined, respectively, by

$$\begin{aligned} \mathcal{L}\Lambda(x,y) = {} & \langle \Lambda'(x), Ax + F(y)\rangle_H \\ & + \frac{1}{2}Tr\{\Lambda''(x)B(y)Q^{1/2}(B(y)Q^{1/2})^*\}, \quad x \in \mathscr{D}(A), \quad y \in H, \end{aligned}$$

and

$$\begin{aligned} \mathcal{L}_n\Lambda(x,y) = {} & \langle \Lambda'(x), Ax + R(n)F(y)\rangle_H \\ & + \frac{1}{2}Tr\{\Lambda''(x)R(n)B(y)Q^{1/2}(R(n)B(y)Q^{1/2})^*\}, \quad x \in \mathscr{D}(A), \quad y \in H. \end{aligned}$$

By analogy with Proposition 1.4.5 and Corollary 1.4.6, we have the following result.

Proposition 4.4.1 *For each $n \in \rho(A)$, the equation (4.4.2) has a unique solution $y_n(t) \in \mathscr{D}(A)$ in the strong sense such that*

$$\lim_{n\to\infty} \mathbb{E}\Big(\sup_{t\in[0,T]} \|y_n(t) - y(t)\|_H^p\Big) = 0 \quad \textit{for any} \quad p > 2, \quad T \geq 0,$$

where y is the mild solution of (4.4.1). Moreover, there exists a subsequence $\{y_{n_k}\}_{k=1}^{\infty}$ of $\{y_n\}$ such that $y_{n_k}(t) \to y(t)$ almost surely as $k \to \infty$ uniformly on any compact set of $[0, \infty)$.

Theorem 4.4.2 *Let $\Lambda : H \to \mathbb{R}_+$ be an Itô functional satisfying*

(i) $\Lambda(x) \geq c_1 \|x\|_H^2$, $x \in H$, for some $c_1 > 0$;
(ii) for any $x \in H$,

$$|\Lambda(x)| + \|x\|_H \|\Lambda'(x)\|_H + \|x\|_H^2 \|\Lambda''(x)\| \leq c_2 \|x\|_H^2 \quad \text{for some} \quad c_2 > 0;$$

(iii) there exist constants $\alpha > 0$, $\lambda \in \mathbb{R}_+$ such that

$$\mathcal{L}\Lambda(x, y) \leq -\alpha \Lambda(x) + \lambda \Lambda(y), \quad x \in \mathscr{D}(A), \quad y \in H.$$

Assume further that

$$\alpha > \lambda, \tag{4.4.3}$$

then there exist numbers $\gamma > 0$ and $C(\varphi) > 0$ such that

$$\mathbb{E}\|y(t, \varphi)\|_H^2 \leq C(\varphi) \cdot e^{-\gamma t}, \qquad \forall t \geq 0.$$

Proof Without loss of generality, we assume that $\lambda \geq 0$. Note that since $\alpha > \lambda$, one can find a positive constant $\varepsilon \in (0, \alpha)$ such that

$$\lambda e^{\varepsilon r} < \alpha - \varepsilon. \tag{4.4.4}$$

Now by applying Itô's formula to the function $v(t, x) = e^{\alpha t}\Lambda(x)$, $t \geq 0$, $x \in H$ and the solution $y_n(t, \varphi)$ of (4.4.2) in the strong sense, we have that for $t \geq 0$,

$$\begin{aligned}
&e^{\alpha t}\Lambda(y_n(t, \varphi)) - \Lambda(y_n(0, \varphi)) \\
&\quad = \alpha \int_0^t e^{\alpha s}\Lambda(y_n(s, \varphi))ds \\
&\qquad + \int_0^t e^{\alpha s}\langle \Lambda'(y_n(s, \varphi)), Ay_n(s, \varphi) + R(n)F(y_n(s - r, \varphi))\rangle_H ds \\
&\qquad + \int_0^t e^{\alpha s}\langle \Lambda'(y_n(s, \varphi)), R(n)B(y_n(s - r, \varphi))dW(s)\rangle_H \\
&\qquad + \frac{1}{2}\int_0^t e^{\alpha s} Tr\Big\{\Lambda''(y_n(s, \varphi))R(n)B(y_n(s - r, \varphi))Q^{1/2} \\
&\qquad\qquad \times \big[R(n)B(y_n(s - r, \varphi))Q^{1/2}\big]^*\Big\}ds.
\end{aligned} \tag{4.4.5}$$

Hence, by virtue of condition (iii) and the martingale property of stochastic integrals, it is easy to get that for $t \geq 0$,

$$
\begin{aligned}
&e^{\alpha t}\mathbb{E}\Lambda(y_n(t,\varphi))\\
&\quad \le \Lambda(y_n(0,\varphi)) + \lambda\int_0^t e^{\alpha s}\mathbb{E}\Lambda(y_n(s-r,\varphi))ds\\
&\quad\quad + \int_0^t e^{\alpha s}\mathbb{E}\Big\{\langle \Lambda'(y_n(s,\varphi)), (R(n)-I)F(y_n(s-r,\varphi))\rangle_H\\
&\quad\quad + \frac{1}{2}Tr\Big[\Lambda''(y_n(s,\varphi))R(n)B(y_n(s-r,\varphi))Q^{1/2}\big(R(n)B(y_n(s-r,\varphi))Q^{1/2}\big)^*\\
&\quad\quad\quad - \Lambda''(y_n(s,\varphi))B(y_n(s-r,\varphi))Q^{1/2}\big(B(y_n(s-r,\varphi))Q^{1/2}\big)^*\Big]\Big\}ds.
\end{aligned}
\tag{4.4.6}
$$

On the other hand, by virtue of Proposition 4.4.1, there exists a sequence $n_k \in \mathbb{N}_+$ in $\rho(A)$ such that $y_{n_k}(t) \to y(t)$ almost surely as $k \to \infty$, uniformly on any compact set of $[0,\infty)$. Consequently, replacing n by n_k and letting $k \to \infty$ in (4.4.6) immediately yields that

$$
\mathbb{E}\Lambda(y(t,\varphi)) \le \Lambda(\varphi(0))\cdot e^{-\alpha t} + \lambda e^{-\alpha t}\int_0^t e^{\alpha s}\mathbb{E}\Lambda(y(s-r,\varphi))ds \quad \text{for all } t \ge 0.
$$

Therefore, for the preceding $\varepsilon > 0$, we have

$$
\begin{aligned}
\int_0^\infty e^{\varepsilon t}\mathbb{E}\Lambda(y(t,\varphi))dt &\le \Lambda(\varphi(0))\int_0^\infty e^{-(\alpha-\varepsilon)t}dt\\
&\quad + \lambda\int_0^\infty e^{-(\alpha-\varepsilon)t}\int_0^t e^{\alpha s}\mathbb{E}\Lambda(y(s-r,\varphi))dsdt\\
&\le \frac{1}{\alpha-\varepsilon}\Lambda(\varphi(0)) + \frac{\lambda}{\alpha-\varepsilon}\int_0^\infty e^{\varepsilon s}\mathbb{E}\Lambda(y(s-r,\varphi))ds.
\end{aligned}
\tag{4.4.7}
$$

On the other hand, it is not difficult to see that

$$
\int_0^\infty e^{\varepsilon s}\mathbb{E}\Lambda(y(s-r,\varphi))ds \le re^{\varepsilon r}\max_{-r\le\theta\le 0}\Lambda(\varphi(\theta)) + e^{\varepsilon r}\int_0^\infty e^{\varepsilon s}\mathbb{E}\Lambda(y(s,\varphi))ds,
$$

which, in addition to (4.4.7), immediately yields

$$
\begin{aligned}
\int_0^\infty e^{\varepsilon t}\mathbb{E}\Lambda(y(t,\varphi))dt &\le \left(\frac{1}{\alpha-\varepsilon} + \frac{\lambda\cdot r\cdot e^{\varepsilon r}}{\alpha-\varepsilon}\right)\max_{-r\le\theta\le 0}\Lambda(\varphi(\theta))\\
&\quad + \frac{\lambda e^{\varepsilon r}}{\alpha-\varepsilon}\int_0^\infty e^{\varepsilon t}\mathbb{E}\Lambda(y(t,\varphi))dt.
\end{aligned}
$$

Hence, in addition to (4.4.4), there exists a positive number $C' = C'(\varphi) > 0$ such that

$$
\int_0^\infty e^{\varepsilon t}\mathbb{E}\Lambda(y(t,\varphi))dt \le C' \tag{4.4.8}
$$

where

$$C' = \frac{1}{1 - \frac{\lambda e^{\varepsilon r}}{\alpha - \varepsilon}} \left[\left(\frac{1}{\alpha - \varepsilon} + \frac{\lambda \cdot r \cdot e^{\varepsilon r}}{\alpha - \varepsilon} \right) \max_{-r \le \theta \le 0} \Lambda(\varphi(\theta)) \right] < \infty.$$

Now by carrying out a similar argument to that at the end of the proof of Theorem 4.3.3 and using (4.4.8) and condition (i), one can obtain for the preceding $\varepsilon > 0$ that

$$\begin{aligned} e^{\varepsilon t}\mathbb{E}\|y(t,\varphi)\|_H^2 &\le 1/c_1 \cdot e^{\varepsilon t}\mathbb{E}\Lambda(y(t,\varphi)) \\ &\le \frac{1}{c_1}\left\{ c_2\|\varphi(0)\|_H^2 + c_2\lambda\left(re^{\varepsilon r} \max_{-r \le \theta \le 0} \|\varphi(\theta)\|_H^2 + e^{\varepsilon r}C' \right) \right\} \\ &:= C(\varphi) < \infty, \end{aligned}$$

that is, the null solution of (4.4.1) is exponentially stable in mean square, and the proof is now complete. □

In an analogous way as in Theorem 3.3.9, we can establish the following almost sure stability result.

Theorem 4.4.3 *Assume that all the hypotheses in Theorem 4.4.2 hold. Then there exist a nonnegative random variable $T(\omega)$ and numbers $K = K(\varphi) > 0$, $\gamma > 0$ such that for all $t \ge T(\omega)$,*

$$\|y(t,\varphi)\|_H \le K(\varphi) \cdot e^{-\gamma t} \qquad a.s.$$

That is, the null solution of (4.4.1) is exponentially stable in the almost sure sense.

4.4.2 Contraction Mapping Method

Consider the following stochastic differential equation with time delay in H,

$$\begin{cases} dy(t) = Ay(t) + F(t, y(t - \rho(t)))dt + B(t, y(t - \tau(t)))dW(t), \quad t \ge 0, \\ y(0) = \phi_0,\ y(\theta) = \phi_1(\theta) \text{ for } \theta \in [-r,0],\ \phi = (\phi_0,\phi_1) \in H \times L^2([-r,0],H), \end{cases} \tag{4.4.9}$$

where A generates a C_0-semigroup e^{tA}, $t \ge 0$, and $F: \mathbb{R}_+ \times H \to H$, $B: \mathbb{R}_+ \times H \to \mathscr{L}_2(K_Q, H)$ are two Borel measurable mappings with $F(t,0) = 0$, $B(t,0) = 0$ for any $t \ge 0$. Here $\rho(\cdot), \tau(\cdot)\colon [0,\infty) \to [0,r]$ are two continuous functions. We also impose the following global Lipschitz condition: for any x, $y \in H$ and $t \ge 0$,

$$\|F(t,x) - F(t,y)\|_H + \|B(t,x) - B(t,y)\|_{\mathscr{L}_2(K_Q,H)} \le \alpha\|x - y\|_H, \tag{4.4.10}$$

for some constant $\alpha > 0$.

Theorem 4.4.4 *Suppose that $\|e^{tA}\| \le Me^{-\gamma t}$, $M > 0$, $\gamma > 0$ for all $t \ge 0$ and (4.4.10) holds. If, further, the following condition is satisfied:*

$$\gamma > \frac{M^2\alpha^2 + \alpha M\sqrt{M^2\alpha^2+8}}{2}, \qquad (4.4.11)$$

then the null solution of (4.4.9) has global stability in mean square, i.e., for any $\phi \in \mathcal{H} = H \times L^2([-r,0],H)$,

$$\lim_{t\to\infty} \mathbb{E}\|y(t,\phi)\|_H^2 = 0.$$

Proof Let $\mathscr{B}$ be the Banach space of all mean square bounded and $\{\mathscr{F}_t\}_{t\ge0}$-adapted process $\varphi(t)\colon [-r,\infty)\times\Omega \to H$ with $\varphi(t) = \phi_1(t)$ for $t \in [-r,0]$, $\varphi(0) = \phi_0$ such that $\varphi(\cdot)$ is mean square continuous and $\mathbb{E}\|\varphi(t)\|_H^2 \to 0$ as $t \to \infty$, equipped with a supreme norm

$$\|\varphi\|_{\mathscr{B}}^2 = \sup_{t\ge -r} \mathbb{E}\|\varphi(t)\|_H^2 \quad \text{for} \quad \varphi \in \mathscr{B}.$$

Let $S\colon \mathscr{B} \to \mathscr{B}$ be a linear operator defined by $Sy(t) = \phi_1(t)$ for $t \in [-r,0]$, $Sy(0) = \phi_0$ and for $t \ge 0$:

$$\begin{aligned} Sy(t) &= e^{tA}\phi_0 + \int_0^t e^{(t-s)A}F(s, y(s-\rho(s)))ds \\ &\quad + \int_0^t e^{(t-s)A}B(s, y(s-\tau(s)))dW(s) \\ &:= I_1(t) + I_2(t) + I_3(t), \qquad y \in \mathscr{B}. \end{aligned}$$

We first want to show that $S(\mathscr{B}) \subset \mathscr{B}$.

For any $y \in \mathscr{B}$, the adaptability of $Sy(t)$ with respect to $\{\mathscr{F}_t\}_{t\ge0}$ is immediate. To proceed further, we first verify the mean square continuity of $Sy(\cdot)$ on $[0,\infty)$ for any given $y \in \mathscr{B}$. To this end, let $y \in \mathscr{B}$, $t \ge 0$ and $\delta > 0$, then

$$\mathbb{E}\|Sy(t+\delta) - Sy(t)\|_H^2 \le 3\sum_{i=1}^3 \mathbb{E}\|I_i(t+\delta) - I_i(t)\|_H^2.$$

It is easy to see by strong continuity of semigroup e^{tA}, $t \ge 0$, that

$$\mathbb{E}\|I_1(t+\delta) - I_1(t)\|_H^2 = \|(e^{(t+\delta)A} - e^{tA})\phi_0\|_H^2 \to 0 \quad \text{as} \quad \delta \to 0.$$

In addition, by the Dominated Convergence Theorem and standard properties of stochastic integrals, we have for any given $t \ge 0$ that

$$
\begin{aligned}
&\mathbb{E}\|I_3(t+\delta)-I_3(t)\|_H^2\\
&\le 2\mathbb{E}\left\|\int_0^t (e^{(t+\delta-s)A}-e^{(t-s)A})B(s,y(s-\tau(s)))dW(s)\right\|_H^2\\
&\quad+2\mathbb{E}\left\|\int_t^{t+\delta} (e^{(t+\delta-s)A}B(s,y(s-\tau(s)))dW(s)\right\|_H^2\\
&\le 2\mathbb{E}\int_0^t \|(e^{(t+\delta-s)A}-e^{(t-s)A})B(s,y(s-\tau(s)))\|^2_{\mathscr{L}_2(K_Q,H)}ds\\
&\quad+2\mathbb{E}\int_t^{t+\delta} \|e^{(t+\delta-s)A}B(s,y(s-\tau(s)))\|^2_{\mathscr{L}_2(K_Q,H)}ds\\
&\to 0 \quad \text{as} \quad \delta\to 0.
\end{aligned}
$$

Similarly, it can be shown that $\mathbb{E}\|I_2(t+\delta)-I_2(t)\|_H^2\to 0$ as $\delta\to 0$. This proves that for any $y\in\mathscr{B}$, $Sy(\cdot)$ is continuous from the right in mean square on $[0,\infty)$. In a similar way, it can be shown that Sy is also continuous from the left in mean square on $[0,\infty)$.

Next, we want to show that for any $y\in\mathscr{B}$, $\mathbb{E}\|Sy(t)\|_H^2\to 0$ as $t\to\infty$. To this end, let $y\in\mathscr{B}$; then we have that for $t\ge 0$,

$$
\begin{aligned}
\mathbb{E}\|Sy(t)\|_H^2 &\le 3\mathbb{E}\|e^{tA}\phi_0\|_H^2+3\mathbb{E}\left\|\int_0^t e^{(t-s)A}F(s,y(s-\rho(s)))ds\right\|_H^2\\
&\quad+3\mathbb{E}\left\|\int_0^t e^{(t-s)A}B(s,y(s-\tau(s)))dW(s)\right\|_H^2\\
&:=J_1(t)+J_2(t)+J_3(t).
\end{aligned}
\tag{4.4.12}
$$

By assumption, it is immediate to see that

$$J_1(t)\le 3M^2e^{-2\gamma t}\cdot\|\phi_0\|_H^2\to 0 \quad \text{as} \quad t\to\infty.$$

On the other hand, by the Hölder inequality and L'Hospital rule, we have for $y\in\mathscr{B}$ that

$$
\begin{aligned}
\lim_{t\to\infty}J_2(t) &= 3\lim_{t\to\infty}\mathbb{E}\left\|\int_0^t e^{(t-s)A}F(s,y(s-\rho(s)))ds\right\|_H^2\\
&\le \lim_{t\to\infty}3M^2\int_0^t e^{-\gamma(t-s)}ds\int_0^t e^{-\gamma(t-s)}\mathbb{E}\|F(s,y(s-\rho(s)))\|_H^2ds\\
&\le \frac{3M^2\alpha^2}{\gamma}\lim_{t\to\infty}\frac{1}{e^{\gamma t}}\int_0^t e^{\gamma s}\mathbb{E}\|y(s-\rho(s))\|_H^2ds\\
&= \frac{3M^2\alpha^2}{\gamma^2}\lim_{t\to\infty}\mathbb{E}\|y(t-\rho(t))\|_H^2=0.
\end{aligned}
\tag{4.4.13}
$$

In a similar way, for any given $y \in \mathscr{B}$, we have by using properties of stochastic integrals and L'Hospital's rule that

$$
\begin{aligned}
\lim_{t\to\infty} J_3(t) &= \lim_{t\to\infty} \mathbb{E}\left\| \int_0^t e^{(t-s)A} B(s, y(s-\tau(s)))dW(s)\right\|_H^2 \\
&= \lim_{t\to\infty} \int_0^t \mathbb{E}\|e^{(t-s)A} B(s, y(s-\tau(s)))\|^2_{\mathscr{L}_2(K_Q,H)} ds \\
&\le M^2 \lim_{t\to\infty} \frac{1}{e^{2\gamma t}} \int_0^t e^{2\gamma s}\mathbb{E}\|y(s-\tau(s))\|_H^2 ds \\
&= \frac{M^2\beta^2}{2\gamma} \lim_{t\to\infty} \mathbb{E}\|y(t-\tau(t))\|_H^2 = 0.
\end{aligned}
\tag{4.4.14}
$$

Therefore, relations (4.4.12) through (4.4.14) together imply that $S(\mathscr{B}) \subset \mathscr{B}$.

Finally, we show that the mapping $S\colon \mathscr{B} \to \mathscr{B}$ is a contraction. To this end, suppose that $x,\ y \in \mathscr{B}$. By the Lipschitz condition (4.4.10), one can easily get

$$
\begin{aligned}
&\sup_{t\ge -r} \mathbb{E}\|Sx(t) - Sy(t)\|_H^2 \\
&\le 2 \sup_{0\le t<\infty} \mathbb{E}\left\| \int_0^t e^{(t-s)A}[F(s, x(s-\rho(s))) - F(s, y(s-\rho(s)))]ds\right\|_H^2 \\
&\quad + 2 \sup_{0\le t<\infty} \mathbb{E}\left\| \int_0^t e^{(t-s)A}[B(s, x(s-\tau(s))) - B(s, y(s-\tau(s)))]dW(s)\right\|_H^2 \\
&\le 2M^2\alpha^2\gamma^{-2}\left(\sup_{t\ge 0}\mathbb{E}\|x(t)-y(t)\|_H^2\right) \\
&\quad + 2M^2\alpha^2(2\gamma)^{-1}\left(\sup_{t\ge 0}\mathbb{E}\|x(t)-y(t)\|_H^2\right) \\
&= 2M^2\left(\alpha^2\gamma^{-2} + \frac{\alpha^2}{2\gamma}\right)\left(\sup_{t\ge 0}\mathbb{E}\|x(t)-y(t)\|_H^2\right).
\end{aligned}
$$

Hence, if

$$
2M^2\left(\frac{\alpha^2}{\gamma^2} + \frac{\alpha^2}{2\gamma}\right) < 1, \tag{4.4.15}
$$

the mapping S is a contraction on $\mathscr{B}$. Note that inequality (4.4.15) is equivalent to

$$
\begin{aligned}
&\gamma > \frac{M^2\alpha^2 + \sqrt{M^4\alpha^4 + 8M^2\alpha^2}}{2} > 0 \\
\text{or}\quad &\gamma < \frac{M^2\alpha^2 - \sqrt{M^4\alpha^4 + 8M^2\alpha^2}}{2} < 0.
\end{aligned}
$$

Since γ is positive, condition (4.4.11) implies that mapping S is a contraction on $\mathscr{B}$. Moreover, by the Banach fixed-point theorem, there exists a unique point y of S, which is a unique mild solution of (4.4.9) with $y(t) = \phi_1(t)$ on $t \in [-r, 0]$, $y(0) = \phi_0$ and $\mathbb{E}\|y(t)\|_H^2 \to 0$ as $t \to \infty$. The proof is thus now complete. □

Note that a condition, such as (4.4.11), is not delicate enough. However, the contraction mapping method does provide a simple and straightforward way to capture stochastic stability in various situations. For example, all the proofs in Theorem 4.4.4 get through when $r = \infty$, i.e., infinite delay, and $t - \rho(t) \to \infty$, $t - \tau(t) \to \infty$ as $t \to \infty$, and the same idea works very well for a time-delay system of neutral type (see, e.g., [157]).

4.4.3 Razumikhin Function Method

First, let us consider the following finite-dimensional stochastic nonlinear differential equation with delay in $\mathbb{R}^n$, $n \geq 1$,

$$\begin{cases} dy(t) = a(y_t)dt + b(y_t)dw(t), \quad t \geq 0, \\ y(0) = \phi_0, \quad y(\theta) = \phi_1(\theta), \quad \theta \in [-r, 0], \\ \qquad\qquad (\phi_0, \phi_1) \in \mathcal{H} = \mathbb{R}^n \times L^2([-r, 0], \mathbb{R}^n), \end{cases} \tag{4.4.16}$$

where $y_t(\theta) = y(t+\theta)$, $\theta \in [-r, 0]$, $t \geq 0$, $a(\cdot)$, $b(\cdot)$ are appropriate nonlinear mappings and $w(\cdot)$ is a standard real Brownian motion. In order to formulate a non–time-delay system, it is natural to define two nonlinear mappings $\mathcal{A}$ and $\mathcal{B}$ by

$$\mathcal{A}(\phi) = (a(\phi_1), \phi_1') \qquad \text{for all} \qquad \phi \in \mathscr{D}(\mathcal{A}),$$

and

$$\mathcal{B}(\phi) = (b(\phi_1), 0) \qquad \text{for all} \qquad \phi \in \mathcal{H}.$$

Then equation (4.4.16) can be lifted up into a nonlinear stochastic system without delay in $\mathcal{H}$,

$$\begin{cases} dY(t) = \mathcal{A}(Y(t))dt + \mathcal{B}(Y(t))dw(t), \quad t \geq 0, \\ Y(0) = (\phi_0, \phi_1) \in \mathcal{H}, \end{cases} \tag{4.4.17}$$

where $Y(t) = (y(t), y_t) \in \mathcal{H}$, $t \geq 0$. To employ the previous theory such as Theorem 3.3.7 or 3.5.2, one needs to find an appropriate Lyapunov function for (4.4.17) on the infinite-dimensional space $\mathcal{H}$, rather than $\mathbb{R}^n$, to handle the stability property of equation (4.4.16). The key idea of Razumikhin's

Lyapunov function method is to develop a Lyapunov function theory for (4.4.16) on the original space $\mathbb{R}^n$.

Consider the strong solution of the following nonlinear autonomous stochastic functional differential equation in V^*,

$$\begin{cases} dy(t) = A(y(t))dt + F(y_t)dt + B(y_t)dW(t), & t \geq 0, \\ y(\theta) = \varphi(\theta), \ \theta \in [-r,0], \ \varphi \in C([-r,0],H), \end{cases} \tag{4.4.18}$$

where $A\colon V \to V^*$ satisfies (1.4.21) and $F\colon C([-r,0],H) \to H$, $B\colon C([-r,0],H) \to \mathscr{L}_2(K_Q,H)$ are Borel measurable mappings with $A(0) = F(0) = B(0) = 0$. Let $\Lambda(\cdot)\colon H \to \mathbb{R}$ be an Itô type of functional, and we define for any $\varphi \in C([-r,0],H)$ with $\varphi(0) \in V$ the following:

$$\begin{aligned} \mathcal{L}\Lambda(\varphi) = {} & \langle\langle \Lambda'_x(\varphi(0)), A(\varphi(0)) + F(\varphi)\rangle\rangle_{V,V^*} \\ & + \frac{1}{2} \cdot Tr[\Lambda''_{xx}(\varphi(0))B(\varphi)Q^{1/2}(B(\varphi)Q^{1/2})^*], \end{aligned} \tag{4.4.19}$$

whenever the right-hand side is meaningful.

Theorem 4.4.5 *Assume that there exist an Itô type of functional $\Lambda\colon H \to \mathbb{R}$ and constants $c_i > 0$, $i = 1, 2$, and $\lambda > 0$ such that*

$$c_1\|x\|_H^2 \leq \Lambda(x) \leq c_2\|x\|_H^2, \qquad x \in H, \tag{4.4.20}$$

and

$$\mathbb{E}\mathcal{L}\Lambda(\bar{\varphi}) \leq -\lambda\mathbb{E}\Lambda(\bar{\varphi}(0)),$$

for any continuous process $\bar{\varphi}\colon [-r,0] \times \Omega \to H$ satisfying $\bar{\varphi}(0) \in V$ almost surely and

$$\mathbb{E}\Lambda(\bar{\varphi}(\theta)) \leq q\mathbb{E}\Lambda(\bar{\varphi}(0)) \quad \text{for all} \quad \theta \in [-r,0], \tag{4.4.21}$$

with some constant $q > 1$. Then the null solution of (4.4.18) is exponentially stable in mean square, i.e., for any $\varphi \in C([-r,0],H)$,

$$\mathbb{E}\|y(t,\varphi)\|_H^2 \leq \frac{c_2}{c_1}\|\varphi\|^2_{C([-r,0],H)}e^{-\gamma t}, \qquad t \geq 0,$$

for some constant $\gamma > 0$.

Proof Let $\gamma \in (0,\lambda)$ be a sufficiently small number such that

$$e^{\gamma r} + \frac{2\gamma}{2-\gamma} \leq q. \tag{4.4.22}$$

Define

$$U(t) = \max_{-r\leq\theta\leq 0}\big[e^{\gamma(t+\theta)}\mathbb{E}\Lambda(y(t+\theta))\big], \qquad t \geq 0. \tag{4.4.23}$$

By the continuity of $\mathbb{E}\Lambda(y(\cdot))$, $U(t)$ is well defined and continuous. We first show

$$D^+U(t) := \overline{\lim_{h\downarrow 0}} \frac{U(t+h)-U(t)}{h} \le 0 \quad \text{for any} \quad t \ge 0. \tag{4.4.24}$$

To this end, for any fixed $t_0 \ge 0$, we put

$$\bar\theta = \bar\theta(t_0) = \max\left\{\theta \in [-r,0]\colon e^{\gamma(t_0+\theta)}\mathbb{E}\Lambda(y(t_0+\theta)) = U(t_0)\right\}.$$

Obviously, $\bar\theta$ is well defined, $\bar\theta \in [-r,0]$, and

$$e^{\gamma(t_0+\bar\theta)}\mathbb{E}\Lambda(y(t_0+\bar\theta)) = U(t_0).$$

If $\bar\theta < 0$, we have

$$e^{\gamma t_0}\mathbb{E}\Lambda(y(t_0)) < e^{\gamma(t_0+\bar\theta)}\mathbb{E}\Lambda(y(t_0+\bar\theta)).$$

By the continuity of $\mathbb{E}\Lambda(y(\cdot))$ again, it is easy to see that for sufficiently small $h > 0$,

$$e^{\gamma(t_0+h)}\mathbb{E}\Lambda(y(t_0+h)) \le e^{\gamma(t_0+\bar\theta)}\mathbb{E}\Lambda(y(t_0+\bar\theta)).$$

Hence, it follows that

$$U(t_0+h) = \max_{-r\le\theta\le 0}\left[e^{\gamma(t_0+h+\theta)}\mathbb{E}\Lambda(y(t_0+h+\theta))\right] \le U(t_0),$$

which immediately implies $D^+U(t_0) \le 0$.

If $\bar\theta = 0$, then

$$e^{\gamma(t_0+\theta)}\mathbb{E}\Lambda(y(t_0+\theta)) \le e^{\gamma t_0}\mathbb{E}\Lambda(y(t_0)) \quad \text{for any} \quad \theta \in [-r,0],$$

which further implies

$$\mathbb{E}\Lambda(y(t_0+\theta)) \le e^{-\gamma\theta}\mathbb{E}\Lambda(y(t_0)) \le e^{\gamma r}\mathbb{E}\Lambda(y(t_0)) \quad \text{for any} \quad \theta \in [-r,0]. \tag{4.4.25}$$

Next, we divide our proofs into two parts by distinguishing the values of $\mathbb{E}\Lambda(y(t_0))$.

(i) If $\mathbb{E}\Lambda(y(t_0)) = 0$, (4.4.25) and (4.4.20) imply that $y(t_0+\theta) = 0$ for all $\theta \in [-r,0]$ almost surely. Since $A(0) = F(0) = B(0) = 0$, this fact, together with the uniqueness of solutions and (4.4.20), implies that $y(t) = 0$ almost surely for all $t \ge t_0 - r$, and thus $D^+U(t_0) = 0$.

(ii) If $\mathbb{E}\Lambda(y(t_0)) > 0$, by the continuity of function $\mathbb{E}\Lambda(y(\cdot))$, there is a sufficiently small $h > 0$ such that for each $t \in [t_0, t_0+h]$,

$$\begin{aligned} e^{\gamma r}\mathbb{E}\Lambda(y(t_0)) &\le e^{\gamma r}\mathbb{E}\Lambda(y(t)) + \frac{\gamma\mathbb{E}\Lambda(y(t_0))}{2}, \\ \mathbb{E}\Lambda(y(t+\theta)) &\le \mathbb{E}\Lambda(y(t_0+\theta)) + \frac{\gamma\mathbb{E}\Lambda(y(t_0))}{2}, \quad \theta \in [-r,0]. \end{aligned} \tag{4.4.26}$$

Hence, by using (4.4.26) and (4.4.25) we have for $t \in [t_0, t_0 + h]$ and $\theta \in [-r, 0]$ that

$$\begin{aligned}\mathbb{E}\Lambda(y(t+\theta)) &\le \mathbb{E}\Lambda(y(t_0+\theta)) + \frac{\gamma\mathbb{E}\Lambda(y(t_0))}{2}\\ &\le e^{\gamma r}\mathbb{E}\Lambda(y(t_0)) + \frac{\gamma\mathbb{E}\Lambda(y(t_0))}{2}\\ &\le e^{\gamma r}\mathbb{E}\Lambda(y(t)) + \gamma\mathbb{E}\Lambda(y(t_0)).\end{aligned} \tag{4.4.27}$$

On the other hand, for any $t \in [t_0, t_0 + h]$, (4.4.26) implies

$$\left(1 - \frac{\gamma}{2}\right)\mathbb{E}\Lambda(y(t_0)) \le \mathbb{E}\Lambda(y(t)),$$

which, in addition to (4.4.27) and (4.4.22), further yields that for $t \in [t_0, t_0+h]$,

$$\mathbb{E}\Lambda(y(t+\theta)) \le \left(e^{\gamma r} + \frac{2\gamma}{2-\gamma}\right)\mathbb{E}\Lambda(y(t)) \le q\mathbb{E}\Lambda(y(t)) \text{ for any } \theta \in [-r, 0]. \tag{4.4.28}$$

Thus, by assumption, (4.4.28) implies

$$\mathbb{E}(\mathcal{L}\Lambda)(y_t) \le -\lambda\mathbb{E}\Lambda(y(t)), \quad \forall t \in [t_0, t_0 + h]. \tag{4.4.29}$$

Applying Itô's formula along the strong solutions $y(\cdot)$ to (4.4.18), we can obtain by using (4.4.29) that for any $t \in [t_0, t_0 + h]$,

$$\begin{aligned}e^{\gamma t}\mathbb{E}\Lambda(y(t)) &\le e^{\gamma t_0}\mathbb{E}\Lambda(y(t_0)) + \int_{t_0}^{t} (\gamma - \lambda)e^{\gamma u}\mathbb{E}\Lambda(y(u))du\\ &\le e^{\gamma t_0}\mathbb{E}\Lambda(y(t_0)).\end{aligned} \tag{4.4.30}$$

Therefore, it follows that $U(t_0 + h) \le U(t_0)$ for sufficiently small $h > 0$, and further $D^+U(t_0) \le 0$. Since $t_0 \ge 0$ is arbitrary, inequality (4.4.24) is shown to hold for all $t \ge 0$, a fact that further implies

$$U(t) \le U(0), \quad \forall t \ge 0.$$

Now it follows from (4.4.23) that

$$e^{\gamma t}\mathbb{E}\Lambda(y(t)) \le U(t) \le U(0) \le c_2\|\varphi\|^2_{C([-r,0],H)}, \quad \forall t \ge 0,$$

which, in addition to (4.4.20), immediately yields

$$\mathbb{E}\|y(t)\|_H^2 \le \frac{c_2}{c_1}\|\varphi\|^2_{C([-r,0],H)}e^{-\gamma t}, \quad \forall t \ge 0.$$

The proof is now complete. □

Now we apply Theorem 4.4.5 to the stochastic delay differential equation

$$\begin{cases} dy(t) = A(y(t))dt + f(y(t-r))dt + g(y(t-r))dW(t), & t \geq 0, \\ y(\theta) = \varphi(\theta) \in C([-r,0], H), \end{cases} \tag{4.4.31}$$

where $A\colon V \to V^*$ is variational, $f(0) = 0$, $g(0) = 0$ and

$$f(\cdot)\colon H \to H \quad \text{and} \quad g(\cdot)\colon H \to \mathscr{L}_2(K_Q, H),$$

are two nonlinear measurable mappings.

Corollary 4.4.6 *Assume that there exist an Itô type of functional* $\Lambda\colon H \to \mathbb{R}$ *and constants* $c_i > 0$, $\lambda_i > 0$, $i = 1, 2$ *such that for any* $x \in H$,

$$c_1\|x\|_H^2 \leq \Lambda(x) \leq c_2\|x\|_H^2 \tag{4.4.32}$$

and

$$\begin{aligned} &\langle\!\langle \Lambda_x'(\varphi(0)), A(\varphi(0)) + f(\varphi(-r))\rangle\!\rangle_{V,V^*} \\ &\quad + \frac{1}{2}Tr\big[\Lambda_{xx}'' g(\varphi(-r))Q^{1/2}(g(\varphi(-r))Q^{1/2})^*\big] \\ &\leq -\lambda_1\Lambda(\varphi(0)) + \lambda_2\Lambda(\varphi(-r)), \end{aligned} \tag{4.4.33}$$

for any $\varphi \in C([-r,0], H)$ *with* $\varphi(0) \in V$. *If* $\lambda_1 > \lambda_2$, *then the null solution of (4.4.31) is exponentially stable in mean square.*

Proof For any $\varphi \in C([-r,0], H)$, define

$$F(\varphi) = f(\varphi(-r)) \quad \text{and} \quad B(\varphi) = g(\varphi(-r)).$$

Let $\bar{\varphi}\colon [-r,0] \times \Omega \to H$ be a continuous process such that $\bar{\varphi}(0) \in V$ almost surely and

$$\mathbb{E}\Lambda(\bar{\varphi}(\theta)) \leq q\mathbb{E}\Lambda(\bar{\varphi}(0)) \quad \text{for all } \theta \in [-r,0]$$

with some $q \in (1, \lambda_1/\lambda_2)$, then by condition (4.4.33), it follows that

$$\begin{aligned} \mathbb{E}\mathcal{L}\Lambda(\bar{\varphi}) &\leq -\lambda_1\mathbb{E}\Lambda(\bar{\varphi}(0)) + \lambda_2\mathbb{E}\Lambda(\bar{\varphi}(-r)) \\ &\leq -(\lambda_1 - q\lambda_2)\mathbb{E}\Lambda(\bar{\varphi}(0)). \end{aligned} \tag{4.4.34}$$

Hence, by virtue of Theorem 4.4.5, the null solution of (4.4.31) is exponentially stable in mean square. The proof is complete. □

Example 4.4.7 Consider a semilinear stochastic partial differential equation with finite time lag $r > 0$,

$$\begin{cases} dy(t,x) = \dfrac{\partial^2}{\partial x^2} y(t,x)dt + g(y(t-r,x))dw(t), & t \geq 0, \quad x \in (0,\pi), \\ y(t,0) = y(t,\pi) = 0, & t \geq 0, \\ y(s,\cdot) = \varphi(s,\cdot) \in H = L^2(0,\pi), & s \in [-r,0], \quad \varphi(\cdot,x) \in C([-r,0],\mathbb{R}), \end{cases} \tag{4.4.35}$$

where $w(t)$ is a standard one-dimensional Brownian motion. Here $g(\cdot)\colon \mathbb{R} \to \mathbb{R}$ is a Lipschitz continuous function with $|g(x)| \leq L|x|$, $L > 0$, for all $x \in \mathbb{R}$.

Let $A = \partial^2/\partial x^2$ with the domain

$$\mathscr{D}(A) = \left\{ u \in L^2(0,\pi)\colon \frac{\partial u}{\partial x}, \frac{\partial^2 u}{\partial x^2} \in L^2(0,\pi),\ u(0) = u(\pi) = 0 \right\},$$

and

$$V = H_0^1(0,\pi), \qquad H = L^2(0,\pi),$$

so it is easy to show that

$$\langle\!\langle u, Au \rangle\!\rangle_{V,V^*} \leq -\|u\|_H^2, \quad u \in V.$$

On the other hand, suppose that

$$\mathbb{E}\|\bar{\varphi}(\theta)\|_H^2 \leq q\mathbb{E}\|\bar{\varphi}(0)\|_H^2, \quad q > 1, \quad \forall \theta \in [-r,0],$$

for a continuous process $\bar{\varphi}\colon [-r,0] \times \Omega \to H$ with $\bar{\varphi}(0) \in V$ almost surely. It is easy to see

$$\mathbb{E}\|g(\bar{\varphi}(-r))\|_H^2 \leq qL^2\mathbb{E}\|\bar{\varphi}(0)\|_H^2.$$

Let $\Lambda(u) = \|u\|_H^2$, then by a direct computation, we have for this process $\bar{\varphi}$ that

$$\mathbb{E}\mathcal{L}\Lambda(\bar{\varphi}) \leq (-2 + qL^2)\mathbb{E}\|\bar{\varphi}(0)\|_H^2.$$

By using Theorem 4.4.5 and letting $q \to 1$, if necessary, it is easy to see that when $L < \sqrt{2}$, the null solution of (4.4.35) is exponentially stable in mean square.

4.4.4 Comparison Method

In this section, we shall present some comparison theorems for stochastic stability. A remarkable feature of this method is that the stability property of stochastic functional differential equations can be handled by considering a similar property of some deterministic functional differential equations.

We say that a function $f(\cdot)\colon C([-r,0],\mathbb{R}) \to \mathbb{R}$ is *nondecreasing* if $f(\xi_1) \le f(\xi_2)$ for any $\xi_1, \xi_2 \in C([-r,0],\mathbb{R})$ satisfying $\xi_1(\theta) \le \xi_2(\theta)$ for each $\theta \in [-r,0]$. Let $T \ge 0$ and consider the following deterministic functional differential equation in $\mathbb{R}$,

$$\begin{cases} dx(t) = f(x_t)dt, & t \in [0,T], \\ x_0 = \xi, & \xi \in C([-r,0],\mathbb{R}), \end{cases} \tag{4.4.36}$$

where $x_t(\theta) = x(t+\theta), \theta \in [-r,0]$.

Definition 4.4.8 Let $\bar{x}(t,\xi)$ be a particular solution of (4.4.36) with initial $\xi \in C([-r,0],\mathbb{R})$ on $[0,T]$. For any other solution $x(t,\xi)$ of (4.4.36), if

$$x(t,\xi) \le \bar{x}(t,\xi), \qquad \forall t \in [0,T],$$

then $\bar{x}(t,\xi)$ is said to be the *maximal solution* of (4.4.36). It is clear that the maximal solution, if it exists, must be unique.

The following lemma establishes the existence of a unique maximal solution of equation (4.4.36) and its proofs are referred to Theorem 6.9.3, p. 36 in [118].

Lemma 4.4.9 *Suppose $f(\cdot)\colon C([-r,0],\mathbb{R}) \to \mathbb{R}$ is a continuous function that is nondecreasing. Then, for arbitrarily given initial function $\xi \in C([-r,0],\mathbb{R})$, there exists some $T \ge 0$ such that equation (4.4.36) admits a unique maximal solution, denote it by $\bar{x}(t,\xi)$, on $[0,T]$.*

Proposition 4.4.10 *In addition to those conditions in Lemma 4.4.9, we assume that*

(a) the function $f(\cdot)$ is concave on $C([-r,0],\mathbb{R})$, i.e.,

$$\frac{f(\xi_1) + f(\xi_2)}{2} \le f\Big(\frac{\xi_1 + \xi_2}{2}\Big) \quad \textit{for any} \quad \xi_1,\ \xi_2 \in C([-r,0],\mathbb{R});$$

(b) there exists an Itô type of functional $\Lambda\colon H \to \mathbb{R}$ such that for any $\varphi \in C([-r,0],H)$ with $\varphi(0) \in V$,

$$\mathcal{L}\Lambda(\varphi) \le f(\Lambda(\varphi(\cdot))),$$

where $\mathcal{L}\Lambda$ is given as in (4.4.19);

(c) the maximal solution $\bar{x}(t,\xi)$ of (4.4.36) exists in $[0,\infty)$, i.e., $T = \infty$.

Further, assume that the inequality

$$\Lambda(\varphi(\theta)) \le \xi(\theta), \quad \forall \theta \in [-r,0], \tag{4.4.37}$$

holds, then the strong solution $y(t) = y(t,\varphi)$, $t \geq 0$, to (4.4.18) with initial $\varphi \in C([-r,0],H)$ satisfies

$$\mathbb{E}\Lambda(y(t,\varphi)) \leq \bar{x}(t,\xi), \quad t \geq 0. \tag{4.4.38}$$

Proof Applying Itô's formula to the function $\Lambda(\cdot)$ along the strong solutions y of (4.4.18) and using (a), (b), and (c), one can easily obtain that for any $t \geq 0$ and $\varepsilon > 0$,

$$\mathbb{E}\Lambda(y(t+\varepsilon)) - \mathbb{E}\Lambda(y(t)) \leq \int_t^{t+\varepsilon} f(\mathbb{E}\Lambda(y(s+\cdot)))ds, \tag{4.4.39}$$

which immediately yields that for any $t \geq 0$,

$$D^+\mathbb{E}\Lambda(y(t)) := \overline{\lim_{\varepsilon\to 0^+}}\, \varepsilon^{-1}\big[\mathbb{E}\Lambda(y(t+\varepsilon)) - \mathbb{E}\Lambda(y(t))\big] \leq f(\mathbb{E}\Lambda(y(t+\cdot))). \tag{4.4.40}$$

Hence, it is not difficult to show (see, e.g., Theorem 8.1.4 in [118]) that

$$\mathbb{E}\Lambda(y(t)) \leq \bar{x}(t,\xi), \quad t \geq 0,$$

whenever $\Lambda(\varphi(\theta)) \leq \xi(\theta)$ for all $\theta \in [-r,0]$. □

A continuous function $g: \mathbb{R}_+ \to \mathbb{R}_+$ is said to be in a class $\mathcal{K}$ if g is strictly increasing with $g(0) = 0$, and in a class $\mathcal{VK}$ if $g \in \mathcal{K}$ and g is convex. A continuous function $g: \mathbb{R}_+ \to \mathbb{R}_+$ is said to be in a class $\mathcal{CK}$ if $g(0) = 0$ and g is concave, strictly increasing. Recall that the null solution of (4.4.36) is called stable if for arbitrarily given $\varepsilon > 0$, there exists $\delta = \delta(\varepsilon) > 0$ such that

$$\|\xi\|_{C([-r,0],\mathbb{R})} < \delta \quad \text{implies} \quad |x(t,\xi)| < \varepsilon \quad \text{for all} \quad t \geq 0. \tag{4.4.41}$$

Theorem 4.4.11 *Under the conditions (a), (b), and (c) in Proposition 4.4.10, if there exist functions $b(\cdot) \in \mathcal{K}$ and $a(\cdot) \in \mathcal{CK}$ such that*

$$b(\|x\|_H) \leq \Lambda(x) \leq a(\|x\|_H), \tag{4.4.42}$$

then the stability in the sense of (4.4.41) of the null solution to (4.4.36) implies stability in probability of the null solution to (4.4.18).

Proof By virtue of (4.4.42), we have

$$0 \leq \mathbb{E}b(\|y(t)\|_H) \leq \mathbb{E}\Lambda(y(t)) \leq \mathbb{E}a(\|y(t)\|_H) \leq a(\mathbb{E}\|y(t)\|_H), \quad t \geq 0. \tag{4.4.43}$$

Now assume that the null solution of (4.4.36) is stable. Let ε and η ($\eta < 1$) be arbitrarily given positive numbers. For the number $\eta b(\varepsilon) > 0$, there exists a number $\delta' = \delta'(\varepsilon,\eta) > 0$ such that $|x(t,\xi)| < \eta b(\varepsilon)$, $\forall t \geq 0$, for $\|\xi\|_{C([-r,0],\mathbb{R})} < \delta'$, and so we have

$$\bar{x}(t,\xi) < \eta b(\varepsilon), \quad \forall t \geq 0. \tag{4.4.44}$$

Since $a \in \mathcal{CK}$, we can choose a proper $\xi \in C([-r,0],\mathbb{R})$ with $0 < \|\xi\|_{C([-r,0],\mathbb{R})} < \delta'$ and a positive number δ small enough such that

$$a(\|\varphi(\theta)\|_H) \le \xi(\theta), \quad \forall \theta \in [-r,0], \tag{4.4.45}$$

for $\|\varphi\|_{C([-r,0],H)} < \delta$. By virtue of (4.4.45) and (4.4.42), we obtain

$$\Lambda(\varphi(\theta)) \le a(\|\varphi(\theta)\|_H) \le \xi(\theta), \quad \theta \in [-r,0].$$

From (4.4.44), Proposition 4.4.10 and the above inequality, we thus get

$$\mathbb{E}\Lambda(y(t,\varphi)) \le \bar{x}(t,\xi) < \eta b(\varepsilon), \quad \forall t \ge 0. \tag{4.4.46}$$

Now we want to show that the null solution of (4.4.18) is stable in probability, i.e.,

$$\mathbb{P}\{\|y(t,\varphi)\|_H \ge \varepsilon\} < \eta, \quad \forall t \ge 0. \tag{4.4.47}$$

Indeed, if (4.4.47) is not true, there will exist some $t^* \ge 0$ such that

$$\mathbb{P}\{\|y(t^*,\varphi)\|_H \ge \varepsilon\} \ge \eta. \tag{4.4.48}$$

Hence, by virtue of (4.4.42), (4.4.45), (4.4.46) and the well-known Chebyshev inequality, we have

$$\begin{aligned}\mathbb{P}\{\|y(t^*,\varphi)\|_H \ge \varepsilon\} &= \mathbb{P}\{b(\|y(t^*,\varphi)\|_H) \ge b(\varepsilon)\}\\ &\le \mathbb{P}\{\Lambda(y(t^*,\varphi)) \ge b(\varepsilon)\}\\ &\le \frac{1}{b(\varepsilon)}\mathbb{E}\Lambda(y(t^*,\varphi)) < \frac{\eta b(\varepsilon)}{b(\varepsilon)} = \eta.\end{aligned} \tag{4.4.49}$$

This contradicts (4.4.48), so (4.4.47) must be true. □

Theorem 4.4.12 *Under the conditions (a), (b) and (c) of Proposition 4.4.10, if there exist functions $b \in \mathcal{VK}$ and $a \in \mathcal{CK}$ such that*

$$b(\|x\|_H^2) \le \Lambda(x) \le a(\|x\|_H^2), \tag{4.4.50}$$

then the stability in the sense of (4.4.41) of the null solution to (4.4.36) implies stability in mean square of the null solution to (4.4.18).

Proof Let ε be an arbitrarily given positive number. By a similar argument to Theorem 4.4.11, one can show that there exists a positive number $\delta_1 = \delta_1(\varepsilon)$ such that if $\|\xi\|_{C([-r,0],\mathbb{R})} < \delta_1$ and $\Lambda(\varphi(\theta)) \le \xi(\theta)$ for each $\theta \in [-r,0]$, then

$$\mathbb{E}\Lambda(y(t,\varphi)) \le \bar{x}(t,\xi) < b(\varepsilon), \quad t \ge 0. \tag{4.4.51}$$

In particular, let

$$\xi(\theta) = a(\|\varphi(\theta)\|_H^2), \quad \forall \theta \in [-r,0].$$

As $a \in \mathcal{CK}$ and for the number $\delta_1 > 0$, we may find $\delta > 0$ such that

$$\xi(\theta) = a(\|\varphi(\theta)\|_H^2) < \delta_1, \quad \forall \theta \in [-r, 0],$$

whenever $\|\varphi\|_{C([-r,0],H)} < \delta$. Hence, by means of (4.4.50) and (4.4.51), we have

$$b(\mathbb{E}\|y(t,\varphi)\|_H^2) < b(\varepsilon), \quad t \geq 0,$$

which further implies

$$\mathbb{E}\|y(t,\varphi)\|_H^2 < \varepsilon, \quad t \geq 0.$$

The proof is now complete. □

4.5 Notes and Comments

In finite-dimensional spaces, there exists a satisfactory theory about stability of functional differential equations, e.g., see Hale [84], Kolmanovskii and Nosov [107], and references therein. For stochastic systems, the reader is referred to Kolmanovskii and Myshkis [106] and Mao [165, 167] for some systematic statements. In an infinite-dimensional setting, Liu [137] is, as far as I know, the only single volume fully dealing with the topic of stochastic stability in the existing literature.

In the first part of this chapter, we introduced two methods, i.e., the Green operator and system lift-up, in handling stochastic stability of linear systems with time delay. The main content of Sections 4.1 and 4.2 is taken from Liu [139, 141], in which the spectrum method plays an essential role. The main ideas of Section 4.3 come from Bierkens [19] and Caraballo [24]. Here we mainly focus on bounded linear operators A_1 and $A_2(\cdot)$ in delay terms. For unbounded operator situations, some stability properties have been considered in Liu [143, 145]. It was first observed by Datko et al. [63] that small delays may ruin exponential stability for a partial differential equation, although the same stability can never be destroyed by small delays in a finite-dimensional setting. More precisely, consider a delay equation in a Hilbert space H,

$$\begin{cases} dy(t) = Ay(t)dt + A_1 y(t-r)dt, \qquad t \geq 0, \\ y(0) = \phi_0 \in H, \; y(t) = \phi_1(t), \; t \in [-r, 0], \; \phi_1 \in L^2([-r,0], H), \end{cases} \tag{4.5.1}$$

where A generates a C_0-semigroup. If the spectrum set of A is unbounded along an imaginary line, it is shown in [12] that for any $r_0 > 0$, one can always find a linear bounded operator A_1 on H and $r \in (0, r_0)$ such that

$A + A_1$ generates an exponentially stable semigroup and meanwhile, system (4.5.1) is not exponentially stable. In fact, it is shown that if A generates a norm continuous C_0-semigroup and the semigroup generated by $A + A_1$ is exponentially stable, then there exists a constant $r_0 > 0$ such that the system (4.5.1) is exponentially stable for all $r \in (0, r_0)$. In Section 4.3.2, we consider a similar sensitivity problem of stability in both the mean square and almost sure sense to small delays for stochastic systems, and Examples 4.3.1 and 4.3.2 show cases where exponential stability is not sensitive to small delays.

Although it is possible to show some best stability results for linear stochastic systems, the results for nonlinear systems are less complete. In the second part, we consider stability of nonlinear stochastic systems by presenting various methods from the existing literature. We introduced in Section 4.4 several useful methods that provide sufficient conditions for stability in one way or another. Among them, the Lyapunov function method is a natural time-delay generalization of those developed in the previous chapters, while contraction mapping and comparison ones are more straightforward, although they may provide less attractive results. Apart from its usefulness in stochastic stability, the contraction mapping method also finds its decent applications in some related topics such as periodic solutions of stochastic evolution equations (see, e.g., Bezandry and Diagana [18]). In finite-dimensional situations, Razumikhin's ideas take advantage of the structure of $\mathbb{R}^n$ while for a Lyapunov function type of approach, one needs to handle difficulties caused by the underlying infinite-dimensional structure. This is clearly illustrated by an example in Levin and Nohel [124]. The basic idea of Razumikhin's method was extended to obtain some stochastic versions of differential equations in Hilbert spaces by many authors, such as Taniguchi [204], Luo and Liu [156], and so on. Because all these methods provide only sufficient conditions, it is important to compare and contrast the results obtained thereby.

There exists some work on stability of second-order stochastic differential equations with delay in abstract spaces, e.g., Ren and Sun [196] or Liang and Guo [128]. To deal with them, one possible method is to convert these kinds of systems into some first-order stochastic systems on proper product spaces, e.g., Bátkai and Piazzera [12], and then employ the associated stochastic stability theory. Another possible way is to introduce some second-order versions such as a cosine family of the usual C_0-semigroups, e.g., Fattorini [72], and then develop a traditional theory to deal with second-order stochastic evolution equations. At the present time, there does not exist a satisfactory theory about this topic.

In this book, we don't present any stability results about stochastic systems with infinite or unbounded delay. This is, of course, an important topic that is, in general, more difficult to handle than finite-delay systems. In this respect, there exist quite a few research works, e.g., Luo and Taniguchi [157]. For stability of stochastic functional differential equations of neutral type, it is informative to see Caraballo et al. [36], Govidan [81], Liu [136], Liu and Xia [152], and Luo [155]. For stability of stochastic equations driven by a jump process or fractional Brownian motion, the reader may find interesting results in Caraballo et al. [25], Cui et al. [44], Liang and Gao [127], and references therein.

5

Some Applications Related to Stochastic Stability

In this chapter, I shall present some selected topics in applications related to stochastic stability of stochastic differential equations in infinite dimensions. The choice of the material reflects my own personal preference and the presentation here is somewhat sketchy. Some material chosen deals with specific stochastic systems that could be regarded as a potential starting point of research in this area. Some material reveals interesting and important relationships between the main topic of this book, stochastic stability, and topics from other branches of science or engineering.

5.1 Applications in Mathematical Biology

In this section, we shall consider a system of two interacting populations in which each population density follows a stochastic partial differential equation. Precisely, consider two populations both living in a bounded domain $\mathcal{O} \subset \mathbb{R}^n$. The evolution of their densities u_i, $i = 1, 2$, defined as the number of individuals per unit volume or area, is a result of three competing factors. Firstly, both populations can migrate in $\mathcal{O}$ according to a macroscopic diffusion described by Laplacian Δ. Secondly, the interaction between these populations is modeled by functions

$$f_i(u_1, u_2) = u_i - a_i u_i^2 \pm b_i u_1 u_2, \tag{5.1.1}$$

where $u_i \in \mathbb{R}$, a_i, $b_i \geq 0$, $i = 1, 2$. This just implies that in the absence of the other population, the considered one will grow in accordance with the logistic law. The term $\pm b_i u_1 u_2$, $i = 1, 2$, describes two typical interactions, namely the *predator–prey* model when the signs in front of b_i are different and the *competition* one when the signs are negative in both terms. Thirdly, it is assumed that each population is randomly disturbed by spatially dependent white noise $\dot{W}_i(t,x)$, $t \geq 0$, $x \in \mathcal{O}$, defined on a complete probability space $(\Omega, \mathscr{F}, \mathbb{P})$ with a state-dependent noise intensity $\sigma_i(u_i)$, $i = 1, 2$. Thus, the system in which we are interested can be modeled as

$$\begin{cases} \partial u_1(t,x)/\partial t = \nu_1 \Delta u_1(t,x) + f_1(u_1(t,x), u_2(t,x)) + \sigma_1(u_1(t,x))\dot{W}_1(t,x), \\ \partial u_2(t,x)/\partial t = \nu_2 \Delta u_2(t,x) + f_2(u_1(t,x), u_2(t,x)) + \sigma_2(u_2(t,x))\dot{W}_2(t,x), \end{cases} \tag{5.1.2}$$

where ν_1, $\nu_2 > 0$, $t \in (0, T]$, $x \in \mathcal{O}$, and $T > 0$ is an arbitrary number. In addition, assume that u_i, $i = 1, 2$, satisfy the initial condition

$$u_i(0,x) = \alpha_i(x) \geq 0, \quad x \in \overline{\mathcal{O}},$$

where $\alpha_i : \overline{\mathcal{O}} \to \mathbb{R}_+$ and $\overline{\mathcal{O}}$ is the closure of $\mathcal{O}$ in $\mathbb{R}^n$. Moreover, u_i, $i = 1, 2$, satisfy some homogeneous Dirichlet boundary condition, i.e.,

$$u_i(t,x) = 0, \quad x \in \partial\mathcal{O}, \quad t \in [0,T],$$

or homogeneous Neumann condition, i.e.,

$$\frac{\partial}{\partial n} u_i(t,x) = 0, \quad x \in \partial\mathcal{O}, \quad t \in [0,T],$$

where n denotes the normal vector. We require that the boundary $\partial\mathcal{O}$ is smooth enough to guarantee that the Green function of $\partial/\partial t - \Delta$ with the corresponding boundary conditions does exist. The formal problem (5.1.2) is a stochastic counterpart of a logistic population growth model with migration (cf. Murray [180]).

Let X be a Banach space and H a separable Hilbert space. Consider a joint stochastic process $(U(t), V(t))$, $t \geq 0$, in the product space $H \times X$ such that its first component U is governed by some stochastic evolution equation in H,

$$\begin{cases} U(t) = e^{tA}U_0 + \int_0^t e^{(t-s)A} F(U(s), V(s))ds \\ \qquad + \int_0^t e^{(t-s)A} B(U(s))dW(s), \quad t \geq 0, \\ U(0) = U_0 \in H, \end{cases} \tag{5.1.3}$$

where e^{tA}, $t \geq 0$, is some strongly continuous semigroup on H with its infinitesimal generator A, and W stands for a Wiener process with its trace class covariance operator Q in a separable Hilbert space K. Here $F: H \times X \to H$ and $B: H \to \mathscr{L}_2(K_Q, H)$ are assumed to be Lipschitz continuous. Furthermore, the process $V(t)$, $t \geq 0$, is assumed to be pathwise continuous in X such that

$$\int_0^T \mathbb{E}\|V(t)\|_X^p dt < \infty \quad \text{for some } p \geq 1 \text{ and each } T \geq 0.$$

Let Λ be an Itô functional and define a mapping $\mathcal{L}$ by

$$(\mathcal{L}\Lambda)(u,v) = \langle Au, \Lambda'(u)\rangle_H + \langle F(u,v), \Lambda'(u)\rangle_H + \frac{1}{2}Tr[\Lambda''(u)B(u)Q^{1/2}(B(u)Q^{1/2})^*]$$

for any $(u,v) \in \mathscr{D}(A) \times X$. Furthermore, let $\Sigma(r)$ be a centered open ball in H with radius $r > 0$. By employing a Yosida approximation type of argument, the proof of the following theorem can be similarly given to Theorem 3.4.3.

Theorem 5.1.1 *Assume that $F(0,v) = 0$ and $B(0) = 0$ for each $v \in X$. Suppose that there exists an Itô functional Λ such that for each $r > 0$ and $u \in H \backslash \Sigma(r)$,*

$$|\Lambda(u)| + \|\Lambda'(u)\|_H + \|\Lambda''(u)\| \le c(r)(1 + \|u\|_H^p), \tag{5.1.4}$$

for some $p \ge 1$, $c(r) > 0$ and

$$(\mathcal{L}\Lambda)(u,v) \le -b(\|v\|_X)\Lambda(u), \quad (u,v) \in \mathscr{D}(A) \times X, \tag{5.1.5}$$

where $b\colon \mathbb{R}_+ \to \mathbb{R}_+$ is continuous and satisfies

$$b(x) \le M(1+x), \qquad x \in \mathbb{R}_+,$$

for some $M > 0$ and

$$\varliminf_{t\to\infty} b(\|V(t)\|_X) > 0 \quad a.s. \tag{5.1.6}$$

Further, suppose that Λ satisfies

$$l_r := \inf_{\|u\|_H > r} \Lambda(u) > 0 \tag{5.1.7}$$

for each $r > 0$ and $\Lambda(0) = 0$. Then the null solution of (5.1.3) is asymptotically stable in sample path,

$$\lim_{t\to\infty} \|U(t)\|_H = 0 \qquad a.s.$$

for each initial $U_0 \in H$.

Next, we use Theorem 5.1.1 to derive some sufficient Lyapunov type conditions for stability of the null solution to system (5.1.2). In particular, we shall consider the following predator–prey case

$$\begin{cases} f_1(u_1,u_2) = u_1 - a_1u_1^2 + b_1u_1u_2, \\ f_2(u_1,u_2) = u_2 - a_2u_2^2 - b_2u_1u_2, \end{cases} \tag{5.1.8}$$

where $u_i \in \mathbb{R}_+$, $a_i, b_i \ge 0$, $i = 1, 2$. The corresponding competition case can be similarly addressed.

Let $X = C(\overline{\mathcal{O}})$, $H = L^2(\mathcal{O})$, $A_i = \nu_i \Delta$ with domains equal either to $H^2(\mathcal{O}) \cap H_0^1(\mathcal{O})$ or to $\{\phi \in H^2(\mathcal{O})\colon \partial\phi(x)/\partial n = 0, x \in \partial\mathcal{O}\}$ in the case of Dirichlet or Neumann boundary conditions, respectively. Furthermore, let

$$(B_i(u)v)(x) = \sigma_i(u(x))v(x), \quad u,\ v \in H, \quad x \in \overline{\mathcal{O}},$$

be the multiplication operators defined by real-valued functions σ_i, $i = 1, 2$. Denote by Q_i the covariance operators of W_i, $i = 1, 2$. The basic idea to obtain stability is a comparison of the solution to (5.1.2) with an essentially simpler system, which is given by

$$\begin{cases} \partial U_1(t,x)/\partial t = A_1 U_1(t,x) + U_1(t,x) + b_1 U_1(t,x) U_2(t,x) \\ \qquad\qquad\qquad + B_1(U_1(t,x))\dot{W}_1(t,x), \qquad t \geq 0, \\ \partial U_2(t,x)/\partial t = A_2 U_2(t,x) + U_2(t,x) + B_2(U_2(t,x))\dot{W}_2(t,x), \qquad t \geq 0, \end{cases} \tag{5.1.9}$$

in the space $H \times X$ with proper initial functions $U_{1,0}$, $U_{2,0}$. For our purpose, we impose the following conditions:

(i) $U_{i,0}\colon \overline{\mathcal{O}} \to \mathbb{R}$ are measurable and there exists a constant $M > 0$ such that

$$0 \leq U_{i,0}(x) \leq M$$

for all $x \in \overline{\mathcal{O}}$, $i = 1, 2$;

(ii) $\sigma_i\colon \mathbb{R} \to \mathbb{R}$, $i = 1, 2$, are globally Lipschitz continuous;

(iii) $\sigma_i(0) = 0$, $i = 1, 2$.

Also, the strongly continuous semigroups e^{tA_i}, $t \geq 0$, defined by their infinitesimal generators A_i are assumed to have the form

$$(e^{tA_i}(t)u)(x) = \int_{\overline{\mathcal{O}}} G_i(t,x,y)u(y)dy, \quad u \in H, \quad x \in \mathcal{O}, \quad t \geq 0,$$

where $G_i(t,x,y) = g(\nu_i t,x,y)$, $i = 1, 2$, and g is the standard Green function to $\partial/\partial t - \Delta$. It can be shown that under the conditions (i) through (iii), the equation (5.1.2) has a pathwise unique nonnegative solution, and so does (5.1.9) since system (5.1.9) represents a particular case of (5.1.2). By establishing a comparison principle (see Manthey and Zausinger [163]), one can carry out a truncation program of solutions to show that for equations (5.1.2) and (5.1.9),

$$\begin{aligned} 0 \leq u_1(t,x) \leq U_1(t,x), \\ 0 \leq u_2(t,x) \leq U_2(t,x), \end{aligned} \tag{5.1.10}$$

almost surely for each $(t,x) \in \mathbb{R}_+ \times \overline{\mathcal{O}}$.

We first present a stability assertion to (5.1.2) that requires stability of the null solution of the second component to (5.1.9).

Theorem 5.1.2 *Suppose that for some nonnegative initial datum* $U_{2,0} \in C(\overline{\mathcal{O}})$,

$$\lim_{t\to\infty} \|U_2(t)\|_{C(\overline{\mathcal{O}})} = 0 \qquad a.s. \tag{5.1.11}$$

Moreover, assume that one of the following conditions (a) or (b) is satisfied:

(a)

$$-\beta_1 + 1 + \frac{1}{2} L_{\sigma_1}^2 Tr(Q_1) < 0, \tag{5.1.12}$$

where $-\beta_1$ *stands for the first eigenvalue of operator* A_1 *and* L_{σ_1} *for Lipschitz constant of* σ_1*;*

(b) $W_1 = w$ *is a standard scalar Brownian motion,*

$$\|\sigma_1(u)\|_H \geq \kappa u, \quad u \geq 0,$$

for some $\kappa \geq 0$ *and*

$$-\beta_1 + 1 + \frac{1}{2} L_{\sigma_1}^2 - \kappa^2 < 0. \tag{5.1.13}$$

Then every solution (u_1, u_2) *of (5.1.2) satisfies that*

$$\lim_{t\to\infty} \|u_1(t,\cdot)\|_H = 0 \qquad a.s.$$

with nonnegative initial $u_1(0,x) = \phi_1 \in L^2(\mathcal{O})$*, and*

$$\lim_{t\to\infty} \|u_2(t,\cdot)\|_{C(\overline{\mathcal{O}})} = 0 \qquad a.s.$$

with nonnegative initial $u_2(0,x) = \phi_2(x) \in C(\overline{\mathcal{O}})$.

Proof According to (5.1.10), it is enough to show that $\|U_1(t)\|_H \to 0$ as $t \to \infty$ almost surely. To this end, we use Theorem 5.1.1 with the function $\Lambda(h) = \|h\|_H^{2\gamma}$ where $\gamma > 0$ will be specified later on. The conditions on Λ required in Theorem 5.1.1 are satisfied trivially except for condition (5.1.5), which is to be checked now with $X = C(\overline{\mathcal{O}})$. In the case (a), letting $\gamma = 1$, we get

$$\begin{aligned}(\mathcal{L}\Lambda)(u,v) = 2(\langle A_1 u, u\rangle_H + \|u\|_H^2 + b_1\langle u\cdot v, u\rangle_H)\\ + Tr[B_1(u)QB_1(u)^*], \quad (u,v) \in \mathscr{D}(A_1) \times X,\end{aligned} \tag{5.1.14}$$

where $u \cdot v$ means the usual multiplication of functions u and v. It thus follows that

$$(\mathcal{L}\Lambda)(u,v) \leq 2\|u\|_H^2\Big(-\beta_1 + 1 + b_1\|v\|_{C(\overline{\mathcal{O}})} + \frac{1}{2}L_{\sigma_1}^2 Tr(Q_1)\Big),$$
$$(u,v) \in \mathscr{D}(A_1) \times X.$$

Thus let

$$b(y) := 2\Big(\beta_1 - 1 - \frac{1}{2}L_{\sigma_1}^2 Tr(Q_1) - b_1 y\Big), \quad y \in \mathbb{R}_+,$$

then we obtain (5.1.5) when (5.1.11) is true. However, conditions (5.1.11) and (5.1.12) trivially yield (5.1.5) in the present case, so the assumptions of Theorem 5.1.1 are verified and we obtain $\|U_1(t)\| \to 0$ as $t \to \infty$ almost surely, which proves case (a).

In the case (b), we may similarly have

$$(\mathcal{L}\Lambda)(u,v) \le 2\gamma \|u\|_H^{2\gamma}\Big(-\beta_1 + 1 + b_1\|v\|_{C(\overline{\mathcal{O}})} + \frac{1}{2}L_{\sigma_1}^2 + (\gamma - 1)\kappa^2\Big),$$

for any $(u, v) \in \mathscr{D}(A_1) \times X$, so we may obtain (5.1.5) with

$$b(y) := 2\gamma\Big(\beta_1 - 1 - \frac{1}{2}L_{\sigma_1}^2 + (1-\gamma)\kappa^2 - b_1 y\Big), \quad y \in \mathbb{R}_+.$$

However, by choosing $\gamma > 0$ small enough and taking into account (5.1.11), it is easy to see that the conditions of Theorem 5.1.1 are verified and, consequently, $\|U_1(t)\|_H \to 0$ as $t \to \infty$ almost surely, which proves the case (b). □

By virtue of Theorem 5.1.2, in order to show stability of the null solution to (5.1.2), it suffices to prove stability of a single equation for U_2 in (5.1.9), where the "drift" is linear. However, the convergence of U_2 in the $C(\overline{\mathcal{O}})$-norm is needed, which is more difficult to prove than the analogous convergence in H. In the following theorem, we present some results on stability in $C(\overline{\mathcal{O}})$ for the second component in (5.1.9). Together with Theorem 5.1.2, this will give sufficient conditions for stability of the original system (5.1.2).

Theorem 5.1.3 *Suppose that $W_2 = w$ is a standard scalar Brownian motion.*

(a) Consider the second equation in (5.1.9) with Neumann boundary condition and assume

$$\kappa u \le \|\sigma_2(u)\|_H, \quad u \ge 0,$$

for some $0 \le \kappa \le L_{\sigma_2}$, and $L_{\sigma_1}^2 < 2\kappa^2 - 2$. Then $\lim_{t\to\infty} \|U_2(t)\|_{C(\overline{\mathcal{O}})} = 0$ almost surely.

(b) If $\sigma_2(u) = L_{\sigma_2} u$ and $2 < L_{\sigma_2}^2$, then $\lim_{t\to\infty} \|U_2(t)\|_{C(\overline{\mathcal{O}})} = 0$ almost surely.

Proof (a) The second equation of (5.1.9) with Neumann boundary conditions and a one-dimensional Wiener process is spatially homogeneous. If we start from a spatially constant initial function, the solution evolves in the space of constant functions. Therefore, taking a constant $C_2 > 0$ such that $C_2 \geq \phi_2(x)$, $x \in \overline{\mathcal{O}}$, a comparison theorem type of argument yields

$$U_2(t,x) \leq v_2(t), \quad x \in \overline{\mathcal{O}},$$

where $v_2(t)$ solves the one-dimensional stochastic differential equation

$$dv_2(t) = v_2(t)dt + \sigma_2(v_2(t))dw(t)$$

with $v_2(0) = C_2 > 0$. A simple one-dimensional version of the proof of Theorem 5.1.2 yields $v_2(t) \to 0$ as $t \to \infty$ almost surely.

In (b), the Neumann boundary condition is no longer demanded. To prove this assertion, note that Itô formula yields

$$U_2(t,x) = (e^{tA_2}\phi_2)(x) \cdot \exp\left[(1 - L_{\sigma_2}^2/2)t + L_{\sigma_2}w(t)\right]$$

for $x \in \overline{\mathcal{O}}$, where e^{tA_2} is the semigroup generated by A_2 in the space $C(\overline{\mathcal{O}})$. Obviously, e^{tA_2} is a contraction semigroup and

$$\exp\left\{\left[(1 - L_{\sigma_2}^2/2) + L_{\sigma_2}w(t)t^{-1}\right]t\right\} \to 0,$$

as $t \to \infty$ almost surely since $w(t)/t \to 0$ almost surely. Hence, $\|U_2(t)\|_{C(\overline{\mathcal{O}})} \to 0$ as $t \to \infty$, which completes this proof. □

The interpretation of the preceding stability results in terms of a predator–prey system is immediate: in case of the prey extinction, the predator would have died out if the prey did not exist and dies out in the presence of the prey as well.

5.2 Applications in Mathematical Physics

This section is devoted to the problem of invariant measures for two classes of important stochastic models in mathematical physics: stochastic reaction-diffusion equations and stochastic Navier–Stokes equations.

5.2.1 Stochastic Reaction-Diffusion Equations

Assume that $\mathcal{O} \subset \mathbb{R}^n$ is a bounded domain with smooth boundary $\partial\mathcal{O}$. Let $A(\cdot)$ denote the second-order elliptic operator in $\mathcal{O}$ given by

$$A(x)\psi(x) = \sum_{i,j=1}^{n} \frac{\partial}{\partial x_i}\left[a_{ij}(x, \nabla\psi)\frac{\partial\psi(x)}{\partial x_j}\right] \quad \text{for any} \quad \psi \in C^2(\mathcal{O}),$$

where $\nabla\psi = \left(\frac{\partial\psi}{\partial x_1}, \ldots, \frac{\partial\psi}{\partial x_n}\right)$ and a_{ij}'s are arbitrarily given functions on $\mathcal{O} \times \mathbb{R}^n$ such that the matrix $(a_{ij})_{n\times n}$ is positive-definite. Let $f(x,\xi,y)$ and $\sigma_i(x,\xi,y)$, $i = 1, 2, \ldots, m$, be real-valued continuous functions in $(x,\xi,y) \in \mathcal{O} \times \mathbb{R} \times \mathbb{R}^n$ and $W_i(t,x)$, for $i = 1, \ldots, m$, be independent Wiener processes in $H = L^2(\mathcal{O})$ with norm $\|\cdot\|_H$ so that $W(t,x) = (W_1(t,x), \ldots, W_m(t,x))$ is a Q-Wiener process in $K := H^m = H \times \cdots \times H$. The kernel matrix of the covariance operator Q of $W(t,x)$ is denoted by $q(x,y) = [\delta_{ij}q_i(x,y)]_{m\times m}$, where δ_{ij} is the Kronecker delta function and q_i is the covariance kernel of $W_i(t,x)$. We assume that q_i's are continuous in $\mathcal{O} \times \mathcal{O}$ such that $q_i(x) := q_i(x,x)$ satisfies

$$\sup_{\substack{x\in\mathcal{O}\\ i=1,\ldots,m}} q_i(x) \le q_0 < \infty \quad \text{for some} \quad q_0 > 0.$$

Now consider the following stochastic parabolic Itô equation of the following form:

$$\begin{cases} dy(t,x) = A(x)y(t,x)dt + f(x,y(t,x),\nabla y)dt + \sum_{i=1}^m \sigma_i(x,y(t,x),\nabla y)dW_i(t,x), \\ y(t,\cdot)|_{\partial\mathcal{O}} = 0, \quad t \ge 0, \\ y(0,x) = y_0(x) \in L^2(\mathcal{O}), \quad x \in \mathcal{O}. \end{cases} \tag{5.2.1}$$

For our purpose, let us impose the following conditions:

(P1) the functions a_{ij} and f are continuous and there exist constants $\alpha > 0$, $p > 1$ and $q = p/(p-1) > 1$ such that

$$\sum_{i,j=1}^n a_{ij}(x,y)y_iy_j + |\xi f(x,\xi,y)| \le \alpha(1 + \|y\|^p_{\mathbb{R}^n} + |\xi|^q),$$

$$\xi \in \mathbb{R}, \; x \in \mathcal{O}, \; y \in \mathbb{R}^n.$$

Then the operator

$$A(v) = A(\cdot)v + f(\cdot,\cdot,\nabla v) \tag{5.2.2}$$

is well defined as a continuous operator from $V = W_0^{1,p}(\mathcal{O}) \cap L^q(\mathcal{O})$ into its dual V^*.

For $v \in V$, we introduce

$$B(v)w = \sum_{i=1}^m \sigma_i(\cdot,\cdot,\nabla v)w_i, \qquad w = (w_1, \ldots, w_m) \in K.$$

Then $B(\cdot)\colon V \to \mathscr{L}_2(K_Q, H)$ if we further assume that

(P2) $\sigma_i(x,\xi,y)$'s are continuous in x and Lipschitz continuous in ξ and y such that for all $(x,\xi,y) \in \mathcal{O} \times \mathbb{R} \times \mathbb{R}^n$,

$$\|\sigma(x,\xi,y)\|^2_{\mathbb{R}^m} \le c(1 + \|y\|^2_{\mathbb{R}^n} + |\xi|^2) \quad \text{for some constant} \quad c > 0.$$

In addition, suppose that the following two conditions hold:

(P3) there exist constants $\beta > 0$, $p > 1$ and θ, $\gamma \in \mathbb{R}$ such that

$$\sum_{i,j=1}^{n} a_{ij}(x,y)y_i y_j - \xi f(x,\xi,y) - \frac{q_0}{2}\|\sigma(x,\xi,y)\|^2_{\mathbb{R}^m} \ge \beta\|y\|^p_{\mathbb{R}^n} + \theta|\xi|^2 + \gamma$$

for any $\xi \in \mathbb{R}$, $x \in \mathcal{O}$ and $y \in \mathbb{R}^n$;

(P4) the operator $A: V \to V^*$ defined by (5.2.2) is monotone in the sense that

$$2\langle\langle u-v, A(u)-A(v)\rangle\rangle_{V,V^*} + q_0\|\sigma(\cdot,\cdot,\nabla u)-\sigma(\cdot,\cdot,\nabla v)\|^2 \le \mu\|u-v\|^2_V,$$

for some constant $\mu \in \mathbb{R}$ and arbitrary u, $v \in V$.

Under conditions (P1) through (P4), it is possible to check that there exists a unique strong solution of (5.2.1) in $W_0^{1,p}(\mathcal{O}) \cap L^q(\mathcal{O})$.

We may proceed to identify a unique invariant measure to the system (5.2.1). For simplicity, assume that $\mathcal{O} \subset \mathbb{R}^3$ is a bounded domain with a smooth boundary $\partial\mathcal{O}$. Let $A(x)$ denote a linear strongly elliptic operator in $\mathcal{O}$,

$$A(x)\varphi(x) = \sum_{i,\,j=1}^{3} \frac{\partial}{\partial x_i}\left[a_{ij}(x)\frac{\partial\varphi(x)}{\partial x_j}\right] \quad \text{for} \quad \varphi \in C^2(\mathcal{O}),$$

where $a_{ij}(\cdot) \in C_b(\mathcal{O})$ and there exists $\alpha_0 > 0$ such that

$$\sum_{i,\,j=1}^{3} a_{ij}(x)\xi_i\xi_j \ge \alpha_0\|\xi\|^2_{\mathbb{R}^3}, \qquad \forall x \in \mathcal{O}, \qquad \xi = (\xi_1,\xi_2,\xi_3) \in \mathbb{R}^3\backslash\{0\}. \tag{5.2.3}$$

Consider the following stochastic reaction-diffusion equation:

$$\begin{cases} \dfrac{\partial y(t,x)}{\partial t} = A(x)y(t,x) - \nu y(t,x) + \displaystyle\sum_{j=1}^{3}\sigma_j(x)\frac{\partial y(t,x)}{\partial x_j}\frac{\partial W_j(t,x)}{\partial t} \\ \qquad\qquad + \sigma_4(x)\dfrac{\partial W_4(t,x)}{\partial t} \quad \text{for all} \quad t \ge 0, \; x \in \mathcal{O}, \\ y(t,\cdot)\big|_{\partial\mathcal{O}} = 0, \quad t \ge 0, \end{cases} \tag{5.2.4}$$

where $\nu > 0$, σ_i's $\in C_b(\mathcal{O})$ and W_is are independent Wiener random fields with covariance function q_is. Here we assume that q_is are continuous functions in $\mathcal{O} \times \mathcal{O}$ such that $q_i(x) = q_i(x,x)$ satisfies

$$\sup_{\substack{x \in \mathcal{O} \\ i=1,\dots,4}} q_i(x) \le q_0 < \infty$$

for some $q_0 > 0$.

Let $H = L^2(\mathcal{O})$, $V = H_0^1(\mathcal{O})$ and $K = \overbrace{H \times \cdots \times H}^{4}$. Let $\Lambda(\cdot) = \|\cdot\|_H^2$. Then it is easy to obtain the following estimate:

$$\mathcal{L}\Lambda(v) \le -(2\alpha_0 - 3\sigma_0^2 q_0)\|v\|_H^2 + q_0\sigma_0^2$$

where $\mathcal{L}$ is given as in (3.5.17) and $\sigma_0 = \sup_{i=1,\dots,4,\ x\in\mathcal{O}} \sigma_i(x)$. By virtue of Corollary 3.5.4, the solution is exponential ultimately bounded if $2\alpha_0 > 3\sigma_0^2 q_0$. On the other hand, for any $u,\ v \in V$, we have

$$\begin{aligned}
2\langle\!\langle u - v, A(u) - A(v)\rangle\!\rangle_{V,V^*} &+ q_0\|\sigma(u) - \sigma(v)\|^2 \\
&\le -2\alpha_0\|u - v\|_V^2 + 2\nu\|u - v\|_H^2 + 3q_0\sigma_0^2\|u - v\|_V^2 \\
&\le -\left(2\alpha_0 - \frac{2\nu}{\lambda_0} - 3q_0\sigma_0^2\right)\|u - v\|_V^2,
\end{aligned}$$

where

$$\lambda_0 = \inf_{\substack{v \in H_0^1, \\ v \ne 0}} \frac{\|v\|_V^2}{\|v\|_H^2} > 0.$$

Hence, when ν and q_0 are so small that

$$2\alpha_0 - \frac{2\nu}{\lambda_0} - 3q_0\sigma_0^2 > 0,$$

the equation (5.2.4) has a unique invariant measure supported in $H_0^1(\mathcal{O})$ according to Theorem 3.7.3 and Corollary 3.7.7.

We may also consider a generalization of equation (5.2.1) where the elliptic operator A is replaced by the nonlinear operator

$$A(v)(x) = A(x)v(x) + A_1(v)(x), \tag{5.2.5}$$

where

$$A_1(v)(x) = \sum_{j=1}^{3} \frac{\partial}{\partial x_j}\left(\left|\frac{\partial v(x)}{\partial x_j}\right|^{p-2} \frac{\partial v(x)}{\partial x_j}\right), \quad p > 2.$$

Then the corresponding stochastic reaction-diffusion equation of (5.2.4) is

$$\begin{cases} \dfrac{\partial y(t,x)}{\partial t} = \displaystyle\sum_{i,j=1}^{3} \frac{\partial}{\partial x_i}\left[\left(a_{ij}(x) + \delta_{ij}\left|\frac{\partial y(t,x)}{\partial x_j}\right|^{p-2}\right)\frac{\partial y(t,x)}{\partial x_j}\right] - \nu y(t,x) \\ \qquad + \displaystyle\sum_{i=1}^{3} \sigma_i(x)\frac{\partial y(t,x)}{\partial x_i}\frac{\partial W_i(t,x)}{\partial t} + \sigma_4\frac{\partial W_4(t,x)}{\partial t}, \quad t \geq 0, \quad x \in \mathcal{O}, \\ y(t,\cdot)\big|_{\partial\mathcal{O}} = 0, \quad t \geq 0. \end{cases} \tag{5.2.6}$$

It is well known that the current nonlinear elliptic operator is coercive and monotone in $W_0^{1,p}(\mathcal{O})$ (see Lions and Magenes [130]). Let $V = W_0^{1,p}(\mathcal{O}) \cap L^4(\mathcal{O})$. Under a similar condition to (5.2.3), one can justify the existence of a unique invariant measure to equation (5.2.6) with support in $W_0^{1,p}(\mathcal{O}) \cap L^4(\mathcal{O})$.

5.2.2 Stochastic Navier–Stokes Equations

The theory of Navier–Stokes equations occupies a central position in the study of nonlinear partial differential equations, dynamical systems, and fluid dynamics. The corresponding theory of randomly perturbed Navier–Stokes equations is rather technical and already very extensive despite being still incomplete. In spite of their intuitive physical background from fluids, these stochastic systems have also been extensively studied without direct reference to the conventional theory of turbulence. As an application of the Lyapunov function approaches, we investigate in this section the invariant measure of a stochastic Navier–Stokes equation that theoretically describes a two-dimensional turbulent flow.

Let $\mathcal{O} \subset \mathbb{R}^2$ be a bounded domain with a smooth boundary $\partial\mathcal{O}$. Let

$$y(t,x) = (y_1(t,x), y_2(t,x)) \in \mathbb{R}^2, \qquad x \in \mathcal{O}, \qquad t \geq 0,$$

be the velocity field and $p(t,x) \in \mathbb{R}$, $x \in \mathcal{O}$, $t \geq 0$, be the pressure field in an incompressible fluid. Then, under a random perturbation by Gaussian white noises, a fluid flow could be governed by the following stochastic Navier–Stokes equation:

$$\begin{cases} dy_i(t,x) = \nu \displaystyle\sum_{j=1}^{2} \frac{\partial^2 y_i(t,x)}{\partial x_j^2}dt - \sum_{j=1}^{2} y_j \frac{\partial y_i(t,x)}{\partial x_j}dt \\ \qquad - \dfrac{1}{\rho}\dfrac{\partial p(t,x)}{\partial x_i}dt + \sigma_i dW_i(t,x), \qquad x \in \mathcal{O}, \ t > 0, \\ \displaystyle\sum_{j=1}^{2} \frac{\partial y_j(t,x)}{\partial x_j} = 0, \quad x \in \mathcal{O}, \quad t > 0, \\ y_i(t,x) = 0, \quad x \in \partial\mathcal{O}, \quad t > 0, \\ y_i(0,x) = \xi_i(x), \quad x \in \mathcal{O}, \quad i = 1, 2, \end{cases} \tag{5.2.7}$$

where $\rho > 0$ is the constant fluid density, $\nu > 0$ kinematic viscosity, σ_i's variance parameters, $\xi(x) = (\xi_1(x), \xi_2(x))$ is a proper initial velocity field, and $W(t,x) = (W_1(t,x), W_2(t,x))$ is a random force with associated covariance operators Q_i, $Tr(Q_i) < \infty$, given by a positive definite kernel

$$q_i(x,y) \in L^2(\mathcal{O} \times \mathcal{O}), \qquad q_i(x,x) \in L^2(\mathcal{O}), \qquad i = 1, 2.$$

In a vectorial notation, the preceding equation takes a more compact form:

$$\begin{cases} dy(t,x) = \nu \Delta y dt - (y \cdot \nabla) y dt - \dfrac{1}{\rho} \nabla p dt + \sigma dW(t,x), & t > 0, \quad x \in \mathcal{O}, \\ \nabla \cdot y = 0, \\ y(t,\cdot)\big|_{\partial\mathcal{O}} = 0, \quad t \geq 0, \\ y(0,x) = \xi(x), \end{cases} \tag{5.2.8}$$

where ∇, $\nabla \cdot$, and Δ are the conventional notations for the gradient, divergence, and Laplace operators, respectively, and

$$\sigma = \begin{pmatrix} \sigma_1 & 0 \\ 0 & \sigma_2 \end{pmatrix}.$$

Let $\Sigma = \{v \in C_0^\infty(\mathcal{O}, \mathbb{R}^2) \colon \nabla \cdot v = 0\}$, and V and H be the closures of Σ in $H^1(\mathcal{O}, \mathbb{R}^2)$ and $L^2(\mathcal{O}, \mathbb{R}^2)$, respectively. It may be shown (see, e.g., [208]) that the norm $\|\cdot\|_V$ on V is equivalent to that one induced by the inner product

$$((u,v)) = \sum_{j=1}^{2} \int_{\mathcal{O}} \nabla u_j(x) \cdot \nabla v_j(x) dx = \langle \nabla u, \nabla v \rangle_{L^2(\mathcal{O}, \mathbb{R}^2)}, \qquad u, v \in V.$$

Let V^* denote the dual of V, then it may be further shown that

$$V \hookrightarrow H \cong H^* \hookrightarrow V^*$$

and the embeddings are compact.

It is known (cf. Vishik and Fursikov [214]) that

$$L^2(\mathcal{O}; \mathbb{R}^2) = H \oplus H^\perp,$$

where $H^\perp$ is characterized by $H^\perp = \{v \colon v = \nabla u \text{ for some } u \in H^1(\mathcal{O}; \mathbb{R}^2)\}$. Let Π be the orthogonal projection operator from $L^2(\mathcal{O}, \mathbb{R}^2)$ unto H, and define for any $v \in \Sigma$,

$$A(v) = \nu \Pi \Delta v - \Pi[(v \cdot \nabla) v].$$

Then $A(\cdot)$ can be extended as a continuous operator, still denote it by $A(\cdot)$, from V to V^*. Hence, let $\xi \in V$, and the equation (5.2.8) can be reformulated as a stochastic differential equation in the form

$$\begin{cases} dy(t) = A(y(t))dt + \sigma dW(t), \\ y(0) = \xi \in V, \quad \xi \in V, \end{cases} \tag{5.2.9}$$

where $W(t)$ is a Q-Wiener process in H. Indeed, let $y \in V$ be a strong solution of (5.2.8). Since

$$\int_{\mathcal{O}} \nabla p \cdot u(x)dx = -\int_{\mathcal{O}} p \operatorname{div} u(x)dx = 0,$$

for any $u \in V$, we have $\Pi\nabla p(t,\cdot) = 0$ for almost all $t \geq 0$. Let

$$-\nu\Pi\Delta H^2(\mathcal{O},\mathbb{R}^2) = \{-\nu\Pi\Delta v\colon v \in H^2(\mathcal{O},\mathbb{R}^2)\},$$

then it is possible to show that $-\nu\Pi\Delta H^2(\mathcal{O},\mathbb{R}^2) = H$. By applying Π to both sides of (5.2.8), one can easily see from the preceding arguments that y is a strong solution to (5.2.9).

On the other hand, by employing a Galerkin approximation technique it can be deduced (cf. p. 347, Vishik and Fursikov [214]) that the problem (5.2.9) has a unique strong solution $\{y(t);\ t \geq 0\}$ satisfying, for any $T \geq 0$,

$$\mathbb{E}\|y(T)\|_H^2 + \nu\mathbb{E}\int_0^T \sum_{i=1}^2 \left\|\frac{\partial y(t)}{\partial x_i}\right\|_H^2 dt \leq \mathbb{E}\|\xi\|_H^2 + \frac{T}{2}Tr(Q).$$

Hence, by using the fact that $\|y(t)\|_V$ is equivalent to $\left(\sum_{j=1}^2 \left\|\frac{\partial y(t)}{\partial x_j}\right\|_H^2\right)^{1/2}$, we obtain

$$\sup_T \frac{1}{T}\int_0^T \mathbb{E}\|y(t)\|_V^2 dt \leq \frac{C}{2\nu}Tr(Q) \quad \text{for some} \quad C > 0.$$

By using Markov's inequality, we further obtain

$$\lim_{R\to\infty} \sup_T \frac{1}{T}\int_0^T \mathbb{P}\{\|y(t)\|_V > R\}dt = 0$$

which, by virtue of Corollary 3.7.2, immediately yields the existence of invariant measures for this stochastic Navier–Stokes equation.

Now let $\mathcal{O}$ be a connected and bounded open set in $\mathbb{R}^3$ with a C^2-smooth boundary $\partial\mathcal{O}$. We denote by H the closure in $L^2(\mathcal{O};\mathbb{R}^3)$ of the set

$$V = \{v \in H_0^1(\mathcal{O};\mathbb{R}^3)\colon \nabla \cdot v = 0 \text{ in } \mathcal{O}\}.$$

Denote by A the so-called Stokes operator defined by

$$Av = -\Pi(\Delta v), \qquad v \in V,$$

where Π is the projection operator from $L^2(\mathcal{O};\mathbb{R}^3)$ unto H. We consider a stochastic three-dimensional Lagrangian averaged Navier–Stokes (3D LANS-α) equation

$$\begin{cases} \partial_t(y-\alpha\Delta y)+\nu(Ay-\alpha\Delta(Ay))+(y\cdot\nabla)(y-\alpha\Delta y)-\alpha(\nabla y)^T\cdot\Delta y+\nabla p \\ \qquad\qquad = F(t,y)+G(t,y)\dot{W}(t) \ \text{ in } \ \mathcal{O}\times(0,\infty), \\ \nabla\cdot y=0 \ \text{ in } \ \mathcal{O}\times(0,\infty), \\ y=0, \ Ay=0 \ \text{ on } \ \partial\mathcal{O}\times(0,\infty), \\ y(0)=y_0 \ \text{ in } \mathcal{O}, \end{cases} \tag{5.2.10}$$

where $y = (y_1, y_2, y_3)$; $(\nabla y)^T$ is the transpose of ∇y; and p are unknown random fields on $\mathcal{O}\times(0,\infty)$, respectively, the large-scale (or averaged) velocity and the pressure, at each point of $\mathcal{O}\times(0,\infty)$, of an incompressible viscous fluid with constant density filling the domain $\mathcal{O}$. The constants $\nu > 0$ and $\alpha > 0$ represent, respectively, the kinematic viscosity of the fluid and the square of the spatial scale at which fluid motion is filtered. The terms $F(t,y)$ and $G(t,y)\dot{W}(t)$ are external forces depending on y where $\dot{W}(t)$ denotes the white noise.

In Caraballo et al. [35], the stability property of the stationary solution to (5.2.10) is considered. More precisely, by formulating a variational equation of (5.2.10) with $G = 0$ one can show that under natural conditions this equation has a unique stationary solution. Moreover, it may be shown that, in addition to some further conditions similar to those in Theorems 3.3.1 and 3.3.3, this unique stationary solution that is an equilibrium state of (5.2.10) has the pathwise exponential stability property. The reader is referred to [35] for more details about this topic and related stabilization results.

5.3 Applications in Stochastic Control

Let $y^u(t)$, $t \geq 0$, be a family of properly defined processes in H with $y^u(0) = y_0$. The parameter u associated with each member is termed a control. Frequently, each u may be identified with a specific member of a given family of functions of $y^u(t)$ that takes values of the form $u(t) = L(t, y^u(t))$ depending on t and $y^u(t)$. This may be developed further by associating to each control u a cost $J(u; y_0)$. The object of control theory is to select a proper control u so that the corresponding process $y^u(\cdot)$ possesses some desired properties, and meanwhile minimizes or maximizes the cost $J(u; y_0)$ with respect to others of a specified class of the so-called "admissible controls."

5.3.1 Optimal Control on an Infinite Interval

Let $(\Omega, \mathscr{F}, \{\mathscr{F}_t\}_{t\geq 0}, \mathbb{P})$ be a complete probability space and H, K^i, $i = 1, 2$, be Hilbert spaces. Consider the controlled stochastic differential equation on H,

$$\begin{cases} dy(t) = (Ay(t) + Bu(t))dt + Gy(t)dW_1(t) + Cu(t)dW_2(t), & 0 \leq t < \infty, \\ y(0) = y_0 \in H, \end{cases} \tag{5.3.1}$$

where A is the infinitesimal generator of a strongly continuous semigroup e^{tA}, $t \geq 0$, on H; $u(t)$ is a control with values in a Hilbert space U; and $B \in \mathscr{L}(U, H)$, $G \in \mathscr{L}(H, \mathscr{L}_2(K^1_{Q_1}, H))$, $C \in \mathscr{L}(U, \mathscr{L}_2(K^2_{Q_2}, H))$, and $W_i(t)$ are independent K^i-valued Wiener processes with covariance operators Q_i, $Tr(Q_i) < \infty$, $i = 1, 2$. For each $u(t)$, it is easy to see that there exists a unique mild solution to (5.3.1) in $C([0, \infty); L^2(\Omega, \mathscr{F}, \mathbb{P}; H))$.

We can consider a quadratic cost functional as follows:

$$J(u; y_0) = \int_0^\infty \mathbb{E}\big\{\langle My(t), y(t)\rangle_H + \langle Nu(t), u(t)\rangle_U\big\}dt, \tag{5.3.2}$$

where $M \in \mathscr{L}^+(H)$ and $N \in \mathscr{L}^+_c(U)$ with inverse $N^{-1} \in \mathscr{L}(U)$. The control problem we are concerned with here is to minimize (5.3.2) over $L^2([0, \infty) \times \Omega; U)$. In finite dimensions, e.g., $H = \mathbb{R}^n$, optimal control problems of this type are well known and given in terms of the solution of a Riccati equation. This is one of the best examples for which dynamic programming gives a complete solution. We shall show that this could be the case in infinite dimensions as well.

To proceed further, we define a *feedback control* by a map $L(t, y)\colon [0, \infty) \times H \to U$ such that the equation given by (5.3.1) with $u(t) = L(t, y(t))$ has a unique mild solution $y(t)$, $t \geq 0$. By *admissible controls*, we mean the class of feedback controls $u = L(t, y)$ such that

(i) $L(t, y)\colon [0, \infty) \times H \to U$ is measurable and for some $c > 0$,

$$\|L(t, y)\|_U \leq c(1 + \|y\|_H), \quad \|L(t, x) - L(t, y)\|_U \leq c\|x - y\|_H,$$

for any $x,\ y \in H$, and

(ii) $\mathbb{E}\|y(t)\|^2_H \to 0$ as $t \to \infty$, where $y(t)$ is the mild solution of (5.3.1) with $u(t) = L(t, y(t))$, $t \geq 0$.

Definition 5.3.1 System (5.3.1) (or (A, B, C, G)) is *stabilizable* if there exists $L \in \mathscr{L}(H, U)$ such that the feedback law $u(t) = -Ly(t)$ yields a L^2-stable mild solution $y(t)$ in mean square, i.e.,

$$\int_0^\infty \mathbb{E}\|y(t,y_0)\|_H^2 dt < \infty.$$

In this case, we also say that $(A - BL, C, G)$ is stable.

Now we introduce two operators $\Gamma(\cdot) \in \mathscr{L}(U)$, $\Delta(\cdot) \in \mathscr{L}(H)$ by

$$\begin{aligned}\langle \Gamma(S)u, u\rangle_U &= Tr(SC(u)Q_2^{1/2}(C(u)Q_2^{1/2})^*), \qquad S \in \mathscr{L}(H), \qquad u \in U,\\ \langle \Delta(S)x, x\rangle_H &= Tr(SG(x)Q_1^{1/2}(G(x)Q_1^{1/2})^*), \qquad S \in \mathscr{L}(H), \qquad x \in H,\end{aligned} \tag{5.3.3}$$

and

$$(\mathcal{L}_u\Lambda)(x) = \langle \Lambda'(x), Ax + Bu\rangle_H + \frac{1}{2}\big\{\langle \Gamma(\Lambda''(x))u, u\rangle_U + \langle \Delta(\Lambda''(x))x, x\rangle_H\big\},$$

for any $x \in \mathscr{D}(A)$, $u \in U$ and Itô's functional $\Lambda(\cdot)$ on H.

If (A, B, C, G) is stabilizable, then the control problem (5.3.1), together with (5.3.2), is meaningful. In particular, we may obtain by analogy with Theorem 2.2.5 the following lemma.

Lemma 5.3.2 *The system (A, B, C, G) is stabilizable if and only if there exists $L \in \mathscr{L}(H, U)$ and $P \in \mathscr{L}^+(H)$ such that*

$$2\langle (A - BL)x, Px\rangle_H + \langle [L^*\Gamma(P)L + \Delta(P)]x, x\rangle_H = -\langle x, x\rangle_H, \quad x \in \mathscr{D}(A).$$

Lemma 5.3.3 *If $(A - BL, C, G)$ is stable, then there exists $P \in \mathscr{L}^+(H)$ such that*

$$2\langle (A-BL)x, Px\rangle_H + \langle (M+L^*NL)x, x\rangle_H + \langle [\Delta(P)+L^*\Gamma(P)L]x, x\rangle_H = 0, \tag{5.3.4}$$

for $x \in \mathscr{D}(A)$.

Proof Let $P_T(t)$, $0 \le t \le T$, be the unique solution of

$$\begin{cases} \dfrac{d}{dt}\langle P(t)x, x\rangle_H + \langle (A - BL)x, P(t)x\rangle_H + \langle (M + L^*NL)x, x\rangle_H \\ \qquad\qquad + \langle [\Delta(P(t)) + L^*\Gamma(P(t))L]x, x\rangle_H = 0, \quad x \in \mathscr{D}(A), \\ P(T) = 0. \end{cases} \tag{5.3.5}$$

By analogy with Proposition 2.2.4, we may show

$$\langle P_T(0)y_0, y_0\rangle_H = \int_0^T \mathbb{E}\big\{\langle My(t), y(t)\rangle_H + \langle NLy(t), Ly(t)\rangle_H\big\}dt,$$

where $y(t)$ is the mild solution of (5.3.1) with $u(t) = -Ly(t)$. Since $(A - BL, C, G)$ is stable, similarly to Theorem 2.2.5 $P_T(0)$ is uniformly bounded in T. But $P_T(0)$ is monotonically increasing and nonnegative. Let $T \to \infty$, then there exists a limit $P \in \mathscr{L}^+(H)$ satisfying (5.3.4). □

Theorem 5.3.4 *Suppose that there exists an admissible control* $\bar{u} = -\bar{L}(y)$ *and an Itô functional* $\Lambda(\cdot)$ *on* H *such that for some constant* $c > 0$ *and*

(a) for any $x \in H$,

$$|\Lambda(x)| \le c\|x\|_H^2;$$

(b) for any $x \in \mathscr{D}(A)$ *and* $u \in U$,

$$-\langle Mx, x\rangle_H = (\mathcal{L}_{\bar{u}}\Lambda)(x) + \langle N\bar{u}, \bar{u}\rangle_U \le (\mathcal{L}_u\Lambda)(x) + \langle Nu, u\rangle_U; \quad (5.3.6)$$

(c) for any $x, y \in H$,

$$\|\bar{L}(x) - \bar{L}(y)\|_U \le c\|x - y\|_H.$$

Then $\bar{u} = -\bar{L}(y)$ *is optimal and* $J(\bar{u}; y_0) = \Lambda(y_0)$.

Proof Let $\bar{y}(t)$, $t \ge 0$, be the mild solution of (5.3.1) with $\bar{u} = -\bar{L}(y)$. By introducing a series of Yosida's approximation solutions to (5.3.1) and applying Itô's formula to $\Lambda(\cdot)$, we can show similarly to Proposition 2.2.4 that

$$\mathbb{E}\Lambda(\bar{y}(t)) - \Lambda(y_0) = -\int_0^t \mathbb{E}\big\{\langle M\bar{y}(s), \bar{y}(s)\rangle_H + \langle N\bar{u}(s), \bar{u}(s)\rangle_U\big\}ds,$$

where $\bar{y}(t)$ is the mild solution of (5.3.1) with $u = \bar{u}$. Note that $|\Lambda(x)| \le c\|x\|_H^2$ and $\mathbb{E}\|\bar{y}(t)\|_H^2 \to 0$ as $t \to \infty$. So

$$\Lambda(y_0) = \int_0^\infty \mathbb{E}\big\{\langle M\bar{y}(t), \bar{y}(t)\rangle_H + \langle N\bar{u}(t), \bar{u}(t)\rangle_U\big\}dt = J(\bar{u}; y_0).$$

In a similar way, for any admissible control u, it is possible to obtain by virtue of (5.3.6) that

$$\Lambda(y_0) \le \int_0^\infty \mathbb{E}\big\{\langle My(t), y(t)\rangle_H + \langle Ny(t), y(t)\rangle_H\big\}dt = J(u; y_0).$$

The proof is thus complete. □

Now we seek a function $\Lambda(x)$ of the form

$$\Lambda(x) = \langle Px, x\rangle_H, \qquad P \in \mathscr{L}^+(H).$$

Then condition (c) yields an algebraic Riccati equation

$$2\langle Ax, Px\rangle_H + \langle\{M + \Delta(P) - PB[N + \Gamma(P)]^{-1}B^*P\}x, x\rangle_H = 0, \quad x \in \mathscr{D}(A), \quad (5.3.7)$$

and the control law

$$u = -Ly, \quad L = [N + \Gamma(P)]^{-1}B^*P. \quad (5.3.8)$$

Hence, to identify the optimal control, we need only establish conditions to ensure the existence of a solution to (5.3.7) and thus the admissible control in (5.3.8). However, if there exist operators $L \in \mathscr{L}(H,U)$ and $P \in \mathscr{L}^+(H)$ satisfying (5.3.4), then it can be shown that Riccati equation (5.3.7) has a solution.

Theorem 5.3.5 *Suppose that there exists $L \in \mathscr{L}(H,U)$ such that $(A - BL, C, G)$ is stable. Then there is a unique solution $P \geq 0$ to the Riccati equation (5.3.7). Moreover, the optimal control for (5.3.1) is the feedback law (5.3.8) and $J(\bar{u}; y_0) = \langle P y_0, y_0\rangle_H$.*

Proof The existence of a solution to (5.3.7) follows from Theorem 5.3.4. Since $\langle P y_0, y_0\rangle_H$ is the minimum cost, it is easy to see that P is unique. The proof is now complete. □

It is possible to generalize the results in this section (see Da Prato and Ichikawa [51]) to deal with some stochastic systems with bounded drift and unbounded diffusion coefficients, i.e., $G \in \mathscr{L}(\mathscr{D}(\sqrt{A}), H)$, $C = 0$, A is self-adjoint and strictly positive, and W_1 is a finite-dimensional Brownian motion in (5.3.1). Also, Flandoli [73] considered a similar case with unbounded drift and bounded diffusion coefficients. The models introduced there cover parabolic equations with boundary control and distributed noise.

5.4 Notes and Comments

The content of Section 5.1 is mainly based on Manthey and Maslowski [161]. The reader is also referred to Manthey and Mittmann [162] for some related material.

The existence and uniqueness of invariant measures for SPDEs have been significantly studied by many researchers over the last several decades. The reader is referred to Da Prato and Zabczyk [52] for a systematic account of this topic. In this chapter, we have presented some applications of Lyapunov function methods to specific stochastic models. Particularly, in connection with the material in this chapter, we mention that the existence of invariant measures for a stochastic Navier–Stokes equation was also studied by asymptotic analysis in Da Prato and Zabczyk [52] and, for periodic boundary conditions in Albeverio and Cruzerio [2] by the Galerkin approximation and the method of averaging. Flandoli and Gątarek [74] considered rather general white noise sources and looked for solutions that are martingales or stationary for a class of stochastic Navier–Stokes equations. Uniqueness and ergodicity questions were also addressed differently from Lyapunov function methods in such work as Flandoli and Maslowski [75]. The existence and uniqueness of invariant measures for stochastic evolution equations with boundary and pointwise noise were also investigated by Maslowski [169].

Much effort has been devoted to the study of optimal control and stabilizability for deterministic PDEs and functional differential equations. Among others, I like to mention Butkovskii [22], Curtain and Pritchard [48], Lasiecka and Triggiani [119, 120], Lions [129], and Wonham [218]. In this chapter, we have formulated stochastic distributed parameter systems as infinite-dimensional stochastic differential equations. This formulation is helpful not only in considering the major concepts of controllability, observability, and stability in system theory, but in posing various problems in optimal control and filtering. We took this approach based on the theory of semigroups. In particular, we employed dynamic programming methods that enable us to determine optimal feedback control and study its connection with the stability property of linear stochastic systems.

There is an important problem that is not discussed in this book, that is, the cost problem for partially observable systems. In this case, admissible controls are those dependent only on observations. It is known that this problem can be decomposed into two parts, filtering and control, and this fact is known as the separation principle; e.g., see Curtain and Ichikawa [47] and Curtain [45]. After reducing problems to those with complete observation, one can use the results with complete observation and obtain an optimal feedback control law on filters. The filtering part can be solved as a dual problem to the deterministic regulator problem. A different formulation is also possible; e.g., see Bensoussan and Viot [17], in which they give necessary and sufficient conditions of optimality for linear stochastic distributed parameter systems with convex differentiable payoffs and partial observation.

Appendix

A Proof of Theorem 4.1.5

We show the existence and uniqueness of solutions for (4.1.2) by the well-known contraction mapping theorem. In association with the delayed term in (4.1.2), we first define a linear operator F on $C([-r,T],X)$ by

$$(Fy)(t) = \int_{-r}^{0} d\eta(\theta)y(t+\theta), \qquad t \in [0,T], \qquad \forall y(\cdot) \in C([-r,T],X). \quad \text{(A.1)}$$

By using Minkowski's inequality, Hölder's inequality, and Fubini's theorem, we obtain that for any $y(\cdot) \in C([-r,T],X)$,

$$\begin{aligned}
\left(\int_0^T \|(Fy)(t)\|_X^2 dt\right)^{1/2} &\le \left(\int_0^T \|A_1\|^2 \|y(t-r)\|_X^2 dt\right)^{1/2} \\
&\quad + \left(\int_{-r}^0 \|A_2(\theta)\|^2 d\theta\right)^{1/2} \left(\int_0^T \int_{-r}^0 \|y(t+\theta)\|_X^2 d\theta dt\right)^{1/2} \\
&\le \left(\int_0^T \|A_1\|^2 \|y(t-r)\|_X^2 dt\right)^{1/2} \\
&\quad + \left(\int_{-r}^0 \|A_2(\theta)\|^2 d\theta\right)^{1/2} \left(\int_{-r}^0 \int_{-r}^T \|y(t)\|_X^2 dt d\theta\right)^{1/2} \\
&\le \left[\|A_1\| + \left(\int_{-r}^0 \|A_2(\theta)\|^2 d\theta\right)^{1/2} \cdot r^{1/2}\right] \left(\int_{-r}^T \|y(t)\|_X^2 dt\right)^{1/2}.
\end{aligned}$$

Since $C([-r,T],X)$ is dense in $L^2([-r,T];X)$, F admits a bounded extension, still denote it by F, from $L^2([-r,T];X)$ to $L^2([0,T];X)$, and moreover

$$\|F\| \le \|A_1\| + \left(\int_{-r}^0 \|A_2(\theta)\|^2 d\theta\right)^{1/2} \cdot r^{1/2}. \quad \text{(A.2)}$$

Given an initial datum $\phi \in \mathcal{X}$, we note that for any $y(\cdot) \in L^2([0,T],X)$, it can be uniquely extended to a function $\tilde{y}(\cdot) \in L^2([-r,T],X)$ on $[-r,T]$ such that $\tilde{y}(0) = \phi_0$,

$\tilde{y}(\theta) = \phi_1(\theta)$ for any $\theta \in [-r, 0]$. We will always identify $y(\cdot) \in L^2([0, T], X)$ with its extension of this kind. In particular, we define a mapping K on $y(\cdot) \in L^2([0, T], X)$ by

$$(Ky)(t) = e^{tA}\phi_0 + \int_0^t e^{(t-s)A} \int_{-r}^0 d\eta(\theta)\tilde{y}(s+\theta)ds, \qquad 0 \le t \le T. \tag{A.3}$$

Then it is easy to see by (A.2) and (A.3) that for any $y(\cdot) \in L^2([0, T], X)$, $(Ky)(t) \in C([0, T], X)$.

We next show that K is a contraction on $L^2([0, \delta], X)$ for small $\delta \in [0, T]$. To this end, let $x(t)$, $y(t) \in L^2([0, \delta], X)$ and note that $\|e^{tA}\| \le Me^{\mu t}$ for some $M \ge 1$, $\mu > 0$ and all $t \ge 0$. Then by virtue of (A.2) and Hölder's inequality, it follows that

$$\begin{aligned}
&\left(\int_0^\delta \|(Kx)(t) - (Ky)(t)\|_X^2 dt\right)^{1/2} \\
&\le \left[\int_0^\delta \left(\int_0^\delta \|e^{(t-s)A}\|\|F(\tilde{x} - \tilde{y})(s)\|_X ds\right)^2 dt\right]^{1/2} \\
&\le M\left\{\frac{1}{2\mu}(e^{2\mu\delta} - 1)\right\}^{1/2} \delta^{1/2}\|F\| \cdot \|x - y\|_{L^2([0,\,\delta],\,X)}.
\end{aligned}$$

Hence, K is a contraction for sufficiently small $\delta > 0$. This proves the local existence and uniqueness of the solution to (4.1.2).

To prove global existence, we shall derive an a priori estimate of this solution. To this end, let $y(t) \in L^2([-r, T], X)$ be a solution of (4.1.2) on some interval $[-r, T]$, $T \ge r$. Then

$$\begin{aligned}
(Fy)(s) &= \int_{-r}^0 d\eta(\theta)y(s+\theta) \\
&= \begin{cases} \displaystyle\int_{-r}^{-s} d\eta(\theta)\phi_1(s+\theta) + \int_{-s}^0 d\eta(\theta)y(s+\theta), & s \in [0, r], \\ \displaystyle\int_{-r}^0 d\eta(\theta)y(s+\theta), & s \in [r, T]. \end{cases}
\end{aligned}$$

We put

$$\begin{aligned}
(Ly)(s) &:= \int_{-r}^{-s} d\eta(\theta)\phi_1(s+\theta) \\
&= A_1\phi_1(s-r) + \int_{-r}^{-s} A_2(\theta)\phi_1(s+\theta)d\theta, \qquad s \in [0, r].
\end{aligned}$$

By a similar argument to that in (A.2), we have that for any $y \in C([0, T], X)$, $\|Ly\|_{L^2([0,r],X)} \le \|F\|\|\phi_1\|_{L_r^2}$. Since $y(t)$ is continuous on $[0, T]$, it is easy to see that

$$\|(Fy - \mathbf{1}_{[0,r]}Ly)(s)\|_X \le Var(\eta) \sup_{0\le\tau\le s} \|y(\tau)\|_X, \qquad s \in [0, T],$$

where

$$Var(\eta) = \|A_1\| + \int_{-r}^0 \|A_2(\theta)\|d\theta < \infty.$$

Hence, by making use of Hölder's inequality, one can immediately obtain

$$\begin{aligned}\sup_{0\le u\le t}\|y(u)\|_X &\le \|e^{tA}\|\|\phi_0\|_X + \int_0^t \|e^{(t-s)A}\|\|(Fy)(s)\|_X ds\\ &\le Me^{\mu t}\|\phi_0\|_X + Me^{\mu t}\|F\|\Big(\int_0^t \|\phi_1(\theta)\|_X^2 d\theta\Big)^{1/2}\\ &\quad + M\cdot Var(\eta)\int_0^t e^{\mu(t-s)}\Big(\sup_{0\le u\le s}\|y(u)\|_X\Big)ds, \qquad t\in[0,T].\end{aligned}$$

Now we can apply Gronwall's inequality to obtain

$$\|y(t)\|_X \le \sup_{0\le u\le t}\|y(u)\|_X \le C\|\phi\|_{\mathcal{X}}e^{\gamma t}, \qquad t\in[0,T],$$

where $\gamma = \mu + M\cdot Var(\eta)\in\mathbb{R}$ and $C>0$ is a constant depending only on η, r, and A. Since $T\ge r$ can be chosen arbitrarily large, the global existence of the solution with the desired estimate is thus proved.

B Proof of Proposition 4.1.7

We denote by $\overline{\mathcal{A}}$ and $\mathscr{D}(\overline{\mathcal{A}})$ the infinitesimal generator of $\mathcal{S}(t)$, $t\ge 0$, and its domain, respectively. We want to show $\overline{\mathcal{A}}=\mathcal{A}$ where $\mathcal{A}$ is given by (4.1.22) and (4.1.23). To this end, let $\phi=(\phi_0,\phi_1)\in\mathscr{D}(\overline{\mathcal{A}})$ and $\overline{\mathcal{A}}\phi=(\psi_0,\psi_1)$. Since the second coordinate of $\mathcal{S}(t)\phi$ is the t-shift $y(t+\cdot)$, it follows by semigroup theory that

$$y(\theta)=\phi_1(\theta)\in W^{1,2}([-r,0],X), \quad \theta\in[-r,0],$$

and

$$\frac{d^+y(\theta)}{d\theta}=\frac{d\phi_1(\theta)}{d\theta}=\psi_1(\theta) \quad \text{in} \quad L^2([-r,0],X), \quad \theta\in[-r,0],$$

where $d^+/d\theta$ denotes the right-hand derivative. By redefining on the set of measure zero, we can suppose that $y(\theta)=\phi_1(\theta)$ is continuous from $[-r,0]$ to X. Since $y(0)=\phi_0$, we have that $\phi_1(0)=\phi_0$, and as $y(\cdot)$ is continuous on $[0,\infty)$, the function $(Fy)(t):=\int_{-r}^0 d\eta(\theta)y(t+\theta)$ is continuous in $t\ge 0$ and satisfies $\lim_{t\downarrow 0}(Fy)(t)=\int_{-r}^0 d\eta(\theta)\phi_1(\theta)$ so that

$$\lim_{t\downarrow 0}\frac{1}{t}\int_0^t e^{(t-s)A}(Fy)(s)ds=\int_{-r}^0 d\eta(\theta)\phi_1(\theta). \tag{B.1}$$

Applying (4.1.2) and (B.1) to the first coordinate of $\overline{\mathcal{A}}\phi=(\psi_0,\psi_1)$, we obtain

$$\begin{aligned}\psi_0 &= \lim_{t\downarrow 0}\frac{1}{t}(y(t)-\phi_0)=\lim_{t\downarrow 0}\frac{1}{t}\Big(e^{tA}\phi_0+\int_0^t e^{(t-s)A}(Fy)(s)ds-\phi_0\Big)\\ &=\lim_{t\downarrow 0}\frac{1}{t}(e^{tA}\phi_0-\phi_0)+\int_{-r}^0 d\eta(\theta)\phi_1(\theta).\end{aligned}$$

Hence, $\lim_{t\downarrow 0} t^{-1}(e^{tA}\phi_0 - \phi_0)$ exists in X, i.e., $\phi_0 \in \mathscr{D}(A)$, and $\psi_0 = A\phi_0 + \int_{-r}^{0} d\eta(\theta)\phi_1(\theta)$. This shows that

$$\mathscr{D}(\overline{\mathcal{A}}) \subseteq \mathscr{D}(\mathcal{A}) \quad \text{and} \quad \mathcal{A}\phi = \overline{\mathcal{A}}\phi \quad \text{for} \quad \phi \in \mathscr{D}(\overline{\mathcal{A}}).$$

Next we show the reverse inclusion. Let $\phi = (\phi_0, \phi_1) \in \mathscr{D}(\mathcal{A})$. For this ϕ, it is not difficult to see from (4.1.2) that $y(\cdot,\phi) \in W^{1,2}([0,T],X)$ for any $T \geq 0$, and so (B.1) is true. Combining this with (4.1.2) and $\phi_0 \in \mathscr{D}(A)$, we see that

$$\lim_{t\downarrow 0} \frac{1}{t}(y(t) - \phi_0) = A\phi_0 + \int_{-r}^{0} d\eta(\theta)\phi_1(\theta).$$

On the other hand, noting that for $\theta \in [-r,0]$,

$$\frac{y(t+\theta) - \phi_1(\theta)}{t} - \phi_1'(\theta) = \frac{y(t+\theta) - y(\theta)}{t} - y'(\theta) = \frac{1}{t}\int_0^t (y'(s+\theta) - y'(\theta))ds,$$

we obtain by using Hölder's inequality that

$$\left\| \frac{1}{t}(y(t+\cdot) - \phi_1(\cdot)) - \phi_1'(\cdot) \right\|^2_{L^2([-r,0],X)} \leq \frac{1}{t}\int_0^t \left[\int_{-r}^{0} \|y'(s+\theta) - y'(\theta)\|_X^2 d\theta \right] ds.$$

This implies that $\lim_{t\downarrow 0} t^{-1}(y(t+\cdot) - \phi_1(\cdot))$ exists in $L^2([-r,0],X)$ and equals ϕ_1'. Thus, we prove $\mathscr{D}(\mathcal{A}) \subset \mathscr{D}(\overline{\mathcal{A}})$ and $\mathcal{A}\phi = \overline{\mathcal{A}}\phi$ for $\phi \in \mathscr{D}(\mathcal{A})$, and hence (4.1.22) and (4.1.23) are shown.

C Proof of Proposition 4.1.10

The equivalence of (i) and (ii) was established in Section 2.1. Thus it suffices to prove the equivalence of (ii), (iii), and (iv). To this end, assume that (iii) is valid, then we intend to prove (ii). First, we note that by analogy with (2.1.4), there exists a constant $\alpha > 0$ such that

$$\int_0^\infty \|G(t)x\|_X^2 dt \leq \alpha \|x\|_X^2, \qquad x \in X. \tag{C.1}$$

On the other hand, we have by using (4.1.12) and Hölder's inequality that

$$\left\| \int_{-r}^{0} G(t+\theta)S\varphi(\theta)d\theta \right\|_X^2 \leq C^2 e^{2\gamma t}\|S\|^2 r \int_{-r}^{0} \|\varphi(\theta)\|_X^2 d\theta, \quad \forall \varphi(\cdot) \in L^2([-r,0],X), \tag{C.2}$$

where $S\colon L^2([-r,0],X) \to L^2([-r,0],X)$ is the structure operator defined in (4.1.15). Let $\beta > \gamma \geq 0$ and by virtue of the quasi semigroup property (4.1.17), we have that for any $t \geq 0$,

$$
\begin{aligned}
\frac{1-e^{-2\beta t}}{2\beta}\|G(t)x\|_X^2 &= \int_0^t e^{-2\beta s}\|G(t)x\|_X^2 ds \\
&\le 2\int_0^t e^{-2\beta s}\|G(s)G(t-s)x\|_X^2 ds \\
&\quad + 2\int_0^t e^{-2\beta s}\left\|\int_{-r}^0 G(s+\theta)[SG_{t-s}x](\theta)d\theta\right\|_X^2 ds \\
&:= 2I_1 + 2I_2.
\end{aligned}
\tag{C.3}
$$

We shall estimate I_1 and I_2 separately. By virtue of (C.1) and (4.1.12), it is easy to see that

$$
I_1 \le \int_0^t e^{-2\beta s}\|G(s)\|^2\|G(t-s)x\|_X^2 ds \le C^2\int_0^t \|G(t-s)x\|_X^2 ds \le C^2\alpha\|x\|_X^2.
\tag{C.4}
$$

On the other hand, by virtue of (C.1), (4.1.12), (C.2), and the fact that $G(t) = 0$ for $t < 0$, we can derive

$$
\begin{aligned}
I_2 &= \int_0^t e^{-2\beta s}\left\|\int_{-r}^0 G(s+\theta)[SG_{t-s}x](\theta)d\theta\right\|_X^2 ds \\
&\le C^2\|S\|^2 r\int_0^t e^{-2\beta s}e^{2\gamma s}\int_{-r}^0 \|G(t-s+\theta)x\|_X^2 d\theta ds \\
&\le C^2\|S\|^2\alpha r^2\|x\|_X^2.
\end{aligned}
\tag{C.5}
$$

Combining (C.3), (C.4), and (C.5), we may obtain, for all $t \ge 0$,

$$
\frac{1-e^{-2\beta t}}{2\beta}\|G(t)x\|_X^2 \le 2C^2\alpha\|x\|_X^2 + 2C^2\|S\|^2\alpha r^2\|x\|_X^2, \qquad x \in X,
$$

which immediately implies that there exists a constant $C_0 > 0$ such that $\|G(t)\| \le C_0$ for all $t \ge 0$. Therefore, it is not difficult to see that for any $\varphi \in L^2([-r,0],X)$, $t > 0$,

$$
\left\|\int_{-r}^0 G(t+\theta)S\varphi(\theta)d\theta\right\|_X^2 \le \|S\|^2\|\varphi\|_{L_r^2}^2\int_{-r}^0 \|G(t+\theta)\|^2 d\theta \le \|S\|^2 r C_0^2\|\varphi\|_{L_r^2}^2.
\tag{C.6}
$$

By virtue of the quasi semigroup property (4.1.17) and (C.6), one may obtain that for any $t > 0$,

$$
\begin{aligned}
t\|G(t)x\|_X^2 &\le \int_0^t \|G(t)x\|_X^2 ds \\
&\le 2\int_0^t \|G(s)G(t-s)x\|_X^2 ds + 2\int_0^t\left\|\int_{-r}^0 G(s+\theta)[SG_{t-s}x](\theta)d\theta\right\|_X^2 ds \\
&\le 2\int_0^t \|G(s)\|^2\|G(t-s)x\|_X^2 ds + 2\|S\|^2 r C_0^2\int_0^t\int_{-r}^0 \|G(t-s+\theta)x\|_X^2 d\theta ds \\
&\le 2C_0^2\alpha\|x\|_X^2 + 2\|S\|^2 r^2 C_0^2\alpha\|x\|_X^2.
\end{aligned}
\tag{C.7}
$$

Therefore, it is easy to see from (C.7) that there exists a constant $C_1 > 0$ such that

$$\|G(t)\| \leq C_1/\sqrt{t} \qquad \text{for all} \quad t > 0. \tag{C.8}$$

Repeating a similar argument as in (C.6) and (C.7) and using (C.8), we can further show that there exists a constant $C_2 > 0$ such that

$$\|G(t)\| \leq C_2/t \qquad \text{for all} \quad t > 0. \tag{C.9}$$

Since the solution $y(t)$ of (4.1.2) is continuous on $[0, \infty)$, it follows that

$$C_3 := \int_0^1 \|y(t)\|_X^2 dt < \infty,$$

which, in addition to (4.1.16), (C.6), and (C.9), implies that

$$\begin{aligned}
&\int_0^\infty \|y(t)\|_X^2 dt \\
&\quad \leq C_3 + 2\int_1^\infty \|G(t)\phi_0\|_X^2 dt + 2\int_1^\infty \left\| \int_{-r}^0 G(t+\theta)(S\phi)(\theta)d\theta \right\|_X^2 dt \\
&\quad \leq C_3 + 2\int_1^\infty \|G(t)\phi_0\|_X^2 dt + 2\|S\|^2 C_2^2 r \int_{-r}^0 \|\phi_1(\theta)\|_X^2 d\theta \int_1^\infty \frac{1}{t^2} dt < \infty.
\end{aligned}$$

Hence, by definition of $e^{t\mathcal{A}}$, $t \geq 0$, we have for any $\phi \in \mathcal{X}$

$$\begin{aligned}
\int_0^\infty \|e^{t\mathcal{A}}\phi\|_{\mathcal{X}}^2 dt &= \int_0^\infty \left(\|y(t)\|_X^2 + \int_{-r}^0 \|y(t+\theta)\|_X^2 d\theta \right) dt \\
&= \int_0^\infty \|y(t)\|_X^2 dt + \int_{-r}^0 \int_\theta^\infty \|y(t)\|_X^2 dt d\theta \\
&\leq \int_0^\infty \|y(t)\|_X^2 dt + r\int_0^\infty \|y(t)\|_X^2 dt + r\int_{-r}^0 \|\phi_1(\theta)\|_X^2 d\theta < \infty.
\end{aligned}$$

This shows the validity of (i), a fact which further implies the validity of (ii).

Now suppose that (ii) is true, then we have by the definition of $G(t)$ that for any $x \in X$, $\phi = (x, 0)$, and $t \geq 0$,

$$\|G(t)x\|_X = \|y(t,\phi)\|_X \leq \|e^{t\mathcal{A}}\phi\|_{\mathcal{X}} \leq Me^{-\mu t}\|\phi\|_{\mathcal{X}} = Me^{-\mu t}\|x\|_X,$$

which implies (iv). The implication (iv) to (iii) is straightforward, and thus the equivalence of (ii), (iii), and (iv) is shown. The proof is thus complete.

D Proof of Proposition 4.1.14

Using the same notations as in the proofs of Theorem 4.1.5, we introduce an operator $Q^t : \mathcal{X} \to X$, $t \geq 0$, defined by

$$Q^t\phi = \int_0^t e^{(t-s)A}(Fy(\phi))(s)ds, \quad \phi \in \mathcal{X},$$

where $y(\phi)$ is the unique solution of system (4.1.2) with initial ϕ and F is defined in (A.1). We may prove that Q^t is compact. Indeed, by using Hölder's inequality, we have

$$\left(\int_0^t \|(Fy)(s)\|_X^2 ds\right)^{1/2} \leq M_1(t)\|\phi\|_{\mathcal{X}},$$

where $M_1(t) = C(t+r)e^{|\gamma|t}$ for some $C > 0$ and $\gamma \in \mathbb{R}$. In order to prove the compactness of Q^t for $t > 0$ under the compactness of e^{tA}, $t > 0$, we define the ε-approximation $Q^t_\varepsilon : \mathcal{X} \to X$ of Q^t for $\varepsilon \in (0,t]$ by

$$Q^t_\varepsilon \phi = e^{\varepsilon A}\int_0^{t-\varepsilon} e^{(t-\varepsilon-s)A}(Fy)(s)ds, \qquad \phi \in \mathcal{X}.$$

Since $e^{\varepsilon A}$ is compact, Q^t_ε is also compact. The compactness of Q^t follows from

$$\begin{aligned}\|(Q^t - Q^t_\varepsilon)\phi\|_X &= \left\|\int_{t-\varepsilon}^t e^{(t-s)A}(Fy)(s)ds\right\|_X \\ &\leq M_2(t)\cdot \varepsilon^{1/2}\cdot \|\phi\|_{\mathcal{X}},\end{aligned}$$

where $M_2(t) = M_1(t)\sup_{s\in[0,t]}\|e^{sA}\|$.

Next let $t > r$ be fixed and define an operator $R^t : \mathcal{X} \to C([t-r,t];X)$ by

$$(R^t\phi)(s) = y(s,\phi), \qquad s \in [t-r,t].$$

Let Σ be a bounded set in $\mathcal{X}$. Since e^{tA} and Q^t are compact for $t > 0$, from equation (4.1.2) it follows that for each $s \in [t-r,t]$, set $\{(R^t\phi)(s) \in X : \phi \in \Sigma\}$ is precompact in X. Now we show that $\{(R^t\phi)(s); s \in [t-r,t], \phi \in \Sigma\}$ is an equicontinuous family of $C([t-r,t],X)$.

Let $0 < \tilde{t} < t-r$, $\phi \in \Sigma$ and $t-r \leq s' < s \leq t$. Then we obtain

$$\begin{aligned}&\|(R^t\phi)(s) - (R^t\phi)(s')\|_X \leq \|e^{sA} - e^{s'A}\|\|\phi_0\|_X + \int_{s'}^s \|e^{(s-u)A}\|\|(Fy)(u)\|_X du \\ &\quad + \int_0^{s'-\tilde{t}} \|e^{(s-u)A} - e^{(s'-u)A}\|\|(Fy)(u)\|_X du \\ &\quad + \int_{s'-\tilde{t}}^{s'} \|e^{(s-u)A} - e^{(s'-u)A}\|\|(Fy)(u)\|_X du \\ &\leq \|e^{sA} - e^{s'A}\|\|\phi_0\|_X + M_2(t)\|\phi\|_{\mathcal{X}}(s-s')^{1/2} \\ &\quad + \sup\left\{\|e^{uA} - e^{u'A}\| : u,u' \in [\tilde{t},t], |u-u'| = |s-s'|\right\}\cdot t^{1/2}\cdot M_2(t)\|\phi\|_{\mathcal{X}} \\ &\quad + 2M_2(t)\|\phi\|_{\mathcal{X}}\cdot \tilde{t}^{1/2}.\end{aligned}$$

Using the fact that e^{tA} is continuous in the uniform operator norm topology of $\mathcal{L}(X)$ and taking $\tilde{t} > 0$ sufficiently small, we have the desired equi continuity. Therefore, by the Arzelá–Ascoli theorem, R^t is compact. Now we introduce a mapping $I^t : C([t-r,t];X) \to \mathcal{X}$ by $I^t y(\cdot) = (y(t), y_t(\cdot))$ for $y \in C([t-r,t];X)$. Clearly, I^t is bounded. Since $e^{t\mathcal{A}}$ can be decomposed as $e^{t\mathcal{A}} = I^t R^t$ for $t > r$, thus $e^{t\mathcal{A}}$ is compact for $t > r$.

References

[1] Ahmed, N. U. *Semigroups Theory with Applications to Systems and Control.* Longman Scientific and Technical, London, (1991).

[2] Albeverio, S. and Cruzerio, A. B. Global flows with invariant (Gibbs) measures for Euler and Navier–Stokes two dimensional fluids. *Comm. Math. Phys.* **129** (1990), 432–444.

[3] Arnold, L. *Stochastic Differential Equation: Theory and Applications.* Wiley, New York, (1974).

[4] Arnold, L. A formula connecting sample and moment stability of linear stochastic systems. *SIAM J. Appl. Math.* **44** (1984), 793–802.

[5] Arnold, L. Stabilization by noise revisited. *Z. Angew. Math. Mech.* **70** (1990), 235–246.

[6] Arnold, L., Crauel, H., and Wihstutz, V. Stabilization for linear systems by noise. *SIAM J. Control Optim.* **21** (1983), 451–461.

[7] Arnold, L., Kliemann, W., and Oeljeklaus, E. *Lyapunov Exponents of Linear Stochastic Systems.* Arnold, L. and Wihstutz, V., Ed. Lecture Notes in Mathematics **1186**, Springer-Verlag, New York, (1984), 85–128.

[8] Arnold, L., Oeljeklaus, E., and Pardoux, E. *Almost Sure and Moment Stability for Linear Itô Equations.* Arnold, L and Wihstutz, V., Ed. Lecture Notes in Mathematics **1186**, Springer-Verlag, New York, (1984), 129–159.

[9] Arnold, L. and Wihstutz, V. Lyapunov Exponents: A Survey. Arnold, L. and Wihstutz, V., Ed. Lecture Notes in Mathematics **1186**, Springer-Verlag, New York, (1984), 1–26.

[10] Bao, J. H., Truman, A., and Yuan, C. G. Almost sure asymptotic stability of stochastic partial differential equations with jumps. *SIAM J. Control Optim.* **49** (2011), 771–787.

[11] Bátkai, A. and Piazzera, S. Semigroups and linear partial differential equations with delay. *J. Math. Anal. Appl.* **264** (2001), 1–20.

[12] Bátkai, A. and Piazzera, S. *Semigroups for Delay Equations.* Research Notes in Mathematics, A. K. Peters, Wellesley, (2005).

[13] Bell, D. R. and Mohammed, S. E. On the solution of stochastic ordinary differential equations via small delays. *Stochastics.* **29** (1989), 293–299.

[14] Benchimol, C. D. Feedback stabilizability in Hilbert spaces. *Appl. Math. Optim.* **4** (1978), 225–248.

[15] Bensoussan, A. and Temam, R. Équations aux dérivées partielles stochastiques non linéaires. *Israel J. Math.* **11** (1972), 95–121.

[16] Bensoussan, A. and Temam, R. Equations stochastiques du type Navier–Stokes. *J. Funct. Anal.* **13** (1973), 195–222.

[17] Bensoussan, A. and Viot, M. Optimal control of stochastic linear distributed parameter systems. *SIAM J. Control.* **13** (1975), 904–926.

[18] Bezandry, P. and Diagana, T. *Almost Periodic Stochastic Processes.* Springer-Verlag, New York, (2011).

[19] Bierkens, J. Pathwise stability of degenerate stochastic evolutions. *Integral Equations and Operator Theory.* **23** (2010), 1–27.

[20] Bierkens, J. and van Gaans, O. Dissipativity of the delay semigroup. *J. Differential Equations.* **257** (2014), 2418–2429.

[21] Brzeźniak, Z., Maslowski, B., and Seidler, J. Stochastic nonlinear beam equations. *Probab. Theory Relat. Fields.* **132** (2005), 119–149.

[22] Butkovskii, A. G. *Theory of Optimal Control of Distributed Parameter Systems.* Elsevier, New York, (1969).

[23] Capiński, M. and Cutland, N. Stochastic Navier–Stokes equations. *Acta. Appl. Math.* **25** (1991), 59–85.

[24] Caraballo, T. Asymptotic exponential stability of stochastic partial differential equations with delay. *Stochastics.* **33** (1990), 27–47.

[25] Caraballo, T., Garrido-Atienza, M. J., and Taniguchi, T. The existence and exponential behavior of solutions to stochastic delay evolution equations with a fractional Brownian motion. *Nonlinear Anal. TMA.* **74** (2011), 3671–3684.

[26] Caraballo, T., Garrido-Atienza, M., and Real, J. Asymptotic stability of nonlinear stochastic evolution equations. *Stoch. Anal. Appl.* **21** (2003), 301–327.

[27] Caraballo, T., Kloeden, P. E., and Schmalfuß, B. Exponentially stable stationary solutions for stochastic evolution equations and their perturbation. *Appl. Math. Optim.* **50** (2004), 183–207.

[28] Caraballo, T. and Langa, J. Comparison of the long-time behavior of linear Itô and Stratonovich partial differential equations. *Stoch. Anal. Appl.* **19** (2001), 183–195.

[29] Caraballo, T., Langa, J., and Taniguchi, T. The exponential behaviour and stabilizability of stochastic 2D–Navier–Stokes equations. *J. Differential Equations.* **179** (2002), 714–737.

[30] Caraballo, T. and Liu, K. Exponential stability of mild solutions of stochastic partial differential equations with delays. *Stoch. Anal. Appl.* **17** (1999), 743–764.

[31] Caraballo, T. and Liu, K. On exponential stability criteria of stochastic partial differential equations. *Stoch. Proc. Appl.* **83** (1999), 289–301.

[32] Caraballo, T. and Liu, K. Asymptotic exponential stability property for diffusion processes driven by stochastic differential equations in duals of nuclear spaces. *Publ. RIMS, Kyoto Univ.* **37** (2001), 239–254.

[33] Caraballo, T., Liu, K., and Mao, X. R. Stabilization of partial differential equations by stochastic noise. *Nagoya Math. J.* **161** (2001), 155–170.

[34] Caraballo, T., Liu, K., and Truman, A. Stochastic functional partial differential equations: existence, uniqueness and asymptotic stability. *Proc. Royal Soc. London A.* **456** (2000), 1775–1802.

[35] Caraballo, T., Márquez-Durán, A., and Real, J. The asymptotic behaviour of a stochastic 3D LANS-α model. *Appl. Math. Optim.* **53** (2006), 141–161.

[36] Caraballo, T., Real, J., and Taniguchi, T. The exponential stability of neutral stochastic delay partial differential equations. *Discrete and Continuous Dynamical Systems: A* **18** (2007), 295–313.

[37] Caraballo, T. and Robinson, J. C. Stabilisation of linear PDEs by Stratonovich noise. *Systems & Control Letters.* **53** (2004), 41–50.

[38] Cerrai, S. Stabilization by noise for a class of stochastic reaction-diffusion equations. *Probab. Theory Relat. Fields.* **133** (2005), 190–214.

[39] Chojnowska-Michalik, A. and Goldys, B. Existence, uniqueness and invariant measures for stochastic semilinear equations on Hilbert spaces. *Probab. Theory Relat. Fields.* **102** (1995), 331–356.

[40] Chow, P. Stability of nonlinear stochastic evolution equations. *J. Math. Anal. Appl.* **89** (1982), 400–419.

[41] Chow, P. *Stochastic Partial Differential Equations.* Chapman & Hall/CRC, New York, (2007).

[42] Chow, P. and Khas'minskii, R. Z. Stationary solutions of nonlinear stochastic evolution equations. *Stoch. Anal. Appl.* **15(5)** (1997), 671–699.

[43] Chueshov, I. and Vuillermot, P. Long time behavior of solutions to a class of stochastic parabolic equations with homogeneous white noise: Itô's case. *Stoch. Anal. Appl.* **18** (2000), 581–615.

[44] Cui, J., Yan, L. T., and Sun, X. C. Exponential stability for neutral stochastic partial differential equations with delays and Poisson jumps. *Statist. Probab. Lett.* **81** (2011), 1970–1977.

[45] Curtain, R. F. Stochastic distributed systems with point observation and boundary control: an abstract theory. *Stochastics.* **3** (1979), 85–104.

[46] Curtain, R. F. Stability of stochastic partial differential equation. *J. Math. Anal. Appl.* **79** (1981), 352–369.

[47] Curtain, R. F. and Ichikawa, A. The separation principle for stochastic evolution equations. *SIAM J. Control Optim.* **15** (1977), 367–383.

[48] Curtain, R. F. and Pritchard, A. *Infinite Dimensional Linear Systems Theory.* Lecture Notes in Control and Information Science, **8**, Springer-Verlag, New York, (1978).

[49] Curtain, R. F. and Zwart, H. J. *Introduction to Infinite Dimensional Linear Systems Theory.* Springer-Verlag, New York, (1985).

[50] Da Prato, G., Gątarek, D., and Zabczyk, J. Invariant measures for semilinear stochastic equations. *Stoch. Anal. Appl.* **10** (1992), 387–408.

[51] Da Prato, G. and Ichikawa, A. Stability and quadratic control for linear stochastic equations with unbounded coefficients. *Boll. Un. Mat. It.* **4** (1985), 987–1001.

[52] Da Prato, G. and Zabczyk, J. *Ergodicity for Infinite Dimensional Systems.* London Mathematical Society Lecture Note Series. **229**, Cambridge University Press, Cambridge, (1996).

[53] Da Prato, G. and Zabczyk, J. *Stochastic Equations in Infinite Dimensions.* Second Edition, Encyclopedia of Mathematics and its Applications, Cambridge University Press, Cambridge, (2014).

[54] Daletskii, L. and Krein, G. *Stability of Solutions of Differential Equations in Banach Spaces.* Trans. Amer. Math. Soc. **43,** Providence, (1974).

[55] Datko, R. Extending a theorem of A. Lyapunov to Hilbert space. *J. Math. Anal. Appl.* **32** (1970), 610–616.

[56] Datko, R. Uniform asymptotic stability of evolutionary processes in a Banach space. *SIAM J. Math. Anal.* **3** (1973), 428–445.

[57] Datko, R. Representation of solutions and stability of linear differential-difference equations in a Banach space. *J. Differential Equations.* **29** (1978), 105–166.

[58] Datko, R. Lyapunov functionals for certain linear delay differential equations in a Hilbert space. *J. Math. Anal. Appl.* **76** (1980), 37–57.

[59] Datko, R. The uniform exponential stability of a class of linear differential-difference equations in a Hilbert space. *Proc. Royal Soc. Edinburgh,* **89A** (1981), 201–215.

[60] Datko, R. An example of an unstable neutral differential equation. *International J. Control.* **20** (1983), 263–267.

[61] Datko, R. The Laplace transform and the integral stability of certain linear processes. *J. Differential Equations.* **48** (1983), 386–403.

[62] Datko, R. Remarks concerning the asymptotic stability and stabilization of linear delay differential equations. *J. Math. Anal. Appl.* **111** (1985), 571–584.

[63] Datko, R., Lagnese, J., and Polis, M. An example on the effect of time delays in boundary feedback of wave equations. *SIAM J. Control Optim.* **24** (1986), 152–156.

[64] Davies, E. B. *Spectral Theory and Differential Operators.* Cambridge Studies in Advanced Mathematics, Vol. **42**, Cambridge University Press, Cambridge, (1996).

[65] Delfour, M., McCalla, C., and Mitter, S. Stability and the infinite-time quadratic cost problem for linear hereditary differential systems. *SIAM J. Control.* **13** (1975), 48–88.

[66] Drozdov, A. Stability of a class of stochastic integro-differential equations. *Stoch. Anal. Appl.* **13(5)** (1995), 517–530.

[67] Duncan, T., Maslowski, B., and Duncan, B. Stochastic equations in Hilbert space with a multiplicative fractional Gaussian noise. *Stoch. Proc. Appl.* **115** (2005), 1357–1383.

[68] Dunford, N. and Schwartz, J. T. *Linear Operators, Part I: General Theory.* Interscience Publishers, Inc., New York, (1958).

[69] Dunford, N. and Schwartz, J. T. *Linear Operators, Part II: Spectral Theory.* Interscience Publishers, Inc., New York, (1963).

[70] Engel, K-J. and Nagel, R. *One-Parameter Semigroups for Linear Evolution Equations.* Graduate Texts in Math. **194**, Springer-Verlag, New York, (2000).

[71] Ethier, S. N. and Kurtz, T. G. *Markov Processes: Characterization and Convergence.* Wiley and Sons, New York, (1986).

[72] Fattorini, H. *Second Order Linear Differential Equations in Banach Spaces.* North Holland Math. Studies Series. **108**, (1985).

[73] Flandoli, F. Riccati equation arising in a stochastic control problem. *Boll. Un. Mat. Ital., Anal. Funz. Appl.* **1** (1982), 377–393.

[74] Flandoli, F. and Gątarek, D. Martingale and stationary solutions for stochastic Navier–Stokes equations. *Probab. Theory Relat. Fields.* **102** (1995), 367–406.

[75] Flandoli, F. and Maslowski, B. Ergodicity of the 2-D Navier–Stokes equations under random perturbations. *Comm. Math. Phys.* **171** (1995), 119–141.

[76] Friedman, A. *Stochastic Differential Equations and Applications.* Vols. I and II, Academic Press, New York, (1975).

[77] Furstenberg, H. Noncommuting random products. *Trans. Amer. Math. Soc.* **108** (1963), 377–428.

[78] Furstenberg, H. A Poisson formula for semi-simple Lie group. *Ann. Math.* **77** (1963), 335–386.

[79] Gawarecki, L. and Mandrekar, V. *Stochastic Differential Equations in Infinite Dimensions with Applications to Stochastic Partial Differential Equations.* Springer-Verlag, New York, (2011).

[80] Gihman, I. J. and Skorokhod, A. V. *Stochastic Differential Equations.* Springer-Verlag, Berlin, New York, (1972).

[81] Govindan, E. Almost sure exponential stability for stochastic neutral partial functional differential equations. *Stochastics.* **77** (2005), 139–154.

[82] Hahn, W. *Stability of Motion.* Springer-Verlag, Berlin, New York, (1967).

[83] Hairer, M. and Mattingly, J. Ergodicity of the 2D Navier–Stokes equations with degenerate stochastic forcing. *Ann. Math.* **164** (2006), 993–1032.

[84] Hale, J. *Theory of Functional Differential Equations.* Springer-Verlag, Berlin, New York, (1977).

[85] Hale, J. and Lunel, S. *Introduction to Functional Differential Equations.* Springer-Verlag, Berlin, New York, (1993).

[86] Has'minskii, R. Z. *Stochastic Stability of Differential Equations.* Sijtjoff and Noordhoff, Alphen, (1980) (translation of the Russian edition, Nauka, Moscow, (1969)).

[87] Haussmann, U. G. Asymptotic stability of the linear Itô equation in infinite dimensional. *J. Math. Anal. Appl.* **65** (1978), 219–235.

[88] Hille, E. and Phillips, R. S. *Functional Analysis and Semigroups.* Amer. Math. Soc., Providence, (1957).

[89] Ichikawa, A. Optimal control of a linear stochastic evolution equation with state and control dependent noise. *Proc. IMA Conference: Recent Theoretical Developments in Control*, Academic Press, Leicester, (1970).

[90] Ichikawa, A. Dynamic programming approach to stochastic evolution equations. *SIAM J. Control Optim.* **17** (1979), 152–174.

[91] Ichikawa, A. Stability of semilinear stochastic evolution equations. *J. Math. Anal. Appl.* **90** (1982), 12–44.

[92] Ichikawa, A. Absolute stability of a stochastic evolution equation. *Stochastics.* **11** (1983), 143–158.

[93] Ichikawa, A. Semilinear stochastic evolution equations: boundedness, stability and invariant measure. *Stochastics.* **12** (1984), 1–39.

[94] Ichikawa, A. Equivalence of L_p stability and exponential stability for a class of nonlinear semilinear semigroups. *Nonlinear Anal. TMA.* **8** (1984), 805–815.

[95] Ichikawa, A. A semigroup model for parabolic equations with boundary and pointwise noise. In: *Stochastic Space-Time Models and Limit Theorems*, D. Reidel Publishing Company, Dordrecht, (1985), 81–94.

[96] Ichikawa, A. Stability of parabolic equations with boundary and pointwise noise. In: *Stochastic Differential Systems*, Lecture Notes in Control and Information Science **69**, Springer-Verlag, Berlin, (1985), 55–66.

[97] Ichikawa, A. and Pritchard, A. J. Existence, uniqueness and stability of nonlinear evolution equations. *J. Math. Anal. Appl.* **68** (1979), 454–476.

[98] Ikeda, N. and Watanabe, S. *Stochastic Differential Equations and Diffusion Processes.* Second Edition, North-Holland, Kodansha, Amsterdam, Tokyo, (1989).

[99] Jahanipur, R. Stability of stochastic delay evolution equations with monotone nonlinearity. *Stoch. Anal. Appl.* **21** (2003), 161–181.

[100] Karatzas, I. and Shreve, S. E. *Brownian Motion and Stochastic Calculus.* Second Edition, Springer-Verlag, Berlin, Heidelberg, New York, (1991).

[101] Kats, I. I. and Krasovskii, N.N. On the stability of systems with random parameters. *Prkil. Met. Mek.* **24**, (1960), 256–270.

[102] Khas'minskii, R. Necessary and sufficient condition for the asymptotic stability of linear stochastic systems. *Theory Probab. Appl.* **12** (1967), 144–147.

[103] Khas'minskii, R. *Stochastic Stability of Differential Equations.* Second Edition, Springer-Verlag, Berlin, Heidelberg, New York, (2012).

[104] Khas'minskii, R. On robustness of some concepts in stability of SDE. *Stochastic Modeling and Nonlinear Dynamics*, W. Kliemann and N.S. Namachchivaya, Eds., Chapman & Hall/CRC, London, New York, (1996), 131–137.

[105] Khas'minskii, R. and Mandrekar, V. On the stability of solutions of stochastic evolution equations. *The Dynkin Festschrift: Markov Processes and their Applications, Progress in Probability*, **34**, Freidlin, M. I., Ed. Birkhäuser, Boston, (1994), 185–197.

[106] Kolmanovskii, V. B. and Myshkis, A. *Introduction to the Theory and Applications of Functional Differential Equations.* Kluwer Academic Publishers, Dordrecht, (1999).

[107] Kolmanovskii, V. B. and Nosov, V. R. *Stability of Functional Differential Equations.* Academic Press, New York, (1986).

[108] Kozin, F. On almost surely asymptotic sample properties of diffusion processes defined by stochastic differential equations. *J. Math. Kyoto Univ.* **4** (1965), 515–528.

[109] Kozin, F. A survey of stability of stochastic systems. *Automatica.* **5** (1969), 95–112.

[110] Kozin, F. *Stability of the Linear Stochastic System.* Lecture Notes in Mathematics **294**, Springer-Verlag, New York, (1972), 186–229.

[111] Krasovskii, N. N. *Stability of Motions.* Stanford University Press, Stanford, (1963).

[112] Kreyszig, E. *Introductory Functional Analysis with Applications.* John Wiley & Sons. Inc., New York (1978).

[113] Krylov, N. V. and Rozovskii, B. L. Stochastic evolution equations. *J. Sov. Math.* **16** (1981), 1233–1277.

[114] Kushner, H. *Stochastic Stability and Control.* Academic Press, New York, (1967).

[115] Kushner, H. *Introduction to Stochastic Control Theory.* Holt, Rinehart and Winston, New York, (1971).

[116] Kwiecińska, A. Stabilization of partial differential equations by noise. *Stoch. Proc. Appl.* **79** (1999), 179–184.

[117] Kwiecińska, A. Stabilization of evolution equations by noise. *Proc. Amer. Math. Soc.* **130** (2001), 3067–3074.

[118] Lakshmikantham, V. and Leela, S. *Differential and Integral Inequalities: Theory and Applications,* Vol. II, Academic Press, New York, (1969).

[119] Lasiecka, I. and Triggiani, R. Feedback semigroups and cosine operators for boundary feedback parabolic and hyperbolic equations, *J. Differential Equations.* **47** (1983), 246–272.

[120] Lasiecka, I. and Triggiani, R. Dirichlet boundary control problem for parabolic equations with quadratic cost: analyticity and Riccati feedback synthesis, *SIAM J. Control Optim.* **21** (1983), 41–67.

[121] Leha, G. and Ritter, G. Lyapunov type of conditions for stationary distributions of diffusion processes on Hilbert spaces. *Stochastics.* **48** (1994), 195–225.

[122] Leha, G., Ritter, G., and Maslowski, B. Stability of solutions to semilinear stochastic evolution equations. *Stoch. Anal. Appl.* **17** (1999), 1009–1051.

[123] Lenhart, S. M. and Travis, C. C. Stability of functional partial differential equations. *J. Differential Equations.* **58** (1985), 212–227.

[124] Levin, J. and Nohel, J. On a nonlinear delay equation. *J. Math. Anal. Appl.* **8** (1964), 31–44.

[125] Li, P. and Ahmed, N. U. Feedback stabilization of some nonlinear stochastic systems on Hilbert space. *Nonlinear Anal. TMA.* **17** (1991), 31–43.

[126] Li, P. and Ahmed, N. U. A note on stability of stochastic systems with unbounded perturbations. *Stoch. Anal. Appl.* **7(4)** (1989), 425–434.

[127] Liang, F. and Gao, H. J. Stochastic nonlinear wave equation with memory driven by compensated Poisson random measures. *J. Math. Phys.* **55** (2014), 033503: 1–23.

[128] Liang, F. and Guo, Z. H. Asymptotic behavior for second order stochastic evolution equations with memory. *J. Math. Anal. Appl.* **419** (2014), 1333–1350.

[129] Lions, J. L. *Optimal Control of Systems Governed by Partial Differential Equations.* Springer-Verlag, Berlin, New York, (1971).

[130] Lions, J. L. and Magenes, E. *Nonhomogeneous Boundary Value Problems and Applications.* Vols. I, II, and III. Springer-Verlag, Berlin, New York, (1972).

[131] Liu, K. On stability for a class of semilinear stochastic evolution equations. *Stoch. Proc. Appl.* **70** (1997), 219–241.

[132] Liu, K. Carathéodory approximate solutions for a class of semilinear stochastic evolution equations with time delays. *J. Math. Anal. Appl.* **220** (1998), 349–364.

[133] Liu, K. Lyapunov functionals and asymptotic stability of stochastic delay evolution equations. *Stochastics.* **63** (1998), 1–26.

[134] Liu, K. Necessary and sufficient conditions for exponential stability and ultimate boundedness of systems governed by stochastic partial differential equations. *J. London Math. Soc.* **62** (2000), 311–320.

[135] Liu, K. Some remarks on exponential stability of stochastic differential equations. *Stoch. Anal. Appl.* **19(1)** (2001), 59–65.

[136] Liu, K. Uniform L^2-stability in mean square of linear autonomous stochastic functional differential equations in Hilbert spaces. *Stoch. Proc. Appl.* **116** (2005), 1131–1165.

[137] Liu, K. *Stability of Infinite Dimensional Stochastic Differential Equations with Applications*. Chapman & Hall/CRC, London, New York, (2006).

[138] Liu, K. Stochastic retarded evolution equations: green operators, convolutions and solutions. *Stoch. Anal. Appl.* **26** (2008), 624–650.

[139] Liu, K. Stationary solutions of retarded Ornstein–Uhlenbeck processes in Hilbert spaces. *Statist. Probab. Lett.* **78** (2008), 1775–1783.

[140] Liu, K. The fundamental solution and its role in the optimal control of infinite dimensional neutral systems. *Appl. Math. Optim.* **60** (2009), 1–38.

[141] Liu, K. A criterion for stationary solutions of retarded linear equations with additive noise. *Stoch. Anal. Appl.* **29** (2011), 799–823.

[142] Liu, K. Existence of invariant measures of stochastic systems with delay in the highest order partial derivatives. *Statist. Probab. Lett.* **94** (2014), 267–272.

[143] Liu, K. On stationarity of stochastic retarded linear equations with unbounded drift operators. *Stoch. Anal. Appl.* **34(4)** (2016), 547–572.

[144] Liu, K. Almost sure exponential stability sensitive to small time delay of stochastic neutral functional differential equations. *Appl. Math. Lett.* **77** (2018), 57–63.

[145] Liu, K. Sensitivity to small delays of pathwise stability for stochastic retarded evolution equations. *J. Theoretical Probab.* **31** (2018), 1625–1646.

[146] Liu, K. Stationary solutions of neutral stochastic partial differential equations with delays in the highest-order derivatives. *Discrete and Continuous Dynamical Systems: B.* **23(9)** (2018), 3915–3934.

[147] Liu, K. and Mao, X. R. Exponential stability of non-linear stochastic evolution equations. *Stoch. Proc. Appl.* **78** (1998), 173–193.

[148] Liu, K. and Mao, X. R. Large time decay behaviour of dynamical equations with random perturbation features. *Stoch. Anal. Appl.* **19(2)** (2001), 295–327.

[149] Liu, K. and Shi, Y. F. Razumikhin-type stability theorems of stochastic functional differential equations in infinite dimensions. In: *Proc. 22nd IFIP TC 7: Systems, Control, Modeling and Optimization*, F. Ceragioli, A. Dontchev, Ed., Springer-Verlag, New York. (2005), 237–247.

[150] Liu, K. and Truman, A. Lyapunov function approaches and asymptotic stability of stochastic evolution equations in Hilbert spaces – a survey of recent developments. In: *Stochastic Partial Differential Equations and Applications*, G. De Prato and L. Dubaro, eds. Lecture Notes in Pure and Applied Math. **27** (2002), 337–371. Dekker, New York.

[151] Liu, K. and Truman, A. Moment and almost sure Lyapunov exponents of mild solutions of stochastic evolution equations with variable delays via approximation approaches. *J. Math. Kyoto Univ.* **41** (2002), 749–768.

[152] Liu, K. and Xia, X. W. On the exponential stability in mean square of neutral stochastic functional differential equations. *Systems & Control Lett.* **37** (1999), 207–215.

[153] Liu, R. and Mandrekar, V. Ultimate boundedness and invariant measures of stochastic evolution equations. *Stochastics.* **56** (1996), 75–101.

[154] Liu, R. and Mandrekar, V. Stochastic semilinear evolution equations: Lyapunov function, stability and ultimate boundedness. *J. Math. Anal. Appl.* **212** (1997), 537–553.

[155] Luo, J. W. Exponential stability for stochastic neutral partial functional differential equations. *J. Math. Anal. Appl.* **355** (2009), 414–425.

[156] Luo, J. W. and Liu, K. Stability of infinite dimensional stochastic evolution equations with memory and Markovian jumps. *Stoch. Proc. Appl.* **118** (2008), 864–895.

[157] Luo, J. W. and Taniguchi, T. Fixed points and stability of stochastic neutral partial differential equations with infinite delays. *Stoch. Anal. Appl.* **27** (2009), 1163–1173.

[158] Luo, Z. H., Guo, B. Z., and Morgul, O. *Stability and Stabilization of Infinite Dimensional with Applications.* Springer-Verlag, London, Berlin, Heidelberg, (1999).

[159] Lyapunov, A. M. Probléme générale de la stabilité du muvement. *Comm. Soc. Math. Kharkov.* **2**, (1892). Reprint *Ann. Math. Studies.* **17**, Princeton University Press, Princeton, (1949).

[160] Mandrekar, V. On Lyapounov stability theorems for stochastic (deterministic) evolution equations. In *Proc. of the NATO-ASI School on: Stochastic Analysis and Applications in Physics*, L. Streit, Ed., Springer-Verlag, New York (1994), 219–237.

[161] Manthey, R. and Maslowski, B. A random continuous model for two interacting populations. *Appl. Math. Optim.* **45(2)** (2002), 213–236.

[162] Manthey, R. and Mittmann, K. On the qualitative behaviour of the solution to a stochastic partial functional differential equation arising in population dynamics. *Stochastics.* **66** (1999), 153–166.

[163] Manthey, R. and Zausinger, T. Stochastic evolution equations in $L^2(2\nu)_p$. *Stochastics.* **66** (1999), 37–85.

[164] Mao, X. R. Almost sure polynomial stability for a class of stochastic differential equations. *Quart. J. Math. Oxford (2).* **43** (1992), 339–348.

[165] Mao, X. R. *Exponental Stability of Stochastic Differential Equations.* Marcel Dekker, New York (1994).

[166] Mao, X. R. Razumikhin-type theorems on exponential stability of neutral stochastic functional differential equations. *SIAM J. Math. Anal.* **28(2)** (1997), 389–401.

[167] Mao, X. R. *Stochastic Differential Equations and Applications.* Second Edition, Woodhead Publishing, Oxford, (2007).

[168] Maslowski, B. Uniqueness and stability of invariant measures for stochastic differential equations in Hilbert spaces. *Stochastics.* **28** (1989), 85–114.

[169] Maslowski, B. Stability of semilinear equations with boundary and pointwise noise. *Annali Scuola Normale Superiore di Pisa*, **IV**, (1995), 55–93.

[170] Maslowski, B., Seidler, J., and Vrkoč, I. Integral continuity and stability for stochastic hyperbolic equations. *Differential Integral Equations.* **6** (1993), 355–382.

[171] Métivier, M. *Semimartingales.* Walter de Gruyter, Berlin, New York, (1982).

[172] Métivier, M. and Pellaumail, J. *Stochastic Integration.* Academic Press, New York, (1980).

[173] Miyahara, Y. Ultimate boundedness of the systems governed by stochastic differential equations. *Nagoya Math. J.* **47** (1972), 111–144.

[174] Miyahara, Y. Invariant measures of ultimately bounded stochastic processes. *Nagoya Math. J.* **49** (1973), 149–153.

[175] Mizel, V. and Trutzer, V. Stochastic hereditary equations: existence and asymptotic stability. *J. Integral Equations.* **7** (1984), 1–72.

[176] Mohammed, S. E. *Stochastic Functional Differential Equations.* Pitman, London, (1984).

[177] Mohammed, S. E. and Scheutzow, M. Lyapunov exponents of linea stochastic functional differential equations driven by semimartingales. Part I. The multiplicative ergodic theory. *Ann. Inst. H. Poincaré Probab. Statist.* **32** (1996), 69–105.

[178] Mohammed, S. E. and Scheutzow, M. Lyapunov exponents of linea stochastic functional differential equations driven by semimartingales. Part II. Examples and case studies. *Ann. Probab.* **25** (1997), 1210–1240.

[179] Morozan, T. Boundedness properties for stochastic systems. In: *Stability of Stochastic Dynamical Systems,* R. F. Curtain, Ed. Lecture Notes in Mathematics **294**, Springer-Verlag, New York, (1972), 21–34.

[180] Murray, J. D. *Mathematical Biology I: An Introduction.* Springer-Verlag, New York, (2001).

[181] Nakagiri, S. Structural properties of functional differential equations in Banach spaces. *Osaka. J. Math.* **25** (1988), 353–398.

[182] Oseledec, V. A multiplicative ergodic theorem Lyapunov characteristic number for dynamical systems. *Trans. Moscow Math. Soc.* **19** (1969), 197–231.

[183] Parthasarathy, K. P. *Probability Measures on Metric Spaces.* Academic Press, New York, (1967).

[184] Pardoux, E. Équations aux dérivées partielles stochastiques non linéaires monotones. Thesis, Université Paris XI, (1975).

[185] Pardoux, E. Stochastic partial differential equations and filtering of diffusion processes. *Stochastics.* **3** (1979), 127–167.

[186] Pazy, A. On the applicability of Lyapunov's theorem in Hilbert space. *SIAM J. Math. Anal. Appl.* **3** (1972), 291–294.

[187] Pazy, A. *Semigroups of Linear Operators and Applications to Partial Differential Equations.* Applied Mathematical Sciences **44**. Springer-Verlag, New York, (1983).

[188] Pinsky, M. A. Stochastic stability and the Dirichlet problem. *Comm. Pure Appl. Math.* **34(4)** (1978), 311–35.

[189] Plaut, R. and Infante, E. On the stability of some continuous systems subject to random excitation. *Trans. ASME J. Appl. Mech.* (1970), 623–627.

[190] Prévôt, C. and Röckner, M. *A Concise Course on Stochastic Partial Differential Equations.* Lecture Notes in Mathematics **1905**, Springer-Verlag, New York, (2007).

[191] Pritchard, A. J. and Zabczyk, J. Stability and stabilizability of infinite dimensional systems. *SIAM Review.* **23** (1981), 25–52.

[192] Razumikhin, B. S. On stability of systems with a delay. *Prikl. Mat. Meh.* **20** (1956), 500–512.

[193] Razumikhin, B. S. Application of Liapunov's methods to problems in stability of systems with a delay. *Automat. i Telemeh.* **21** (1960), 740–749.

[194] Real, J. Stochastic partial differential equations with delays. *Stochastics.* **8** (1982), 81–102.

[195] Reed, M. and Simon, B. *Methods of Modern Mathematical Physics I: Functional Analysis.* Revised and Enlarged Edition, New York, Academic Press, (1980).

[196] Ren, Y. and Sun, D. Second-order neutral impulsive stochastic evolution equations with delay. *J. Math. Phys.* **50** (2009), 102709: 1–12.

[197] Revuz, D. and Yor, M. *Continuous Martingales and Brownian Motion.* Third Edition, Springer-Verlag, Berlin Heidelberg, (1999).

[198] Rogers, L. and Williams, D. *Diffusion, Markov Processes and Martingales.* Second Edition. Vol. I, New York, Wiley, (1994)

[199] Rozovskii, B. L. *Stochastic Evolution Systems: Linear Theory and Applications to Non-Linear Filtering.* Kluwer Academic Publishers, Dordrecht, (1990).

[200] Scheutzow, M. Stabilization and destabilization by noise in the plane. *Stoch. Anal. Appl.* **11** (1993), 97–113.

[201] Shiga, T. Ergodic theorems and exponential decay of sample paths for certain interactive diffusion systems. *Osaka J. Math.* **29** (1992), 789–807.

[202] Skorokhod, A. V. *Asymptotic Methods in the Theory of Stochastic Differential Equations.* American Mathematical Society, Providence, (1989).

[203] Tanabe, H. *Equations of Evolution.* Monographs and Studies in Mathematics, **6**, Pitman, London, (1979).

[204] Taniguchi, T. Asymptotic stability theorems of semilinear stochastic evolution equations in Hilbert spaces. *Stochastics.* **53** (1995), 41–52.

[205] Taniguchi, T. Moment asymptotic behavior and almost sure Lyapunov exponent of stochastic functional differential equations with finite delays via Lyapunov Razumikhin method. *Stochastics.* **58** (1996), 191–208.

[206] Taniguchi, T. The existence and asymptotic behaviour of energy solutions to stochastic 2D functional Navier–Stokes equations driven by Lévy processes. *J. Math. Anal. Appl.* **385** (2012), 634–654.

[207] Taniguchi, T., Liu, K., and Truman, A. Existence, uniqueness and asymptotic behavior of mild solutions to stochastic functional differential equations in Hilbert spaces. *J. Differential Equations.* **181** (2002), 72–91.

[208] Temam, R. *Navier–Stokes Equations, Theory and Numerical Analysis.* Third Revised Edition, North-Holland, Amsterdam, (1984).

[209] Temam, R. *Infinite Dimensional Dynamical Systems in Mechanics and Physics.* Second Edition, Springer-Verlag, New York, (1988).

[210] Travis, C. C. and Webb, G. F. Existence and stability for partial functional differential equations. *Trans. Amer. Math. Soc.* **200** (1974), 395–418.

[211] Travis, C. C. and Webb, G. F. Existence, stability and compactness in the α-norm for partial functional differential equations. *Trans. Amer. Math. Soc.* **240** (1978), 129–143.

[212] Tubaro, L. An estimate of Burkholder type for stochastic processes defined by the stochastic integral. *Stoch. Anal. Appl.* **2** (1984), 187–192.

[213] Vinter, R. B. Filter stability for stochastic evolution equations. *SIAM J. Control Optim.* **15** (1977), 465–485.

[214] Vishik, M. J. and Fursikov, A. V. *Mathematical Problems in Statistical Hydromechanics.* Kluwer Academic Publishers, Dordrecht (1988).

[215] Walsh, J. B. An Introduction to stochastic partial differential equations. In: *École d'eté de Probabilité de Saint Flour XIV,* P. L. Hennequin, Ed., Lecture Notes in Mathematics. **1180**, (1984), 265–439 Springer-Verlag, New York.

[216] Wan, L. and Duan, J. Q. Exponential stability of non-autonomous stochastic partial differential equations with finite memory. *Statist. Probab. Lett.* **78** (2008), 490–498.

[217] Wang, P. K. On the almost sure stability of linear stochastic distributed parameter dynamical systems. *Trans. ASME J. Appl. Mech.* (1966), 182–186.

[218] Wonham, W. M. Random differential equations in control theory. In *Probabilistic Methods in Applied Mathematics.* **2**, A. Bharucha-Reid, Ed., Academic Press, New York, (1970), 131–212.

[219] Wonham, W. M. Lyapunov criteria for weak stochastic stability. *J. Differential Equations.* **2** (1966), 195–207.

[220] Wu, J. H. *Theory and Applications of Partial Functional Differential Equations.* Appl. Math. Sci. **119**, Springer-Verlag, New York, (1996).

[221] Xie, B. The moment and almost surely exponential stability of stochastic heat equations. *Proc. Amer. Math. Soc.* **136** (2008), 3627–3634.

[222] Yavin, Y. On the stochastic stability of a parabolic type system. *International J. Syst. Sci.* **5** (1974), 623–632.

[223] Yavin, Y. On the modelling and stability of a stochastic distributed parameter system. *International J. Syst. Sci.* **6** (1975), 301–311.

[224] Yosida, Y. *Functional Analysis.* Sixth Edition, Springer-Verlag, New York, (1980).

[225] Zabczyk, J. Remarks on the control of discrete-time distributed systems. *SIAM J. Control Optim.* **12** (1974), 721–735.

[226] Zabczyk, J. A note on C_0 semigroups. *Bull. Polish Acad. Sci. (Math.)* **162** (1975), 895–898.

[227] Zabczyk, J. On optimal stochastic control of discrete-time systems in Hilbert space. *SIAM J. Control Optim.* **13** (1975), 1217–1234.

[228] Zabczyk, J. On stability of infinite dimensional stochastic systems. Probab. Theory, Z. Ciesislski, Ed., Banach Center Publications, **5**, Warsaw, (1979), 273–281.

[229] Zabczyk, J. A comment on the paper "Stability and stabilizability of infinite dimensional systems" by A.J. Pritchard and J. Zabczyk, SIAM Review, **23** (1981), 25–52. *Bull. Polish Acad. Sci. (Math.)*. Private Communication.
[230] Zakai, M. On the ultimate boundedness of moments associated with solutions of stochastic differential equations. *SIAM J. Control.* **5** (1967), 588–593.
[231] Zakai, M. A Lyapunov criterion for the existence of stationary probability distributions for systems perturbed by noise. *SIAM J. Control.* **7** (1969), 390–397.

Index

admissible control, 240, 242
algebraic Riccati equation, 242
asymptotic stability, 36
  almost sure, 38
  in probability, 37
  in pth moment, 37

Bochner integral, 17
Borel–Cantelli lemma, 81
boundary noise, 82
bound (growth, spectral), 51
Burkholder–Davis–Gundy's inequality, 25

Chapman–Kolmogorov equation, 19
comparison method, 219
contraction mapping method, 210
covariance operator, 22
cross quadratic variation, 21

decay, decayable
  with rate $\lambda(t)$ in the almost sure sense, 39
  with rate $\lambda(t)$ in the pth moment, 39
dissipative (of an operator), 9
distribution (of a random variable), 16
Doob's inequality, 21

exponential stability, 39
  almost sure, 39
  in pth moment, 39
exponential ultimately bounded in mean square, 146

feedback control, 240
Feller property (strong), 19
filtration, 16
  normal, 16
  satisfying the usual conditions, 16
fractional power (of an operator), 15
fundamental solution, 179

Gaussian measure, 22
Green operator, 179

Hahn–Banach theorem, 4
Hille–Yosida theorem, 8

invariant measure, 150
Itô's formula, 26, 35
Itô type of functional, 26, 35

linear growth condition, 28
Lipschitz continuous condition, 28, 100
Lumer–Phillips theorem, 9
Lyapunov
  equation, 56
  exponent, 45
  exponent method, 45
  function, 42, 54, 56, 117
  second (direct) method, 42

Markov process, 19
Markov property (strong), 19
martingale, 20
  continuous L^p, 20
  local, 20
measurable
  mapping, 16
  space, 16
  stochastic process, 17

modification of a stochastic process, 18

operator (linear), 3
 compact, 6
 Hilbert–Schmidt, trace-class, 6
 resolvent, 7
 self-adjoint, symmetric, 5
 variational, 131
optimal control, 240

predator–prey model, 226
probability
 measure, 16
 space, 16
 space, complete, 16
process
 adapted, 18
 continuous, 20
 integrable, 20
 progressively measurable, 18

quadratic cost functional, 240
quadratic variation, 21
Q-Wiener process, 22

random variable, 16
Razumikhin function method, 214
Riesz representation theorem, 4

semigroup (strongly continuous or C_0), 8
 analytic, (eventually) compact, 13, 14
 bounded, norm continuous, contraction, 8
Sobolev space, 2
solution in the strong sense, 28
solution, mild or strong, 27, 34
spectrum (retarded), 7, 189
 continuous, point, residual (retarded), 7, 189
spectrum-determined growth condition, 51
submartingale, 20
supermartingale, 20
stability, 36
 almost sure (strong), 39
 in probability (strong), 37
 in the global (strong), 36
 in the pth moment, 37
stabilizable (of a controllable system), 240
stabilization by noise, 168
stable C_0-semigroup, 47
stationary solution, 88, 89
stochastic
 control, 239
 delay differential equation, 196, 206, 218
 differential equation, 27
 Navier–Stokes equation, 236
 reaction-diffusion equation, 232
stochastic convolution, 24
stochastic Fubini theorem, 25
stochastic integral, 23
stopping time, 17
Stratonovich integral, 177

tensor (cross) quadratic variation, 21
trajectory of a stochastic process, 18
transition probability function, 19
 homogeneous, 19

ultimate boundedness in the pth moment, 39

weakly convergent, 151

Yosida approximation, 9, 30
Young's inequality, 115